铅元素人为流动

毛建素 等 著

科 学 出 版 社
北 京

内 容 简 介

本书研究铅元素在人类活动圈的流动，共分四篇：第一篇铅元素人为循环，介绍人为循环流动的概念、系统组成，分析物质数量在各股流动间的分配和不同"库"间的转移与蓄积；第二篇人为迁移与转化，着重分析人为循环流动过程中铅元素形态和服务功能方面的变化；第三篇外部效应与评估，分别阐述铅元素人为流动对外部社会、经济和资源、环境系统的影响，评估其生态效率和风险水平；第四篇废铅管理与政策，侧重分析废铅形成规律，剖析我国铅流变化的原因，梳理我国现行涉铅管理政策。

本书可为地学、工程与材料等领域的科研工作者提供研究方法和案例参考，也可为各级管理人员提供借鉴。

图书在版编目（CIP）数据

铅元素人为流动/毛建素等著. —北京：科学出版社，2016.6
ISBN 978-7-03-048028-6

Ⅰ.①铅… Ⅱ.①毛… Ⅲ.①铅－有害元素－人为因素－社会流动－研究 Ⅳ.①F062.1 ②X24

中国版本图书馆 CIP 数据核字（2016）第 071643 号

责任编辑：张 震 孟莹莹 / 责任校对：李 影
责任印制：张 倩 / 封面设计：无极书装

科学出版社 出版
北京东黄城根北街 16 号
邮政编码：100717
http://www.sciencep.com

北京通州皇家印刷厂 印刷
科学出版社发行 各地新华书店经销

*

2016 年 6 月第 一 版 开本：720×1000 1/16
2016 年 6 月第一次印刷 印张：24 1/2
字数：478 000

定价：130.00 元
（如有印装质量问题，我社负责调换）

前　言

伴随着人口规模与需求的增长，大量物质被从自然资源中开采出来，经过一系列生产活动被转化成具有特定使用功能的工业产品，由此形成了物质在人类社会经济复合系统中的流动，简称物质的人为流动。这一流动不仅改变着物质自身属性和赋存状态、影响资源储备和环境质量，而且，长期作用下还衍生出物质在地表人类活动圈内迁移转化新组分，因此，它深刻反映人类发展与地表交互作用关系，对未来人类持续发展和地表演变具有深远意义。

本书以铅元素为例，借助笔者十五年来的研究成果，集成了发表的多篇中英文学术论文，围绕铅元素在人类社会活动圈内流动过程中发生数量分配、性能与时空变化等属性，设置了铅元素人为循环、人为迁移与转化、外部效应与评估和废铅管理四个篇章，系统地介绍铅元素人为循环、迁移与转化等基本概念，阐述其人为循环流动和转化过程的研究方法；结合我国和全球52个国家的铅金属应用技术状况，详尽地展示铅元素人为流动水平、历史演变过程、铅元素人为使用（in-use）累积和环境释放物的环境累积清单等重要结果；描述铅元素人为流动基本规律。在此基础上，本书借助追踪铅元素人为流动过程所连接的资源、环境、社会、经济间的内在关系，阐述了铅元素人为流动的评估方法，展示了该过程对外部铅矿资源环境系统和社会经济系统的可能影响，并阐述了废铅形成基本规律，论述了废铅和含铅环境释放物的源头综合管理方法和措施，以期帮助读者形成对铅元素人为流动的系统全面的认识，为我国重金属环境污染物的源头综合管理提供新思路。同时，笔者也希望本书能进一步丰富物质地表循环、迁移转化基本理论，成为研究人类活动圈内物质流动的基础。

笔者铅的人为流动研究过程曾得到东北大学陆钟武院士、美国耶鲁大学 Thomas E. Graedel 院士的悉心指导，获得美国 2004 年 Henry Luce Foundation "Industrial Ecology in Asia"、中国国家自然基金项目（面上）"铅元素人为迁移转化及其释放物环境蓄积定量解析"（41171361）等资金支持，也得到多个企业界人士的帮助，在此表示衷心感谢！也感谢研究小组成员马兰、梁静、曾润、孙梦瀛、廖珍梅等不畏艰难、勇于探索作出的重要贡献！更感谢亲人们的殷殷期待和不懈支持与鼓励！

尽管希望本书能全面、系统、科学地展示铅元素人为流动整个过程，并对管理有所借鉴。但由于笔者水平所限，书中仍难免存在诸多错误与不足，真诚欢迎广大读者提出宝贵意见。

毛建素

2015 年 12 月于北京

目　录

前言

第一篇　铅元素人为循环

第 1 章　多尺度铅元素人为循环分析：方法
The Multilevel Cycle of Anthropogenic Lead：I. Methodology ········· 3
   1.1　Introduction ········· 3
   1.2　Framing the Lead Cycle ········· 4
   1.3　The Mining and Processing of Lead ········· 5
   1.4　Lead Fabrication and Product Manufacture ········· 7
   1.5　Lead Flow at the Use Stage ········· 12
   1.6　Lead Flow at the Waste Management & Recycling Stage ········· 13
   1.7　Data Quality and Uncertainty ········· 14
   1.8　Summary ········· 15
   References ········· 17

第 2 章　多尺度铅元素人为循环分析：结果与讨论
The Multilevel Cycle of Anthropogenic Lead：Ⅱ. Results and Discussion ········· 20
   2.1　Introduction ········· 20
   2.2　The Framework of Multilevel Lead Cycle Characterization ········· 21
   2.3　Multilevel Lead Flows and Cycles ········· 22
      2.3.1　The Lead Cycle for Selected Countries ········· 22
      2.3.2　Regional Lead Cycles ········· 26
      2.3.3　A Detailed Regional-Level Lead Cycle ········· 31
      2.3.4　The Global Lead Cycle ········· 32
   2.4　Comparing Lead Flows in Countries and Regions ········· 35
   2.5　Discussion ········· 38
   References ········· 40

第 3 章　铅元素使用蓄积动力分析
Lead In-Use Stock：A Dynamic Analysis ········· 42
   3.1　Introduction ········· 42

3.2 Methodology ··· 44
   3.2.1 "Top-Down" Computation of In-Use Stocks ··············· 44
   3.2.2 Considerations for Specific Lead Product Groups ············ 46
   3.2.3 Types of Stocks and Losses ······································ 47
3.3 Lead In-Use Stock in the 20th Century ································· 47
   3.3.1 Additions to Stock in Year 2000 ································ 47
   3.3.2 Global In-Use Stock，1900–2000 ································ 49
3.4 In-Use Lead Stocks in Regions and Countries ······················ 52
   3.4.1 The Disaggregation Methodology ································ 52
   3.4.2 The Results of Disaggregation ···································· 53
3.5 Discussion ··· 55
References ··· 58

第 4 章 中国铅元素人为释放动力分析及其环境累积
A Dynamic Analysis of Environmental Losses from Anthropogenic Lead Flow and Their Accumulation in China ·· 60

4.1 Introduction ··· 60
4.2 Methodology ··· 61
   4.2.1 The Model of Lead Emissions in Anthropogenic Cycle ········ 61
   4.2.2 Intensity and Accumulative Equations of Lead Emissions ····· 65
   4.2.3 Values of the Parameters and the Data Sources ················ 66
4.3 Results and Discussions ·· 69
   4.3.1 The Intensity of Lead Emissions ································ 69
   4.3.2 The Accumulative Lead Emissions ······························ 70
   4.3.3 Data Uncertainty and Perspectives ······························ 72
4.4 Conclusions ··· 72
References ··· 73

第 5 章 铅酸电池系统的铅流分析
The Lead Flow Analysis for Lead-Acid Battery Systems ··············· 76

5.1 研究方法 ·· 76
   5.1.1 铅酸电池系统及其铅的流动 ·· 76
   5.1.2 评价指标 ·· 77
   5.1.3 铅酸电池生命周期铅流图 ·· 77
5.2 铅流基本规律 ··· 78
   5.2.1 资源效率 ·· 78
   5.2.2 环境效率 ·· 79

5.2.3　环境效率与资源效率的关系 ·· 79
5.3　铅酸电池系统中铅的流动（实例分析）·· 80
5.3.1　中国铅酸电池系统的铅流 ·· 80
5.3.2　结果与讨论 ··· 81
5.4　结论 ··· 84
参考文献 ··· 84

第 6 章　中国铅流分析
The Industrial Flow of Lead in China ·· 86
6.1　Introduction ·· 86
6.2　Theoretical Study ·· 88
6.2.1　The IFL Model ··· 88
6.2.2　Evaluation Indices：External Indices ································· 91
6.2.3　Primary Regulation of the System ····································· 92
6.2.4　The Evaluation Indices：Internal Indices ···························· 94
6.3　Case Study：the Industrial Flow of Lead in China ························ 95
6.3.1　Brief Description of Lead Flow in China ···························· 95
6.3.2　Data Sources ··· 97
6.3.3　Evaluation of the IFL in China ······································· 98
6.3.4　Analysis of Causes and Proposed Improvements ···················· 101
6.4　Conclusions ·· 105
References ··· 106

第 7 章　2005 年北京市铅的使用蓄积研究
Lead in Use Stock of Beijing in 2005 ··· 109
7.1　研究方法 ··· 109
7.1.1　研究对象的确定 ··· 109
7.1.2　自下而上（bottom-up）法 ·· 110
7.1.3　铅酸蓄电池的产品种类 ·· 110
7.1.4　铅酸电池使用量及其数据来源 ·· 111
7.1.5　单元铅酸电池含铅量 ·· 112
7.2　结果与讨论 ·· 112
7.2.1　蓄积量及其构成 ··· 112
7.2.2　与国外部分城市对比 ·· 114
7.2.3　结果的不确定性 ··· 115
7.3　结论 ·· 115
参考文献 ··· 115

第 8 章　我国耗散型铅的使用现状及趋势分析
Tread Analysis on Dissipative Uses of Pb in China ·············· 117
- 8.1　概述 ·············· 117
- 8.2　我国典型耗散型铅制品的现状及分析 ·············· 117
  - 8.2.1　使用状况 ·············· 117
  - 8.2.2　相关政策 ·············· 118
  - 8.2.3　相关标准 ·············· 118
- 8.3　国外同类产品的现状及分析 ·············· 119
  - 8.3.1　国外使用状况 ·············· 119
  - 8.3.2　国外相关政策 ·············· 120
  - 8.3.3　国外相关标准 ·············· 120
- 8.4　我国相关政策标准趋势分析 ·············· 121
- 8.5　改善建议与展望 ·············· 122
- 参考文献 ·············· 122

第二篇　人为迁移与转化

第 9 章　矿物资源服务归趋：概念、内涵与议题
The Flows of Mineral Resources to Provide Human Service: Concepts, Connotation and Contents ·············· 127
- 9.1　概念 ·············· 128
  - 9.1.1　服务归趋的概念 ·············· 128
  - 9.1.2　服务归趋研究框架 ·············· 130
  - 9.1.3　物质服务归趋与环境归趋的对比 ·············· 131
- 9.2　科学意义 ·············· 132
  - 9.2.1　科学内涵 ·············· 132
  - 9.2.2　学科意义 ·············· 133
- 9.3　核心议题 ·············· 135
  - 9.3.1　关键问题 ·············· 135
  - 9.3.2　工作内容 ·············· 135
- 9.4　展望与结语 ·············· 136
- 参考文献 ·············· 137

第 10 章　重金属人为迁移转化：概念、内涵与内容
Anthropogenic Transfer & Transformation of Heavy Metals in Anthrosphere: Concepts, Connotations and Contents ·············· 139
- 10.1　Introduction ·············· 139

10.2　Concepts ··· 140
　　10.2.1　Anthropogenic Transfers ··· 140
　　10.2.2　Anthropogenic Transformations ··· 143
　　10.2.3　Essential Characteristics ··· 145
10.3　Scientific Connotations ··· 145
10.4　Core Issues ··· 148
　　10.4.1　Core Issues ··· 148
　　10.4.2　Main Research Contents ··· 149
10.5　Prospects and Conclusions ··· 151
References ··· 152

第 11 章　铅元素人为服务归趋中的变化：功能、形态与位置
Changes in the Functions, Species and Locations of Lead during Its Anthropogenic Flows to Provide Services ··· 154

11.1　Introduction ··· 154
11.2　Analogy between Anthropogenic Flows and Environmental Flows ··· 156
　　11.2.1　Factors That Influence Environmental Lead Flows ··· 156
　　11.2.2　The Relationships between the Anthropogenic and Biogeo-Chemical Lead Cycles ··· 157
　　11.2.3　Comparing the Anthropogenic and Environmental Flows of Lead ··· 157
11.3　Factors That Influence Anthropogenic Lead Flows ··· 159
　　11.3.1　Tracing Flows of Lead Through Its Life Cycle ··· 159
　　11.3.2　The Factors That Influence the Flows of Lead to Providing Human Services ··· 161
11.4　Results ··· 164
　　11.4.1　Changes in the Functions of Lead ··· 164
　　11.4.2　Changes in the Forms of Lead ··· 166
　　11.4.3　Changes in the Locations of Lead ··· 166
　　11.4.4　Characteristics of the Anthropogenic Flows to Providing Service ··· 168
11.5　Conclusions ··· 168
References ··· 169

第 12 章　中国铅元素的人为迁移与转变
Lead Anthropogenic Transfer and Transformation in China ··· 172

12.1　Introduction ··· 172
12.2　Methodology ··· 173
　　12.2.1　Basic Concepts ··· 173
　　12.2.2　Analysis of Lead Anthropogenic Transfer and Transformation ··· 175
　　12.2.3　Data Sources ··· 176

12.3　Determination of Lead Flows and Species ··· 177
　　12.3.1　Lead Flow Quantities in Anthropogenic Cycle ································ 177
　　12.3.2　Lead Transformation at Production Stage ······································ 178
　　12.3.3　Lead Transformation at F&M Stage ··· 179
　　12.3.4　Lead Transformation at Use and WMR Stage ································ 179
12.4　Results and Discussion ·· 180
　　12.4.1　Anthropogenic Transfer in Anthropogenic Lead Cycle ··················· 180
　　12.4.2　Implications on Resources and Environment ································· 181
　　12.4.3　Data Uncertainty ··· 183
12.5　Conclusions ··· 183
References ·· 184

第 13 章　铅元素人为循环环境释放物形态分析
Speciation Analysis of Lead Losses from Anthropogenic Flow in China ············ 187

13.1　材料与方法 ·· 188
　　13.1.1　环境释放物研究框架 ·· 188
　　13.1.2　生命周期各阶段铅环境释放物形态分析 ·· 189
13.2　结果与讨论 ·· 192
　　13.2.1　各阶段铅释放物形态比例 ·· 192
　　13.2.2　生命周期铅释放物形态构成 ·· 193
　　13.2.3　生命周期铅释放物来源构成 ·· 194
　　13.2.4　讨论 ··· 196
13.3　结论 ·· 196
参考文献 ·· 196

第 14 章　全球铅元素人为释放物源头数量与形态分析
Source Analysis of Global Anthropogenic Lead Emissions：Their Quantities and Species ·· 199

14.1　Introduction ·· 199
14.2　Methodology ·· 200
　　14.2.1　Estimation of Lead Emissions ·· 200
　　14.2.2　Lead Emissions Species Analysis ··· 205
　　14.2.3　Data Sources ··· 206
14.3　Results and Discussion ·· 207
　　14.3.1　Global Lead Emissions ··· 207
　　14.3.2　Species Present in the Global Lead Emissions ································ 209
　　14.3.3　Discussion and Uncertainty ·· 211

14.4　Conclusions⋯⋯⋯⋯⋯⋯⋯⋯⋯⋯⋯⋯⋯⋯⋯⋯⋯⋯⋯⋯⋯⋯⋯⋯⋯⋯⋯⋯ 213
References⋯⋯⋯⋯⋯⋯⋯⋯⋯⋯⋯⋯⋯⋯⋯⋯⋯⋯⋯⋯⋯⋯⋯⋯⋯⋯⋯⋯⋯⋯⋯ 213

第 15 章　多尺度铅元素人为循环的环境排放
Losses to the Environment from the Multilevel Cycle of Anthropogenic Lead⋯⋯⋯ 217

15.1　Introduction⋯⋯⋯⋯⋯⋯⋯⋯⋯⋯⋯⋯⋯⋯⋯⋯⋯⋯⋯⋯⋯⋯⋯⋯⋯⋯⋯ 217
15.2　Losses of Lead and Potential Environmental Risk⋯⋯⋯⋯⋯⋯⋯⋯⋯⋯ 218
　　15.2.1　Lead Loss in Tailings⋯⋯⋯⋯⋯⋯⋯⋯⋯⋯⋯⋯⋯⋯⋯⋯⋯⋯⋯ 219
　　15.2.2　Lead Loss in Slag⋯⋯⋯⋯⋯⋯⋯⋯⋯⋯⋯⋯⋯⋯⋯⋯⋯⋯⋯⋯⋯ 219
　　15.2.3　Lead Losses from Fabrication & Manufacture⋯⋯⋯⋯⋯⋯⋯⋯⋯ 219
　　15.2.4　Lead Losses from the In-Use Stage⋯⋯⋯⋯⋯⋯⋯⋯⋯⋯⋯⋯⋯ 220
　　15.2.5　Lead Losses from the Waste Management & Recycling Life Stage⋯⋯ 221
　　15.2.6　Lead Losses Not Treated in This Study⋯⋯⋯⋯⋯⋯⋯⋯⋯⋯⋯ 221
　　15.2.7　Evaluation of Environmental Hazard of Lead Emissions⋯⋯⋯⋯⋯ 221
15.3　Results⋯⋯⋯⋯⋯⋯⋯⋯⋯⋯⋯⋯⋯⋯⋯⋯⋯⋯⋯⋯⋯⋯⋯⋯⋯⋯⋯⋯ 222
　　15.3.1　Lead Loss at the Global Level⋯⋯⋯⋯⋯⋯⋯⋯⋯⋯⋯⋯⋯⋯⋯ 222
　　15.3.2　Lead Loss at the Regional Level⋯⋯⋯⋯⋯⋯⋯⋯⋯⋯⋯⋯⋯⋯ 224
　　15.3.3　Comprehensive Lead Emission Patterns for Countries and Regions⋯⋯ 225
15.4　Incorporating Potential Environmental Hazard into the Life Cycle⋯⋯⋯ 228
15.5　Discussion⋯⋯⋯⋯⋯⋯⋯⋯⋯⋯⋯⋯⋯⋯⋯⋯⋯⋯⋯⋯⋯⋯⋯⋯⋯⋯⋯ 229
References⋯⋯⋯⋯⋯⋯⋯⋯⋯⋯⋯⋯⋯⋯⋯⋯⋯⋯⋯⋯⋯⋯⋯⋯⋯⋯⋯⋯⋯⋯⋯ 230

第三篇　外部效应与评估

第 16 章　铅酸电池中铅的生态效率
The Eco-efficiency of Lead in China's Lead-acid Battery System⋯⋯⋯⋯⋯⋯⋯ 235

16.1　Introduction⋯⋯⋯⋯⋯⋯⋯⋯⋯⋯⋯⋯⋯⋯⋯⋯⋯⋯⋯⋯⋯⋯⋯⋯⋯⋯ 235
　　16.1.1　Background⋯⋯⋯⋯⋯⋯⋯⋯⋯⋯⋯⋯⋯⋯⋯⋯⋯⋯⋯⋯⋯⋯ 235
　　16.1.2　The Present Study⋯⋯⋯⋯⋯⋯⋯⋯⋯⋯⋯⋯⋯⋯⋯⋯⋯⋯⋯ 236
16.2　Primary Regulations⋯⋯⋯⋯⋯⋯⋯⋯⋯⋯⋯⋯⋯⋯⋯⋯⋯⋯⋯⋯⋯⋯ 238
　　16.2.1　Methodology：the Lead-flow Diagram in the LAB System⋯⋯⋯⋯ 238
　　16.2.2　Results and Discussion⋯⋯⋯⋯⋯⋯⋯⋯⋯⋯⋯⋯⋯⋯⋯⋯⋯⋯ 241
16.3　A Case Study：the Eco-efficiency of Lead in China's LAB System⋯⋯⋯ 243
　　16.3.1　Brief Description of Lead Flow in the LAB System⋯⋯⋯⋯⋯⋯ 243
　　16.3.2　Data Sources⋯⋯⋯⋯⋯⋯⋯⋯⋯⋯⋯⋯⋯⋯⋯⋯⋯⋯⋯⋯⋯⋯ 244
　　16.3.3　Results and Discussion⋯⋯⋯⋯⋯⋯⋯⋯⋯⋯⋯⋯⋯⋯⋯⋯⋯⋯ 245

16.4 Conclusions ··· 249
References ··· 250

第 17 章 关于中国铅的资源效率的研究
Study on the Resource Efficiency of Lead for China ··· 252
17.1 铅的资源效率及其变化规律 ··· 252
17.1.1 铅的资源效率的定义 ··· 252
17.1.2 铅的资源效率变化规律 ··· 252
17.1.3 资源效率估算式 ··· 253
17.2 实例应用——中国铅的资源效率分析 ··· 254
17.2.1 中国铅的资源效率现状 ··· 254
17.2.2 原因分析 ··· 254
17.2.3 提高铅的资源效率的途径及建议 ··· 256
17.3 结论 ··· 257
参考文献 ··· 257

第 18 章 铅元素人为循环释放物的风险评价
Risk Assessment of Lead Emissions from Anthropogenic Cycle ··· 258
18.1 Introduction ··· 258
18.2 Methodology ··· 259
18.2.1 Framework for Lead Risk Assessment ··· 259
18.2.2 Model for Lead Risk Assessment ··· 260
18.2.3 Risk Assessment ··· 263
18.3 Results and Discussions ··· 264
18.3.1 Human Health Risk Assessment ··· 264
18.3.2 Ecological Risk Assessment ··· 265
18.3.3 Total Risk Assessment ··· 266
18.3.4 Uncertainty Analysis ··· 268
18.4 Conclusions ··· 268
References ··· 269

第 19 章 物质的循环流动与价值循环流动
The Material Circular Flow and Value Circular Flow ··· 271
19.1 元素 M 的循环流动 ··· 271
19.2 元素 M 的价位变化 ··· 273
19.2.1 元素 M 的价位 ··· 273
19.2.2 元素 M 的价位变化 ··· 273

19.2.3　价值的循环流动 ·· 274
19.3　结语 ··· 276
参考文献 ··· 276

第 20 章　物质循环流动对价值源强的影响
The Influence of Recycling of Materials on Value Source Intensity ············ 278
20.1　Introduction ·· 278
20.1.1　Background ··· 278
20.1.2　This Study ··· 279
20.2　Primary Rules（Theoretical Study）··· 280
20.2.1　Methodology ·· 280
20.2.2　Results and Discussion ·· 284
20.3　A Case Study：the VSI for the Production Stages in the Lead-acid Battery System ··· 286
20.3.1　Methodology ·· 286
20.3.2　Results and Discussion ·· 287
20.4　Conclusions ··· 289
References ·· 290

第 21 章　工业物质循环的若干收益
Several Benefits From Recycling of Industrial Materials ·························· 292
21.1　Introduction ·· 292
21.1.1　Background ··· 292
21.1.2　The Present Study ··· 293
21.2　Methodology ·· 294
21.2.1　CFIM and Its Material Benefits ·· 294
21.2.2　The CFV and Its Economic Benefits ·· 297
21.3　Results and Discussion ··· 300
21.3.1　Results on Various Benefits ·· 300
21.3.2　Discussion on Various Benefits ·· 300
21.4　A Case Study：the Benefit From the Circular Flow of Lead in Lead-acid Battery System ·· 302
21.4.1　Methodology ·· 302
21.4.2　Results and Discussion ·· 304
21.5　Conclusions ··· 305
References ·· 306

第四篇　废铅管理与政策

第 22 章　论工业中的废金属资源
On Metal Scrap Resource for Industry ········· 311
- 22.1　引言 ········· 311
- 22.2　理论研究 ········· 312
  - 22.2.1　若干基本概念 ········· 312
  - 22.2.2　废金属指数基本规律 ········· 313
- 22.3　实例分析：中国大陆若干金属的废金属指数 ········· 316
  - 22.3.1　估算方法 ········· 316
  - 22.3.2　估算结果 ········· 317
  - 22.3.3　讨论 ········· 318
- 22.4　结论 ········· 319
- 参考文献 ········· 320

第 23 章　论铅业的废铅资源
On Lead Scrap Resource for Lead Industry ········· 321
- 23.1　分析方法 ········· 321
  - 23.1.1　废铅的来源 ········· 321
  - 23.1.2　废铅指数 ········· 322
  - 23.1.3　铅产量变化与废铅指数之间的关系 ········· 322
  - 23.1.4　废铅实得率与废铅指数之间的关系 ········· 325
- 23.2　实例——中国、美国、瑞典三国废铅指数的估算 ········· 326
  - 23.2.1　铅产量的变化情况 ········· 326
  - 23.2.2　废铅实得率的变化情况 ········· 326
  - 23.2.3　废铅指数的估算 ········· 327
  - 23.2.4　小结 ········· 328
- 23.3　讨论 ········· 329
  - 23.3.1　中国 ········· 329
  - 23.3.2　美国 ········· 329
  - 23.3.3　瑞典 ········· 329
- 23.4　结论 ········· 329
- 参考文献 ········· 330

第 24 章　关于我国废铅实得率低下的原因的研究
A Study on the Reasons of Low Lead Recycling Rate in China ········· 331
- 24.1　废铅实得率 ········· 331

24.2　铅的生命周期流动示意图 ··· 332
24.3　废铅实得率影响因数的估算 ··· 333
  24.3.1　估算方法 ··· 333
  24.3.2　估算结果 ··· 334
  24.3.3　讨论 ·· 336
24.4　改善建议 ··· 336
24.5　结论 ·· 337
参考文献 ·· 337

第 25 章　中国铅流变化的定量分析
Quantitative Analysis on the Changes in Anthropogenic Lead Flows of China ········ 339

25.1　研究方法 ··· 340
  25.1.1　铅元素人为流动分析 ·· 340
  25.1.2　对比研究 ··· 341
  25.1.3　数据来源和计算说明 ·· 342
25.2　结果与讨论 ··· 342
  25.2.1　2010 年我国铅流分析结果 ·· 342
  25.2.2　指标对比分析 ·· 343
25.3　结论 ·· 345
参考文献 ·· 345

第 26 章　中国铅流改变原因分析
The Reasons for the Changes in Anthropogenic Lead Flows of China ··············· 347

26.1　影响铅流的因素分析 ··· 347
  26.1.1　铅流分析框架 ·· 347
  26.1.2　影响指标 ··· 348
26.2　原因分析 ··· 351
  26.2.1　终端消费对铅消费的拉动作用 ·· 351
  26.2.2　铅消费对铅生产的拉动作用 ··· 352
  26.2.3　铅制品国内消费率增大便于铅循环率提高 ·································· 353
  26.2.4　先进技术的采用 ··· 353
  26.2.5　宏观管理加强 ·· 354
26.3　结论 ·· 355
参考文献 ·· 355

第 27 章　中国铅的使用政策现状分析
Analysis of Current Policies on Lead Usage in China ·· 357

27.1　Introduction ··· 357
27.2　Methodology ··· 358

- 27.2.1 Lead Product Life Cycle ·········· 358
- 27.2.2 Policy Classification ·········· 359
- 27.3 Existing Policies and Analysis ·········· 359
  - 27.3.1 Lead Production ·········· 363
  - 27.3.2 Lead Products Fabrication and Manufacturing ·········· 365
  - 27.3.3 Lead Products Use ·········· 367
  - 27.3.4 Waste Management and Recycling ·········· 368
  - 27.3.5 Lead Trade ·········· 369
- 27.4 Discussion and Recommendations ·········· 370
  - 27.4.1 Discussion ·········· 370
  - 27.4.2 Recommendations ·········· 371
- References ·········· 371

第一篇

铅元素
人为循环

第1章 多尺度铅元素人为循环分析：方法
The Multilevel Cycle of Anthropogenic Lead: I. Methodology*

1.1 Introduction

Lead is one of the few elements known to the ancient world, and was used in the manufacture of products, such as pipe and tableware in the time of the Roman Empire and earlier. Indeed, it appears that lead was the first metal to be mined and worked by humans (Gale and Stos-Gale, 1981). Soft and ductile, lead is easily worked, and sediment chemical analyses record a history of several millennia of anthropogenic processing (Shotyk et al., 1998).

The rate of use of lead has varied considerably over the centuries. The first experimental data showed that Earth's atmosphere had been extensively polluted by industrial lead emissions from smelters and automobiles over the past century (Settle and Patterson, 1982; Patterson and Settle, 1987). In a reconstruction by Nriagu (1998), its first major phase was a byproduct of silver-lead mining for the production of silver coinage. After the fall of the Roman Empire, lead production diminished sharply, but rose again in the 18th century as a consequence of the Industrial Revolution. From 1923 until nearly the present day, lead's use in gasoline additives was the impetus for major increases in extraction. The modern industrial application of lead is heavily concentrated in the battery industry, but pigments, compounds, rolled or extruded products, alloys, ammunition, and so on, each employ a few percent of annual lead use (Table 1.1).

Table 1.1 Global average percentages of lead use in principal products (ILZSG, 2005)

Products	Percent of Total*
Batteries	75.5
Pigments/Compounds	8.4

---

* Mao J S, Dong J, Graedel T E. 2008. The Multilevel Cycle of Anthropogenic Lead: I. Methodology. Resource Conservation & Recycling, 52 (52): 1058-1064.

Products	Percent of Total*
	Continued
Rolled and extruded products	5.9
Miscellaneous	3.1
Alloys	2.8
Ammunition	2.4
Cable sheathing	1.3
Gasoline additives	0.5

* The data refer to more developed countries only.

Lead differs from other extensively used metals in that it is highly toxic. A cumulative poison, it is absorbed into the blood stream where it interferes with the production of hemoglobin (NRC, 1993). As a result, uses of lead in which human or animal ingestion is likely – gasoline additives, paint pigments, fishing weights, etc. – are gradually being phased out.

The creation of comprehensive material flow cycles for lead has substantial policy potential (Hendriks et al., 2000; Uihlein et al., 2006). Such an effort requires that lead uses be well characterized, and that each stage of its life cycle be understood in some detail. Recovery and refining efficiencies and other model parameters must be quantified, and the total of the information formed into a coherent framework for analysis. In this chapter, we describe such an effort. In companion papers (Mao et al., 2008), we use this information to generate lead cycles for countries, regions, and the planet, and explore lead emissions in detail.

1.2 Framing the Lead Cycle

The methodology used to characterize the multilevel cycles of lead draws on the pioneering work of Lohm and associates (Lohm et al., 1994; Sörme et al., 2001), and is based on the simplified framework shown in Fig. 1.1. The cycle consists of four stages: Production, Fabrication and Manufacturing (F&M hereinafter), Use, and Waste Management and Recycling (WMR hereinafter). Earth's lithosphere serves as the initial source of lead, and environmental repositories (landfills, sediments, road construction, etc.) as the eventual sink. The lead accumulated in society during use is referred to as in-use stock.

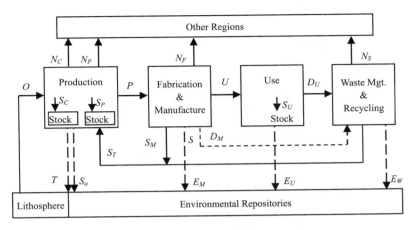

Fig. 1.1 A simplified schematic diagram of a comprehensive lead cycle,
with successive life stages plotted from left to right

For each life cycle stage, the relationships among inflows, outflows, and changes to stock obey the mass conservation law (Kleijn, 2000). The lead flows between stages are constructed by continuity of lead flow. Thus, missing information can often be resolved by difference calculations.

A realistic lead cycle is, of course, much more complex than it appears in Fig. 1.1, as has been shown previously for copper (Spatari et al., 2002). We discuss below the challenges this complexity presents for data acquisition and resolution, and demonstrate how those issues may be resolved.

1.3 The Mining and Processing of Lead

The production of lead is classified into two parts: primary lead production and secondary lead production. The former is chiefly related to lead ore extraction, milling and concentration processes, whereas the latter addresses recycling of lead scrap. Fig. 1.2 gives the framework of lead flow at the Production stage.

The data that were used for the Production stage of the life cycle were collected, estimated, or calculated as follows:

The data for mine production, lead production (primary and secondary), use of refined lead, and lead in trade of concentrate and refined lead on the country level are collected directly from lead and zinc statistical database (ILZSG 2007).

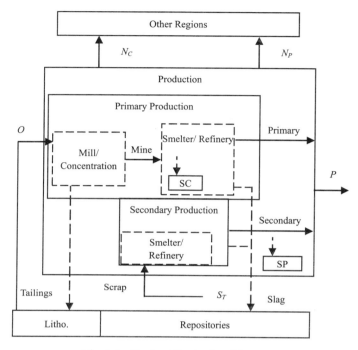

Fig. 1.2 The lead Production life stage

At the primary production facilities, the ore extraction is calculated as the concentrate production divided by the recovery rate in milling and concentrating, which mainly depends upon the grade of lead ore and the technology employed in different countries. We estimate that this recovery rate varies within a range of 0.75 to 0.91, based on data from China Nonferrous Metals Industry Yearbook (2001) and BGRIMM (Beijing General Research Institute of Mining and Metallurgy) (2000). Data for the grade of lead ore is available from RMG (Raw Material Group) (2005). The lead contained in the ore but not entering concentration becomes the lead in tailings. Similarly, some losses occur at the smelter/refinery facility; we estimate that the primary refinery rate varies from 0.89 to 0.99 based on data from the China Nonferrous Metals Industry Yearbook (2001). Lead not recovered is discarded as part of the slag.

At the secondary production facilities, the scrap entering secondary production is estimated as the secondary refined lead divided by the recovery rate at the secondary refinery; this ratio ranges from 0.86 to 0.985 (Li et al., 1999; Jiang, 2000; Ma, 2000; Ma and Yang, 2001). The lead in secondary slag is the difference between the scrap entering secondary refineries and the secondary refined lead produced. Lead scrap is mainly from obsolete lead products, with a small portion from F&M residues.

The sum of the primary and secondary refined lead is available for export or for use in fabrication and manufacturing.

The estimated recovery rates for selected countries, regions, and the world are listed in Table 1.2.

Table 1.2 Estimated values for mill and refinery recovery and recycling rates for selected countries, world regions, and the planet in 2002

	Region							
	AF	AS	CIS	EU	LAC	ME	NA	OC
Lead grade in production[1]/%	2.5	3.0	2.9	3.4	3.1	2.7	5.1	8.8
Recovery rate at concentrate[2]	0.75	0.80	0.78	0.84	0.82	0.82	0.89	0.91
Primary refinery rate[3]	0.89	0.94	0.95	0.98	0.95	0.94	0.99	0.98
Secondary refinery rate[4]	0.86	0.96	0.93	0.985	0.93	0.90	0.985	0.98
	Country and world							
	FR	IT	JP	US	GB	IN	World	
Lead grade in production[1]/%	3.4	3.4	3	5.1	3.4	3.0	3.47	
Recovery rate at concentrate[2]	0.84	0.84	0.82	0.89	0.84	0.79	0.84	
Primary refinery rate[3]	0.98	0.98	0.98	0.98	0.99	0.94	0.95	
Secondary refinery rate[4]	0.99	0.99	0.99	0.99	0.99	0.87	0.99	

[1] Tiscali SpA, 2005.
[2] BGRIMM, 2000; RMG, 2005.
[3] Karlsson, 1999; China Nonferrous Metals Industry Yearbook, 2001.
[4] Jiang, 2000; Ma, 2000; China Nonferrous Metals Industry Yearbook, 2001; Mao et al., 2006.

1.4 Lead Fabrication and Product Manufacture

The fabrication and manufacturing (F&M) life stage is designed to transform refined lead into semi-finished or finished lead products. Lead products may be thought of as falling into three types of use: ①Uses that are economically and technologically compatible with recycling under present prices and regulations; ②Uses that are not economically compatible with recycling but where recycling is technically feasible; ③Uses for which recycling is inherently not feasible (Ayres, 1994). Most lead use falls into the first class, making high recycling efficiencies possible. For example, in the United States 85%–95% of lead scrap is obsolete batteries, and most of the remainder is new scrap from F&M (Smith, 2000; USGS, 2005). Li et al. (1999) reported that

2%–5% of lead scrap is customarily from obsolete pipe or sheet. Lead in compounds is mainly used as paste in battery grids, pigments and other paint additives, additives in glasses, glazes, poly (vinyl chloride) stabilizers, and so on. This lead has low economic value, and is generally difficult to recover and reuse. The same is true of lead in ammunition and gasoline additives, which are effectively emitted to the environment whenever they are put to use.

By combining the life spans of lead products with the potential for recycling, we categorized the products into five groups: batteries, sheet & pipe, cable sheathing, alloys, and others, a category principally composed of lead chemicals. We include semi-products in the sheet & pipe group. Thus, using data for 52 countries reported by ILZSG (2005), the global use pattern of lead in the year 2000 can be expressed as follows: batteries, 74.5%; sheet & pipe, 4.8%; cable sheathing, 2.4%; alloys, 3.8%; and other 14.9%. The designation of these five groups produces a detailed framework of lead flow at the F&M life stage as shown in Fig. 1.3.

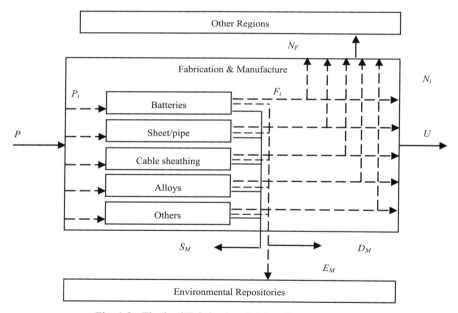

Fig. 1.3　The lead Fabrication & Manufacture life stage

The proportions of the refined lead that enter the five product groups vary according to the relative demand for the products and with the technology employed. In our study, we estimated these proportions by a two-step process. First, we multiplied the total lead use for each selected country by the geographically-specific product group distribution of uses. The use patterns for 29 countries among the selected 52 countries was calculated according to the data of ILZSG（2005）；those for other 23 countries were set equal to that of data-rich countries in the same region or estimated according to a country's development status. The use pattern for regions and the world are generated by aggregating the use patterns for the countries. The product group patterns for 6 selected countries（i.e., the United States，United Kingdom，Italy，Japan，France and India），8 regions and the world are given in Table 1.3.

Table 1.3　Estimation of the relative use patterns of lead in the world，regions，and selected countries in 2000（%）. See text for details

	Region							
	AF	AS	CIS	EU	LAC	ME	NA	OC
Batteries	75.3	73.2	72.2	60.4	77.3	74.5	84.6	67.8
Pipe & sheet	2.4	1.7	6.5	12.4	1.3	6.9	2.6	14.8
Cable sheathing	5.9	2.5	1.3	3.6	1.8	1.6	0.3	3.3
Solder alloys	4.2	7.8	2.8	3.9	4.8	2.8	1.1	4.2
Other	12.2	14.8	17.2	19.6	14.9	14.1	11.4	10.0
	Country							
	US	FR	IT	GB	JP	IN	World	
Batteries	86.3	73.0	74.5	33.7	72.7	72.9	74.5	
Pipe & sheet	3.1	6.8	3.3	33.0	1.1	1.5	4.8	
Cable sheathing	0.2	3.7	0.6	3.0	1.5	5.0	2.4	
Solder alloys	0.9	2.3	1.1	7.2	3.4	6.9	3.5	
Other	9.6	14.1	20.5	23.0	21.4	13.7	14.9	

Data source：Based on ILZSG，2005. World data generated by aggregating country data.

The second step in allocating lead to the product groups within F&M involves multiplying the product group patterns by coefficients termed the fabrication efficiencies，which refer to the lead retained in a specific finished product as a fraction of the corresponding F&M input. Fabrication efficiencies for specific products have been estimated by Jiang（2000），Karlsson（1999），Mao and Lu （2003a，b，c，d），and Mao et al.（2006，2007），and are listed in Table 1.4 for the world's regions and the selected countries.

Table 1.4 Fabrication efficiencies for world regions and selected countries

	World	AF	AS	CIS	EU	LAC	ME	NA	OC
Batteries	0.91	0.89	0.91	0.92	0.93	0.91	0.91	0.93	0.93
Pipe & sheet	0.89	0.87	0.89	0.88	0.91	0.89	0.89	0.91	0.91
Cable sheathing	0.87	0.85	0.87	0.86	0.89	0.87	0.87	0.89	0.89
Solder alloys	0.88	0.86	0.88	0.87	0.90	0.88	0.88	0.90	0.90
Other	0.96	0.94	0.96	0.95	0.98	0.96	0.96	0.98	0.98
	US	FR	GB	IT	JP	IN			
Batteries	0.93	0.93	0.93	0.93	0.93	0.91			
Pipe & sheet	0.91	0.91	0.91	0.91	0.91	0.88			
Cable sheathing	0.89	0.89	0.89	0.89	0.89	0.86			
Solder alloys	0.90	0.90	0.90	0.90	0.90	0.87			
Other	0.98	0.98	0.98	0.98	0.98	0.95			

References are given in the text.

Lead that is not retained as part of finished products becomes prompt scrap (S_M), discards to waste management (D_M), or emissions (E_M). The division among these three flows largely depends on the level and policies of environmental management, and thus may vary greatly among countries and regions. Few such data are available. In our approach, these three flows were roughly as 50%, 30%, 20%, respectively, of the total lead not entering finished products. For countries with F&M capacity but without secondary production capacity, lead entering F&M but not leaving F&M in products is treated entirely as emissions.

Data for the trade of semi- or finished products at F&M stage were collected from United Nations statistics (UNSD, 2000). These data, however, record the number of products transferred from one repository to another, not the lead contained therein. The estimated lead contents in the different products that were used in our calculations are listed in Table 1.5. Batteries must be treated both as products and as product constituents, because batteries are traded by themselves and also as components of vehicles.

Table 1.5 Conversion factors for the lead content of commercial products

Items	Name	Unit-conversion factor $/kg commodities		Lead-content conversion factor kg Pb/kg commodities
		Import	Export	
Batteries	Lead-acid batteries for vehicles	1.35	1.35	0.58
	Lead-acid batteries for other than vehicles	2.19	2.60	0.58

Continued

Items	Name	Unit-conversion factor $/kg commodities		Lead-content conversion factor kg Pb/kg commodities
		Import	Export	
Sheet &pipe	Lead plates, sheets, strip, foil, powders and flakes	0.77	1.16	0.95
	Lead tubes, pipes and fittings	1.73	1.03	0.95
	Articles of lead nes	1.54	3.33	0.95
Cable sheathing	Lead bars, rods, profiles and wire	3.63	1.60	0.95
Alloys	Lead alloys, unwrought	0.79	0.50	0.98 for Sb-Pb, 0.4 for Sn-Pb*
Other	Lead oxides; red lead and orange lead.	0.68	0.67	0.8
	TV picture tubes, CRT, etc	2.26	3.92	0.21
	Anti-knock preparations based on lead compounds	1.80	2.72	0.00033
	Drinking glasses of lead crystal	5.26	7.06	0.23
	Glassware except kitchen, table ware, of lead crystal	11.9	17.4	0.23

* Thornton et al., 2001.

Most of the principle uses of lead are such that the products become obsolete within only a few years (Mao and Lu, 2003a). Table 1.6 shows the range of life spans of the principal lead products. In several cases, we list the life-span as zero, by which we mean that lead in those uses is effectively lost to the environment as soon as it enters into use, because recovery is unlikely or impossible.

Table 1.6　The range of life spans for lead products

Lead product	Life span/year
SLI batteries[a]	2.5–5
Stationary batteries	8–15
Pipe	10–50
Sheet	20–100
Cable sheathing	20–50
Semi-manufactures	5–40
Ammunition	0[b]
Gasoline additives	0[b]
Solder alloys	0[b]
Others	0[b]

a SLI (starting, lighting, and ignition) batteries are used primarily for vehicles.
b In these uses, lead is essentially lost to the environment as soon as it enters into service.

1.5 Lead Flow at the Use Stage

At the use stage, the flows of interest are lead entering Use phase, annual addition to (or subtraction from) in-use stock, lead leaving Use through product discard, and lead emissions. The flow diagram is shown in Fig. 1.4.

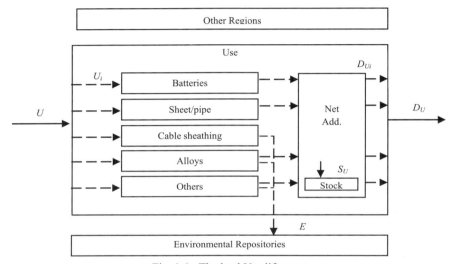

Fig. 1.4 The lead Use life stage

Strictly speaking, lead contained in products purchased by potential users is regarded as having entered the Use stage. Most of this material that in vehicle batteries, for example, is immediately put into service. Small amounts may be purchased and retained for future use (e.g., ammunition); we designate this material as part of in-use stock, recognizing that in some cases it may be discarded without ever actually having been used. The quantities subject to this situation are very small on a percentage basis.

S_U, the annual additions to in-use stock, result from more lead entering the Use phase than leaving it. We can use mass closure around the Use stage to compute the annual additions to in-use stock (S_U):

$$S_U = U - D_U - E_U \qquad (1.1)$$

As noted in Fig. 1.4, the lead leaving the Use phase may become end-of-life discards or may be emitted to environmental repositories. The end-of-life discards (D_U) come from lead products with non-zero life span that entered Use one life span ago, i.e., $D_U = \sum D_{Ui}$. The product lifetime used for this calculation, and the percentage of each product group that entered WMR, are given in Table 1.7.

Table 1.7 Data used in estimating end-of-life discards

Lead product	Life span/year	Potential recycling percentage/%
SLI battery	4	100
Stationary battery	12	100
Pipe/sheet	30	100
Alloys	12	40
Others	12	50

Lead leaving the Use phase but can not going to WMR will become emissions to environmental repositories, i.e., $E_U = \sum E_{Ui}$. Emissions at the Use phase include cable sheathing, lead alloys and others. Cable sheathing, which is employed to provide integrity for the cable interior and for protection against cable operating failure due to marine and land animal bites, constitutes "hibernating stock": not finally disposed of, but not being utilized. Because a very large fraction of these sheathed cables are deployed on the ocean floor or underground, it rarely makes economic sense to recover these cables when they are no longer being used. Thus lead in sheathed cables is nearly always lost to the lead cycle once deployed. We therefore treat lead flows into cable sheathing as special emissions to environmental repositories, i.e., hibernating emissions, even though those emissions may take place over long time periods. The second emission from the Use phase is alloys, for which we estimate that 60% is dissipative so far as the lead cycle is concerned. The third component of emission at the Use phase is the category "others", half of which can become end-of-life discards, and the remainder dissipative uses. Thus, 50% of end-lead uses in "others" was treated as emissions. The sum of the above three flows constitute the total emissions at the Use phase.

Because U has been estimated in the F&M stage, and E_U and D_U as described above, the value for S_U is calculated by equation (1.1).

1.6 Lead Flow at the Waste Management & Recycling Stage

The lead flows in Waste Management & Recycling (WMR) are shown in Fig. 1.5. There are two main sources for lead entering this life stage. One is obsolete lead products from the Use stage. The other is discards from F&M, the so-called "industrial waste". The data for these two lead flows have been estimated as discussed above. The lead flow in import and export trade at the WMR stage is mainly lead scrap. These data are available from United States Statistics Division (UNSD, 2000). We estimate that lead scrap is 90% lead by weight, based on private discussions with experts in secondary refineries. The outflow from WMR is divided into scrap returned to Production and that emitted to the environment. The former can be estimated as the difference between the total scrap returned to production minus the scrap from F&M, i.e., $S_T - S_M$, and the latter (E_D) by the lead balance at the WMR stage:

$$E_D = D_U + D_M - N_S - S \qquad (1.2)$$

where $S = S_T - S_M$.

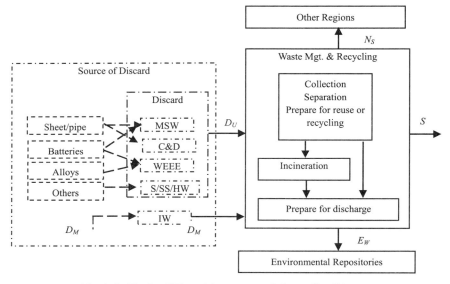

Fig. 1.5 The lead Waste Management & Recycling life stage

1.7 Data Quality and Uncertainty

In assembling the data for the multilevel lead cycle, we have been required to use numerous data sources of varying quality, mostly available at the country level. (The situation is not restricted to lead, but holds generally for the cycles of all metals [Gordon et al., 2003]). The utilization of import/export data and end-of-life flow information is particularly problematic but important in the analysis. Especially in this context, we have used some empirical information, such as the range of life-span, the percentages that become end-of-life discards, lead use patterns, the lead content in different semi- or finished lead products, present and historical efficiencies in fabricating and manufacturing of specific lead products, and so on.

In general, data for the early stages of the lead life cycle (extraction, concentration, refining) is widely available and of good quality, largely because it is important to global financial markets. The manufacturing stage, at which the lead is widely distributed and assembled into a spectrum of lead products, is reasonably well characterized in the public literature for the main use categories in around 30 countries, which has been discussed at section 4. Such data do not exist for most less-developed countries, and are estimated on the basis of the countries state of development. For the fabrication efficiencies at manufacturing stage, we generally estimated it according to the reported data from China, and take a relative low value for less developed countries and higher values for developed countries. The losses to the environment, especially at the Use stage, are also not reported systematically, although some information is available for specific emissions, especially for the lead in gasoline additives. For Waste Management & Recycling, data for scrap return to production is of good quality because it is closely related to publicly reported secondary lead production. However, lead losses from this life stage have high uncertainty, being largely derived from less than rigorous data generated by the US Environmental Protection Agency (EPA, 2005).

Information is highly variable in availability and quality from one geographical region to another. The most consistent data tend to be at the country level; data heterogeneity arises in scaling up (to continental or global levels) or scaling down (to city or watershed levels, for example). When scaling up, we are generally forced to use a few countries to represent a regional lead use pattern. When scaling down, the challenge is to accurately subdivide data so as to appropriately treat a specific region.

Additional uncertainty arises from the complex dynamics of the lead cycle itself, which is never in steady state because of increasing or decreasing lead production, technology development, and transformations in the development level of the economy and society.

Overall, our judgment is that the data completeness and accuracy in our study are sufficient for the approximate characterization of the lead cycle on our global, regional, and country levels, though parts of the quantity are problematic. Our results should be adequate to guide engineering decision-making, resource, environment and policy perspectives so far as the major lead flows are concerned.

1.8　Summary

We have constructed a comprehensive lead cycle, and collected and summarized a large body of data pertaining to the flows of lead in various forms. In a companion paper (Mao et al., 2008), we use this framework to generate the quantitative lead cycle for the year 2000 for 52 countries, 8 regions, and the world.

The comprehensive lead cycle that results from the consideration discussed above is shown in Fig. 1.6. In a companion paper (Mao et al., 2008a), we use this frameworks to generate the quantitative lead cycle for year 2000 for 52 countries, 8 regions, and the world.

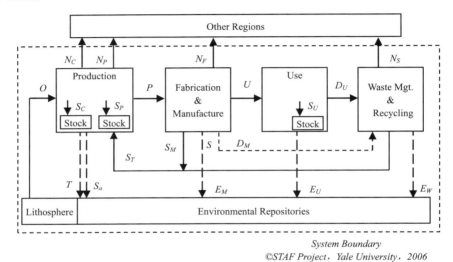

Fig. 1.6 The comprehensive lead cycle

O	Symbols and definition	S_a	Lead in slag
P	Refined lead	S_P	Lead entering stock of refined lead
U	Lead entering Use phase	S_C	Lead entering concentrate stock
D_U	End-of-life lead to waste management & recycling	S_U	Lead entering in-use stock
D_M	Lead in F&M discards to WMR	S	Lead in scrap from WMR
N	Lead in net export (export minus import)	S_T	Lead in total scrap to production
N_P	Lead in net export of refined lead	S_M	Lead in new scrap from Manufacture
N_C	Lead in net export of concentrate or mine	E	Lead in emissions to the environment
N_S	Lead in net export of scrap	E_M	Lead in F&M emissions
N_F	Lead in net export of semi- or finished products	E_U	Lead in Use emissions
T	Lead in tailings	E_W	Lead in WMR emissions

We have described in this chapter the availability and use of data for the flows of lead within the technological society, and methods for analyzing and assembling these data into lead cycles. In a number of cases, the challenge comes from the gathering of information from a wide variety of heterogeneous sources. In other cases, data are quite fragmentary and it is necessary to proceed on the basis what appears to be representative information. As a consequence, we do not claim that the technological lead cycles that can be assembled with information currently available will be highly accurate. Nonetheless, we conclude that information is sufficiently available to permit quantitative cycles to be developed at national, regional, and global scales. Refinement of the quantitative information can be expected as comprehensive lead flow information becomes integrated into national data sets. The utility of the resulting cycles as information frameworks for resource management, environmental science, and policy analysis is likely to be substantial, as the cycles will provide hitherto unavailable perspectives on the approximate level of stocks in various reservoirs, the flows among them, and the potential ways in which stocks and flows might be modified should that prove desirable.

Acknowledgements

This work is part of the Stocks and Flows Project at Yale University's Center for Industrial Ecology. The project is supported by the US National Science Foundation under Grant BES 0329470. The participation of J. S. Mao was made possible by the Henry Luce Foundation of the USA and the "National Basic Research (973) Program" Project (No. 2005CB724204) of the Ministry of Science and Technology of China.

References

Ayres R U. 1994. Industrial metabolism: theory and policy. In The Greening of Industrial Ecosystems. Washington D.C: National Academy Press.

BGRIMM (Beijing General Research Institute of Mining and Metallurgy). 2000. China Investigation Report on the Exploitation and Utilization of Lead-Zinc Mineral Resource.

Editorial Staff of The Yearbook of The Nonferrous Metals Industry. 2001. China nonferrous metals industry yearbook (1990-2001) (in Chinese). Beijing: China Nonferrous Metals Industry Yearbook Press.

Gale N H, Stos-Gale Z. 1981. Lead and silver in the ancient Aegean. Scientific American, 244 (244): 176-192.

Gordon R B, Graedel T E, Bertram M, et al. 2003. The characterization of technological zinc cycles. Resources Conservation & Recycling, 39 (2): 107-135.

Hendriks C, Obernosterer R, Müller D, et al. 2000. Material flow analysis: A tool to support environmental policy decision making. Case-studies on the city of Vienna and the Swiss lowlands. Local Environment the International Journal of

Justice & Sustainability, 5 (3): 311-328.

ILZSG (International Lead and Zinc Study Group). 2005. Principal Uses of Lead and Zinc. London: ILZSG (International Lead and Zinc Study Group), 7: 21-31.

ILZSG (International Lead and Zinc Study Group). 2007. Lead and Zinc in the World. Interactive statistical database. http://www.ilzsg.org/ (accessed November.16, 2007).

Jiang S. 2000. The development and utilization of secondary nonferrous resource in China (in Chinese). Resource Saving and Comprehensive Utilization, 2 (1): 18-21.

Karlsson S. 1999. Closing the technospheric flows of toxic metals-modeling lead losses from a lead acid battery system for Sweden. Journal of Industrial Ecology, 3 (1): 23-40.

Kleijn R. 2000. In = out: The trivial central paradigm of MFA? Journal of Industrial Ecology, 3 (2/3): 8-10.

Li F Y, Li S S, Wang J. 1999. The present production status of recycled lead and its future in domestic and overseas markets (in Chinese). World Nonferrous Metals, 5: 26-30.

Lohm U, Anderberg S, Bergbäck B. 1994. Industrial metabolism at the national level: a case study on chromium and lead pollution in Sweden, 1880-1980. In: Ayres RU, Simonis U E, editors. Industrial metabolism. Tokyo: United Nations University Press, p. 103-18.

Ma YG. 2000. Some suggestions on changing present state of secondary lead refining in China and its related policy and regulations (in Chinese). Resource Saving and Comprehensive Utilization, 3 (1): 15-19.

Ma YG, Yang HY. 2001. The recovery of obsolete lead-acid batteries and secondary lead refining (in Chinese). Environ. Herald, 1: 52-53.

Mao J S, Lu Z W. 2003a. Resource-service efficiency of lead in lead-acid battery (in Chinese). Journal of Northeastern University (Natural Science), 24: 1173-1176.

Mao J S, Lu Z W. 2003b. A study on causes of low recovery of scrap lead (in Chinese). World Nonferrous Metals, 11: 24-32.

Mao J S, Lu Z W. 2003c. Study on the resource efficiency of lead for China (in Chinese). Research of Environmental Sciences, 17 (3): 78-80.

Mao J S, Lu Z W. 2003d. On the lead scrap resource in the lead industry (in Chinese). World Nonferrous Metals, 7: 10-14.

Mao J S, Lu Z W, Yang Z F. 2006. The eco-efficiency of lead in China's lead-acid battery system. Journal of Industrial Ecology, 10 (1-2): 185-197.

Mao J S, Yang Z F, Lu Z W. 2007. Industrial flow of lead in China. Transactions of Nonferrous Metals Society of China, 17 (2): 400-411.

Mao J S, Dong J, Graedel T E. 2008a. The multilevel cycle of anthropogenic lead: II. Results and Discussion. Resources Conservation & Recycling, 52 (s 8-9): 1050-1057.

Mao J S, Cao J, Graedel T E. 2009. Losses to the environment from the multilevel cycle of anthropogenic lead. Environmental Pollution, 157 (10): 2670-2677.

NRC(National Research Council). 1993. Measuring Lead Exposure in Infants, Children, and Other Sensitive Populations. Washington, DC: National Academy Press.

Nriagu J O. 1998. Tales told in lead. Science, 281: 1622-1623.

Patterson C C, Settle D M. 1987a. Review of data on eolian fluxes of industrial and natural lead to the lands and seas in remote regions on a global scale. Marine Chemistry, 22 (2-4): 137-162.

Patterson C C, Settle D M. 1987b. Magnitude of lead flux to the atmosphere from volcanoes. Geochimica Et Cosmochimica Acta, 51 (3): 675-681

RMG (Raw Material Group). 2005. Raw materials data- the mining database. Stockholm: RMG (Raw Material Group): http://www.rmg.se/RMG2005/pages/database.htm (accessed August 17, 2005).

Settle D, Patterson C. 1982. Magnitudes and sources of precipitation and dry deposition fluxes of industrial and natural leads to the North Pacific at Enewetak. Journal of Geophysical Research Atmospheres. 87 (C11): 8857-8869.

Shotyk W, Weiss D, Appleby P G, et al. 1998. History of atmospheric lead deposition since 12, 370 ^{14}C yr BP from a peat bog, Jura mountains, Switzerland. Science, 281 (281): 1635-1640.

Smith G R. 2000. Lead. http://minerals.usgs.gov/minerals/pubs/commodity/lead/380400.pdf.(accessed October 10, 2005).

Smith G R. 2004. Lead recycling in the United States in 1998. Reston, VA: Circular 1196-F, U.S. Geological Survey.

Spatari S, Bertram M, Fuse K et al. 2002. The contemporary European copper cycle: 1-year stocks and flows. Ecological Economics, 42 (S1-2): 27-42.

Sörme L, Bergbäck B, Lohm U. 2001. Goods in the anthroposphere as a metal emission source a case study of Stockholm, Sweden. Water Air & Soil Pollution Focus, 1 (1): 213-227.

Thornton I, Rautiu R. Brush S. 2001. Lead, the Facts. IC Consultants Ltd. London, UK.

Tiscali SpA, 2005. Lead ore: 2005. http://www.tiscali.co.uk/reference/encyclopaedia/hutchinson/m0006412.html (accessed Oct. 20, 2005).

Uihlein A, Poganietz W-R, Schebek L. 2006. Carbon flows and carbon use in the German anthroposphere: an inventory. Resources Conservation & Recycling, 46: 410-429.

UNSD (United States Statistics Division). 2000. UN Commodity Trade Statistics Database (UN Comtrade), http://unstats.un.org/unsd/comtrade/ (accessed October 14, 2005).

US Environmental Protection Agency. 2005. United States Environmental Protection Agency Lead and Copper Rule Workshop: Lead in Plumbing Fittings and Fixtures. http://www.epa.gov/safewater/lcrmr/pdfs/summary_lcmr_review_school_plumbing.pdf (accessed July 26, 2005).

USGS (U.S. Geological Survey). 2005. Mineral Commodity Summaries. http://minerals.usgs.gov/minerals/pubs/commodity/lead/lead_mcs05.pdf. (accessed January. 20, 2006).

Woodbury W D, Edelstein D, Jasinski S M. 1993. Lead materials flow in the United States, 1940-1998. Unpublished report. Reston, VAUSGS: U.S. Geological Survey.

第 2 章　多尺度铅元素人为循环分析：结果与讨论
The Multilevel Cycle of Anthropogenic Lead：Ⅱ. Results and Discussion*

2.1 Introduction

A study of the multilevel cycle of anthropogenic lead can enhance our perspective on a variety of related topics，including lead ore availability and utilization，recycling potential，environmental impacts of dissipative use，and environmental policy. In the first paper of this series（Mao et al.，2008a；hereafter "Paper I"），we presented the methodology needed to characterize lead cycles：the division of lead uses into five categories，the specification of stocks and flows for the four principal life stages（Processing，Fabrication & Manufacture（F&M），Use，and Waste Management & Recycling（WMR）），and estimates of product life spans and recovery and refining efficiencies. In this chapter，we use that information to prepare lead cycles at three discrete geographical levels—country，region，and the world.

As with other metals，lead trades freely around the world. A detailed understanding of its use thus requires that its entire cycle be characterized. To date，few attempts have been made to develop comprehensive cycles for lead. Those of which we are aware are summarized in Table 2.1. None is at the regional or global level，and thus is not comprehensive enough to indicate the complex multi-scale situations involving lead ore extraction，lead emissions，and so on. Those cycles at a national level tend to be focused on specific product systems or restricted to portions of the life cycle. It is clear that a broader-based and geographically more complete approach would be useful.

Table 2.1 Existing contemporary lead cycles

Region	Year	Reference	Comments
Unteres Bünztal, Switzerland	1986	Brunner et al., 1994	Recycling not considered, but other aspects carefully treated

---

* Mao J S*，Dong J，Graedel T E. 2008. The Multilevel Cycle of Anthropogenic Lead：Ⅱ. Results and Discussion. Methodology. Resource Conservation & Recycling，（52）：1050-1057.

Continued

Region	Year	Reference	Comments
Vienna	1991	Obernosterer and Brunner, 2001	Urban flows and stocks were quantified, fabrication & manufacturing, losses, and recycling were not
USA	1993–1994	Socolow and Thomas, 1997	Thorough treatment of all aspects of the lead cycle
Sweden	1993–1995	Karlsson, 1999	Lead flow in battery systems was analyzed, no analysis for the whole system
Stockholm	1995	Bergbäck et al., 2001	Detailed treatment of losses, but no analysis of fabrication & manufacturing, or of recycling
USA	1998	Smith, 2004	Lead supply and demand were emphasized, losses were not evaluated
Japan	1990s	NIMS-EMC, 2003	Lead flows by industrial sectors were described, but only the fabrication & manufacture stage was evaluated.
China	1999	Mao et al., 2006a, b	Lead flow in battery systems was analyzed, no analysis for the whole system
UK	2001	Thornton et al., 2001	Lead use and its relationship to resources and the environment was emphasized, no system contemporary flow
Europe and Netherlands	2004	Elshkaki and van der Voet, 2004a, b	A stock model was provided and the lead stocks and flows for Europe and the Netherlands were researched. Specific lead uses, lead in reuse or recycling, and lead emissions were not evaluated

The multilevel lead cycle resulting from our work is the seventh to be completed by our research group, the others being copper (Graedel et al., 2004), zinc (Graedel et al., 2005), silver (Johnson et al., 2005), chromium (Johnson et al., 2006), iron (Wang et al., 2007), and nickel (Rech et al., 2008). All of the cycles have followed a common framework, and thus begin to comprise a composite set of information for the cycles of the non-ferrous metals.

2.2 The Framework of Multilevel Lead Cycle Characterization

Paper I presented the main relationships among the four life cycle stages. We chose 2000 as the reference year and thus constructed a "snapshot" of the multilevel lead cycle for that year. Our basic unit of study is the country, for which information on lead cycling is most readily available. The approach is to study a small group of important countries in detail, and then to extend this perspective to countries for which information is less extensive. For this purpose, we selected the United States, France, Italy, United

Kingdom, Japan, and India as diverse but information-rich examples of lead cycle and flow. From this base, lead cycles for a total of 52 countries, representing 96.6%, 97.1% and 90.0% of the total concentrate, lead production, and lead use of the world, respectively, were chosen for study. Where data for specific countries were not readily available, we proceeded by utilizing information from similar countries or regions. Not only does this group of countries enable us to study virtually the entire lead situation, it will also permit us to investigate at a later date the relationships of the lead cycles to those of other metals.

The information for the 52 countries can be aggregated into 8 regions, as defined in previous studies by our group and designated as Africa (AF), Asia (AS), Commonwealth of Independent States (CIS), Europe (EU), Latin America and the Caribbean (LAC), Middle East (ME), North America (NA), and Oceania (OC).

2.3 Multilevel Lead Flows and Cycles

2.3.1 The Lead Cycle for Selected Countries

As examples of the fully-characterized lead cycles, we illustrate in Fig. 2.1 the cycles for the six countries we selected for detailed study. Complete lead flow information for the remaining 46 countries is contained in the Supplementary Material for this chapter.

The United States cycle[Fig. 2.1(a)] is typical of a country that mines, processes, and recycles large amounts of lead. Note that almost 2/3 of the lead flow to F&M comes from recycled scrap (mostly batteries). The United States has a net import flow of refined and semi- or finished lead products, but a net export flow of lead concentrate and scrap. About 10% of the flow of lead into Use is accumulated as additional in-use stock. Losses to the environmental repositories from WMR are significantly larger than those from Production, F&M, and Use.

In contrast to the United States lead cycle, the cycles of France, the United Kingdom, and Italy[Fig. 2.1(b)-Fig. 2.1(d)] are typical of countries with little or no domestic lead extraction. As a result, their lead production depends completely upon the importation of lead, largely as concentrate, but partially as refined lead they recycle quantities of lead essentially equal to their imports, and these flows may be considered domestic "mining" from previously extracted material. Losses are mainly from the WMR and Use life stages.

The lead cycle for Japan[Fig. 2.1（e）] is largely similar to that of Italy，but differs in an interesting way：Japan receives more than half its Production lead from domestic recycling，and has thus developed its domestic "mines" to a greater percentage extent than the other developed countries in our study.

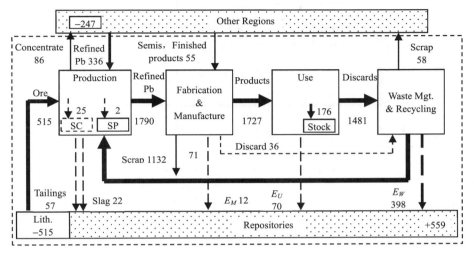

Lead，*2000*
(a) United States

System Boundary "United States of America（US）"
©STAF Project，*Yale University*，*2006*

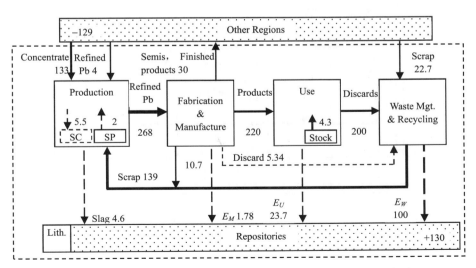

Lead，*2000*
(b) France

System Boundary "France（FR）"
©STAF Project，*Yale University*，*2006*

| 铅元素人为流动 |

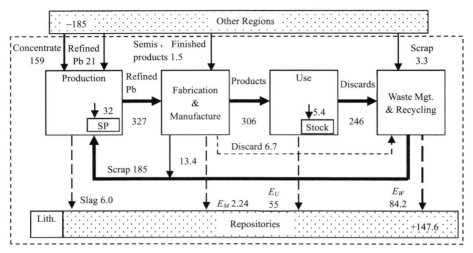

Lead, 2000
(c) United Kingdom

System Boundary "United Kingdom (GB)"
©STAF Project, Yale University, 2006

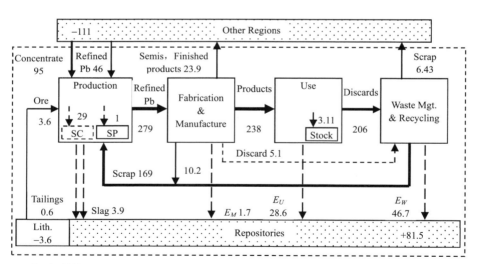

Lead, 2000
(d) Italy

System Boundary "Italy (IT)"
©STAF Project, Yale University, 2006

| 第 2 章 | 多尺度铅元素人为循环分析：结果与讨论

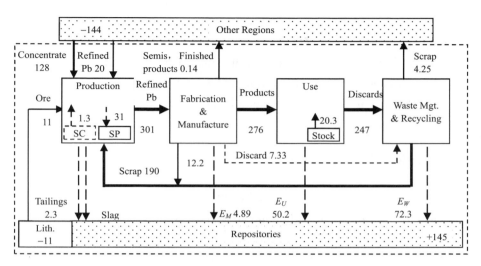

(e) Japan

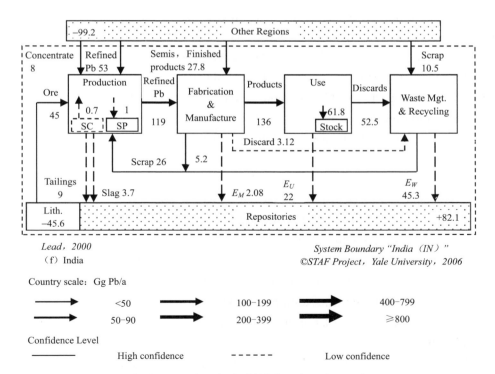

(f) India

Fig. 2.1 Lead cycles in 2000 for selected countries

The final cycle in Fig. 2.1 is that for India. India receives only about 22% of its Production lead from domestic recycling, the remainder being about equally divided between domestic ore extraction and the importation of refined lead. Such a pattern is typical of a developing country in which the historic use of lead, and thus its availability for recycling, has been modest. It is also significant that a net of almost 45% of the flow into Use is added to in-use stock, a much higher percentage than those of the other countries studied here.

2.3.2 Regional Lead Cycles

As noted above, regional-level lead cycles can be produced by aggregating the appropriate country flows. The eight regional lead cycles appear in Fig. 2.2.

The cycle for Africa is shown in Fig. 2.2 (a). Africa mines a modest amount of lead, processes it, and exports almost half in either concentrate or refined form. The input from scrap to processing is only about 30%. We calculate that stock in Use is actually decreasing, or a consequence of the removal from use without replacement of lead pipe and sheet. The relatively low technological level of the processing is such that lead losses in tailings and slag easily exceed those from later life stages.

In contrast to the lead cycle for Africa, that for Asia[Fig. 2.2 (b)] demonstrates that it is a major extractor, producer, and manufacturer. Asia's development is recent enough that it imports roughly the same amount lead for Production as it gets from recycling. Addition to stock in Use is a net 17% of input. There is a net export of semi- or finished lead products. Losses are high at almost all life stages but that at WMR are the highest.

The CIS lead cycle[Fig. 2.2 (c)] is largely self-contained, and most flows are small. Stock in Use is decreasing due to losses of lead without replacement for domestic uses (ILZSG, 1992, 2005, 2007). Imports outweigh exports, except at Production. Losses at WMR stage are the highest.

Europe[Fig. 2.2 (d)] has quite modest lead extraction. It is especially strong in secondary production (i.e., production from scrap), and it is a net importer of lead in various lead-based products. Addition to stock in Use is 2%. Europe's lead flows to F&M and Use are relatively large. Its losses are predominantly from both WMR and Use life stage.

Latin America and the Caribbean[Fig. 2.2 (e)] have relatively modest flows throughout the cycle. It is a net exporter of concentrate and refined lead and a net importer of lead in finished products and scrap. Addition to stock in Use is about 25% of input flow, caused by the sharp increase of lead uses there in the last 2 or 3 decades (ILZSG, 1992, 2005, 2007).

The Middle East[Fig. 2.2（f）] is a small extractor and producer of lead. Recycling provides about half of the input to Production. Loss at the Use stage is the same as that at WMR stage, and is high relative to flows. Addition to in-use stock is about 19% of the input flow to Use.

The lead cycle for North America[Fig. 2.2（g）] seems the most balanced within its boundaries of any region, but the region nonetheless is a large importer of refined lead and lead in products. It is a major extractor, producer, manufacturer and user. About 14% of the flow into Use phases added as stock.

Oceania, the final region[Fig. 2.2(h)] is a large lead extractor and a major exporter. The flows within the region are small, and recycling is efficient.

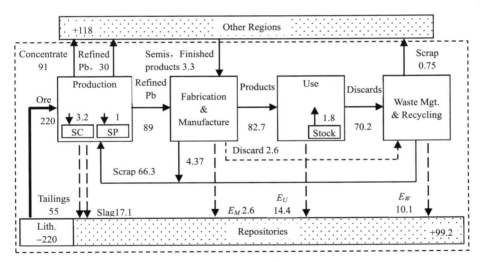

Lead, 2000　　　　　　　　　　　　　　　System Boundary "Africa（AF）"
（a）Africa　　　　　　　　　　　　　　　©STAF Project, Yale University, 2006

| 铅元素人为流动 |

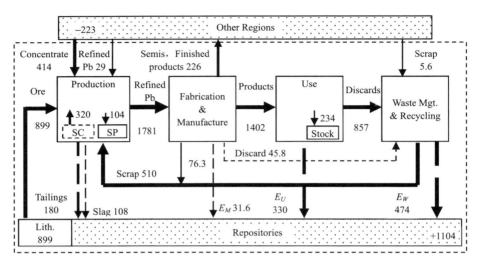

Lead,2000
(b) Asia

System Boundary "Asia (AS)"
©*STAF Project,Yale University,2006*

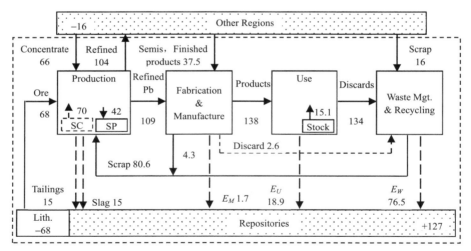

Lead,2000
(c) Commonwealth of Independent States

System Boundary "Commonwealth of Independent States (CIS)"
©*STAF Project,Yale University,2006*

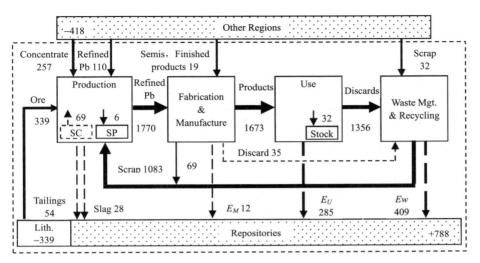

Lead, 2000
(d) Europe

System Boundary "Europe (EU)"
©STAF Project, Yale University, 2006

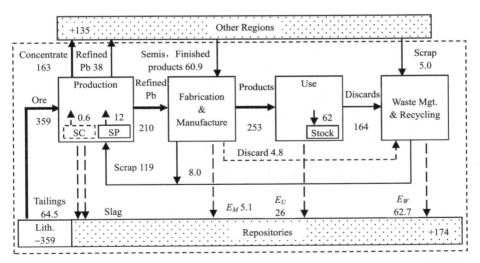

Lead, 2000
(e) Latin America and the Caribbean

System Boundary "Latin America and the Caribbean (LAC)"
©STAF Project, Yale University, 2006

| 铅元素人为流动 |

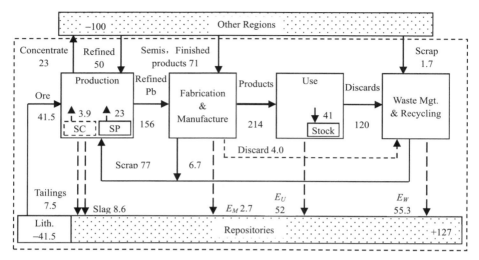

Lead, 2000
(f) Middle East

System Boundary "Middle East (ME)"
©STAF Project, Yale University, 2006

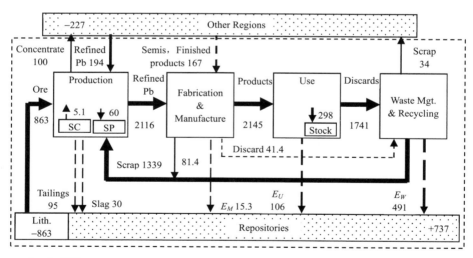

Lead, 2000
(g) North America,

System Boundary "North America (NA)"
©STAF Project, Yale University, 2006

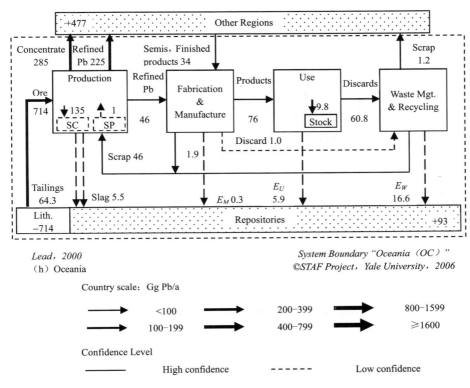

Fig. 2.2　Lead cycle in 2000 for the world's regions

2.3.3　A Detailed Regional-Level Lead Cycle

Because our cycles utilize information on the five major groups of lead products (batteries, sheep/pipe, cable sheathing, alloys, and "other"), it is possible to generate more detailed cycles than those shown in Fig. 2.1 and Fig. 2.2. These expansive cycles can provide additional insight on product flows, lead losses, and recycling potential. We illustrate such a cycle in Fig. 2.3, for North America.

Several features of this expanded lead cycle are worth noting, in contrast to the compact cycles shown previously. The most obvious is the overwhelming dominance of batteries: some 84.6% of all the lead employed in North America. As a result, lead losses and lead recycling are similarly dominated by the battery sector. The diagram also is explicit about imports and exports of lead at the several different life stages. As seen, these are not one-way flows, but markets in which refined lead and lead in products flows both ways across the regional boundary. In the case of semi-finished and finished products, import and export flows are shown for each of the five product groups.

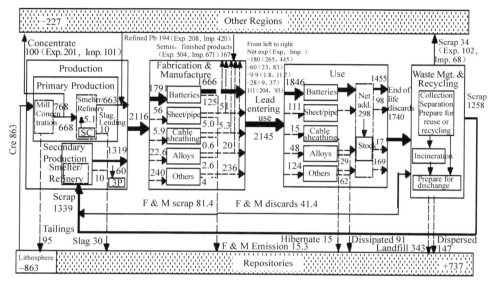

Fig. 2.3 The detailed lead cycle for North America in 2000. The units are Gg Pb/year

2.3.4 The Global Lead Cycle

The global-level lead cycle is constructed as the aggregate of the lead cycles for the eight regions, with minor adjustments to compensate for lead concentrate, production, Use, and trade not captured in the country set that was utilized (Paper I). The resulting global cycle is displayed in Fig. 2.4a.

The major features of the global cycle can be briefly summarized. About 6300 Gg Pb enter Production. Of this amount, about 57% is from secondary sources. About 11% of the flow into Use is retained as additional in-use stock. An amount of lead 19% less than the amount extracted from the lithosphere is returned to the environmental repositories in the form of various waste flows.

A true global lead cycle has, of course, no import or export flows. In our simple addition of the data from lower levels, various inaccuracies introduce import/export discrepancies of the order of 1%–2%. In a speculative vein, we can construct a completely closed global cycle if we make a few reasonable assumptions, following the approach of Graedel et al. (2004). Beginning with the schematic cycle diagram of Fig. 2.4, we first constrain the lead flows at the four life-cycle stages to be closed by balancing the outgoing flows with the incoming flows. This gives:

$$O + S_T = P + T_S \qquad (2.1)$$

$$P = U + S_M + D_M + E_M \qquad (2.2)$$

$$U = D_U + S_U + E_U \qquad (2.3)$$

$$D_U + D_M = S_T - S_M + E_W \qquad (2.4)$$

where the variables are defined in Fig. 1.3 of Paper I.

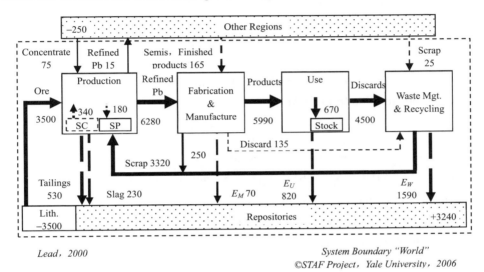

Fig. 2.4　The global anthropogenic lead cycle for 2000 produced by aggregating the regional cycles, the units are Gg Pb/year

We then impose several relationships that are supported by data or reasonable inference.

From Smith (2000) we know that about 4.16% of the scrap input to production is from F&M scrap. Thus,

$$\frac{S_M}{S_T} = 0.0416 \qquad (2.5)$$

We assume that the numeric values for O, P and T_S are correct, i.e., $O=3500$, $P=6280$ and $T_S=540+230=770$, as those data are relatively well characterized. Thus there remain 9 unknown values: S_T, U, S_M, D_M, E_M, D_U, S_U, E_U and E_W, requiring four additional independent equations. We adopt the following ratios, representing the median values from the best-characterized country-level cycles:

Utilization efficiency: Manufacturing output minus F & M discards and losses:

$$\psi = 1 - \frac{E_M + D_M}{U} = 0.966 \qquad (2.6)$$

Accumulation ratio: Addition of in-use stock as a fraction of lead entering Use phase:

$$\alpha = \frac{S_U}{U} = 0.112 \qquad (2.7)$$

End of life recycling ratio: Scrap from WMR back to production as a fraction of Waste Management & Recycling input:

$$\zeta = \frac{S_T - S_M}{D_U + D_M} = 0.662 \qquad (2.8)$$

Discard-management ratio: F&M discards entering WMR as a fraction of the sum of discards and emission from F&M:

$$\mu = \frac{D_M}{D_M + E_M} = 0.659 \qquad (2.9)$$

The above 9 equations give: $S_T = 3540$, $U = 5930$, $S_M = 147$, $D_M = 134$, $E_M = 69$, $D_U = 4989$, $S_U = 663$, $E_U = 278$ and $E_W = 1730$. We allocate the lead emissions in tailings and slag in the same proportion as in Fig. 2.4, where the lead losses in tailings and slag are 530 and 230 kt Pb/ year, respectively. The resulting "best estimate" global cycle is given in Fig. 2.5.

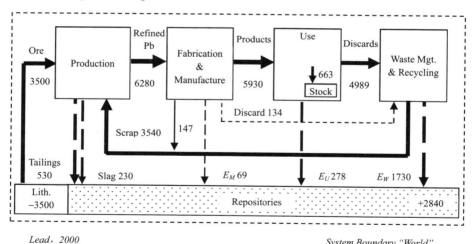

Lead, 2000

System Boundary "World"
©STAF Project, Yale University, 2006

Fig. 2.5 The "global best estimate anthropogenic lead cycle" for 2000. Closure in not precisely achieved for each reservoir due to rounding. In both diagrams the units are Gg Pb/year

The closed cycle analysis confirms the Fig. 2.4 results, showing that virgin lead made up 50% of the total lead production. Most lead passes quickly through the in-use life stage, only a net 11% being accumulated as in-use stock. The end of life recycling ratio for the world is around 66%. Lead escapes from the cycle because of discards with low lead content, high dispersion, or with complex composition, such as the lead in glass, ammunition, gasoline, PVC, CRTs, and the like. The major lead losses occur at the Production and WMR stages (27% and 61% of the total, respectively).

The largest difference between Fig. 2.4 and Fig. 2.5 is that E_U, the emissions of lead to the environment from the Use stage, has been significantly decreased. This primarily due to our assumed recycling and discard management ratios, which we feel are reasonable choices. The result is that much of the flow characterized as emissions in Fig. 2.4 has been redistributed into enhanced discard, recycling, and landfill flows, which we regard as somewhat more realistic. Additional field studies of lead losses would help confirm this result.

From the above facts on multilevel lead cycles, we found the lead cycle can be generally characterized three ratios: accumulation ratio, end of life recycling ratio, and secondary supply ratio (the scrap supplied to production as a fraction of refined lead input to F&M). The former two ratios have been defined and can be estimated according to equation (2.7) and equation (2.8), respectively. The third is given by:

$$\theta = \frac{S_T}{P} \quad (2.10)$$

These three ratios reflect the relationships between lead flow and stock, lead discard and recycling, and lead recycling and lead production, respectively.

2.4　Comparing Lead Flows in Countries and Regions

In the electronic supplement to this chapter, we give the cycle diagrams for each of the 52 countries in our data set except for the 6 selected countries shown here. In Fig. 2.6 we show that, of the six countries we have studied in detail, the United States leads in ore extraction, flow of lead into Use, and lead emissions. France, the United Kingdom, Italy, and Japan extract no lead ore, and their flows into use are similar to each other. India performs a small amount of lead ore extraction. Its flow into use is less than that of any of the other five countries, but its emissions level is nearly the same that of Italy, France, and Japan, largely because of a very low secondary supply ratio.

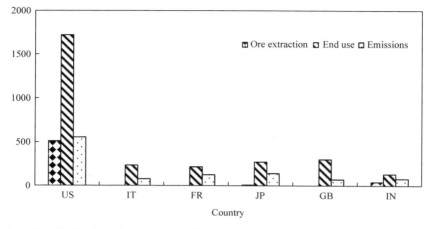

Fig.2.6 The distribution of annual rates of lead ore extraction, end use, and emissions to the environment to selected countries in 2000. The units are Gg Pb/year

The lead cycles of countries can also be compared by examining the values of the three characteristic ratios: the accumulation ratio, the end of life recycling ratio, and the secondary supply ratio. For countries and regions, for which import and export must be included, equation (2.8) is reversed to be:

$$\zeta' = \frac{S_T - S_M + N_S}{D_U + D_M} \quad (2.11)$$

Here we treat the net export of lead scrap as WMR scrap back to production, even though the scrap is not used domestically.

We list the characteristic ratios in Table 2.2 for our six selected countries; values for all 52 countries are included in the Supplementary Material. Of the six countries, the United States and Japan have the highest domestic scrap supply with a secondary supply ratio of 0.63. India has the highest accumulation ratio of 0.45, but the lowest end of life recycling ratio and secondary supply ratio, with values of 0.37 and 0.22, respectively. The end of life recycling ratio and secondary supply ratio for Italy, France, Japan, and United Kingdom are nearly at the same level, with values of 0.62–0.75, and 0.52–0.63, respectively. The two lowest accumulation ratios occur for Japan and France, with negative values of 0.07 and 0.02, reflecting trends of reduction in lead uses for these countries.

Table 2.2 Characteristic ratios for lead cycles in 2000

	Region							
	AF	AS	CIS	EU	LAC	ME	NA	OC
Accumulation ratio	−0.02	0.17	−0.11	0.02	0.25	0.19	0.14	0.13
End of life recycling ratio	0.86	0.47	0.44	0.71	0.63	0.55	0.72	0.73
Secondary supply ratio	0.74	0.29	0.74	0.61	0.57	0.49	0.63	1.00
	Country and world							
	US	IT	FR	GB	JP	IN	World*	
Accumulation ratio	0.10	0.01	−0.02	0.02	−0.07	0.45	0.11	
End of life recycling ratio	0.74	0.78	0.51	0.67	0.72	0.19	0.66	
Secondary supply ratio	0.63	0.60	0.52	0.57	0.63	0.22	0.57	

Note: the values for the world refer to the "best estimate" lead cycle. Others are calculated from equation (2.11).

In Fig. 2.7 we compare for the eight regions the three lead flows that were identified for countries in Fig. 2.6. The flows vary greatly among the regions: lead ore is mainly extracted in Asia, North America and Oceania, lead flows into Use chiefly in North America, Europe, and Asia, and lead emissions are largest in Asia, followed by North America and Europe.

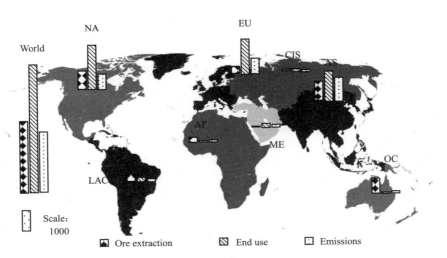

Fig. 2.7 The distribution of annual rates of lead ore extraction, end use, and emissions to the environment in the world's regions in 2000

The numerical data are given in the Fig. 2.2 diagrams. The units are Gg Pb/year.

2.5 Discussion

Once the lead cycles are established, it is of interest to attempt to relate the flows to potential explanatory variables. We begin the attempt with Fig. 2.8(a), which explores the relationship between country-level lead use and gross domestic product. Because the United States' use of lead is three times that of the next largest country, we omit it from the plot in order to better explore the overall correlation. Of the remaining countries, China's rate of use is by far the highest, with Germany and the United Kingdom following. No obvious correlation emerges.

An alternative approach is to plot lead use as a function of per capita GDP, which is done in Fig. 2.8 (b). Again no obvious relationship is present, but it is interesting that for the same level of per capita income the lead use can differ by a factor of ten. Because vehicle batteries are the major lead reservoir, the per capita number and size of vehicles appear likely to be the major cause of this discrepancy.

Our final exploration of cause and effect is to plot country level per capita lead use as a function of the human development index (HDI), a composite index that measures a country's status in three basic aspects of human development: longevity, knowledge, and standard of living (HDRO, 2003). The result is shown in Fig. 2.8 (c). Just as with per capita GDP, lead use can vary by a factor of about ten for identical values of HDI. Thus, high levels of lead use do not appear to be necessary in order to achieve a high standard of living.

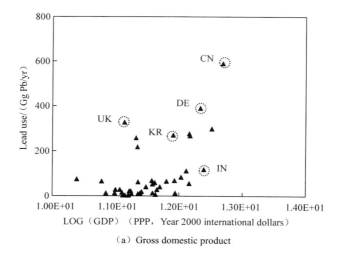

(a) Gross domestic product

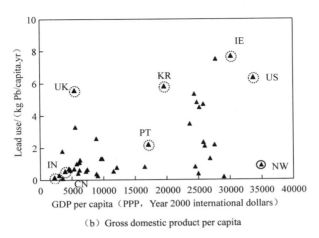

(b) Gross domestic product per capita

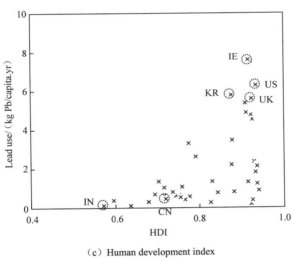

(c) Human development index

Fig. 2.8 Lead use as a function of potential correlative parameters in 2000

As is the case with all other material flow analysis, the existence of data limitations and deficiencies means that our cycles of lead will neither be completely correct nor entirely complete. Nevertheless, the results in this chapter present a consistent and useful picture of the multilevel lead cycle as of 2000. Lead's toxicity distinguishes its cycle from those of most of the other metals, and renders the emissions of particular interest. We will discuss the lead emissions in detail in future papers.

Acknowledgements

This work is part of the Stocks and Flows Project at Yale University's Center for Industrial Ecology. The project is supported by the US National Science Foundation under Grant BES-0329470. The participation of J.S. Mao. was made possible by the Henry Luce Foundation of the USA and the "National Basic Research（973）Program" Project（No. 2005CB724204）of the Ministry of Science and Technology of China.

References

Brunner P H, Daxbeck H, Baccini P. 1994. Industrial metabolism at the regional and local level: A case study on a Swiss region. In Industrial Metabolism: Restructuring for Sustainable Development, R.U. Ayres and U.E. Simonis, Eds. Tokyo: United Nations University Press, 163-193.

Bergbäck B, Johansson K, Mohlander U. 2001. Urban metal flows – A case study of Stockholm. Water, Air, & Soil Pollution Focus, 1（3）: 3-24.

Elshkaki A, van der Voet E, Van Holderbeke M, et al. 2004a. Development of a dynamic model for substance flow analysis: Part 1 – General stock model. Delft: Delft University. http: //www.emis.vito.be/index.cfm?PageID=393.（accessed Jan. 13, 2006）.

Elshkaki A, van der Voet E, Van Holderbeke M, et al. 2004b. Development of a dynamic model for substance flow analysis: Part 2 – Integration of stock and flow model. Delft: Delft University. http: //www.emis.vito.be/index.cfm?PageID=393, accessed Jan. 13, 2006.

Graedel T E, van Beers D, Bertram M, et al. 2004. Multilevel cycle of anthropogenic copper. Environmental Science & Technology, 38（4）: 1242-1252.

Graedel T E, van Beers D, Bertram M, et al. 2005. The multilevel cycle of anthropogenic zinc. Journal of Industrial Ecology, 9（3）: 67-90.

HDRO（Human Development Report Office）. 2003. Human Development Reports. http: //hdr.undp.org/statistics/data/,（accessed Oct. 20, 2005）.

ILZSG（International Lead and Zinc Study Group）. 1992. Principal Uses of Lead and Zinc 1960-1990. London.

ILZSG（International Lead and Zinc Study Group）. 2005. Lead and Zinc: End Use Industry Statistical Supplement, 1994-2005. London.

ILZSG（International Lead and Zinc Study Group）. Lead and Zinc in the World. Interactive Statistical Database. http: //www.ilzsg.org/（accessed November 16, 2007）.

Johnson J, Jirikowic J, Bertram M, et al. 2005. Contemporary anthropogenic silver cycle: A multilevel analysis. Environmental Science & Technology, 39（12）: 4655-4665.

Johnson J, Schewel L, Graedel T E. 2006. Contemporary anthropogenic chromium cycle: A multilevel analysis.

Environmental Science & Technology,40:7060-7069.

Karlsson S. 2008. Closing the technospheric flows of toxic metals–Modeling lead losses from a lead acid battery system for Sweden. Journal of Industrial Ecology,3(1):23-40.

Mao J S,Lu Z W,Yang Z F. 2006a. The eco-efficiency of lead in China's lead-acid battery system. Journal of Industrial Ecology,10(1-2):185-197.

Mao J S,Lu Z W,Yang Z F. 2006b. Lead flow analysis for lead acid battery system(In Chinese). Environmental Science,27(3):442-447.

Mao J S,Dong J,Graedel T E. 2008a. The multilevel cycle of anthropogenic lead:I. Methodology. Resources Conservation & Recycling,52(52):1058-1064.

NIMS-EMC(National Institute of Material Science,Ecomaterial Center). MDE Report No. 2:Fundamental Survey for Lead Material Flow in Japan, http://www.nims.go.jp/ecinateruak/emc_eng/mde-report-en.html.(accessed Oct. 11,2005).

Obernosterer R,Brunner P H. 2001. Urban metal management:The example of lead. Water,Air,& Soil Pollution:Focus,1(3):241-253.

Reck B,Müller D B,Rostkowski K,et al. 2008. The anthropogenic nickel cycle:insights into use,trade,and recycling. Environmental Science & Technology,42(9):3394-3400.

Smith G R. Lead:2000. http://minerals.usgs.gov/minerals/pubs/commodity/lead/380400.pdf(accessed October 10,2005).

Smith G R. Lead recycling in the United States in 1998. In:Sibley,Scott F.,editor. Flow studies for recycling metal commodities in the United States. Version 2.0 [revised 17-July-2006] Washington:U.S. Geological Survey;2004. http://pubs.usgs.gov/circ/2004/1196am/. 73-85,accessed December 12,2007.

Socolow R,Thomas V. 1997. The industrial ecology of lead and electric vehicles. Journal of Industrial Ecology,1(1):13-36.

Thornton I,Rautiu R,Brush S. 2001. Lead the facts. London,UK:IC Consultants Ltd.

Wang T,Müller D B,Graedel T E. 2007. Forging the anthropogenic iron cycle. Environmental Science & Technology,41(14):5120-9.

第 3 章 铅元素使用蓄积动力分析
Lead In-Use Stock: A Dynamic Analysis*

3.1 Introduction

Lead in-use stock refers to the lead currently providing services of various kinds to humanity. It results when more material enters use over time than leaves use. The amount of lead in-use stock for a region or country reflects that region's standard of living under currently available technology; therefore, lead in-use stock is an indicator of the amount of metal that less-developed regions may need to put in place to attain comparable material services. Additionally, knowledge of the magnitude of lead in-use stock is useful in estimations of the amount of lead that will be available for future recycling.

The growing concern regarding the toxicity of lead has resulted in a series of lead replacement activities in recent years: the replacement of lead pipes in water supply infrastructure (DCWASA 2007), the use of lead-free electrical components in consumer applications (AKTINA 2008), and the development of lead-free solders (Nakamura et al. 2008), for example. The results of these activities are likely to include a decline in lead demand and, over time, a decrease in lead in use stocks. Thus, a study of lead in-use stock has the potential to guide resource conservation, environmental management, and industrial and governmental policy (e.g., van der Voet et al. 2002). In a demonstration of this potential utility, Gordon and colleagues (2006) have linked metal stocks with sustainability and discussed the in-use stocks of copper, zinc, platinum, silver, and nickel. The result showed that providing today's developed-country level of services for copper worldwide (as well as for zinc and perhaps platinum) appears to require conversion of essentially all the ore in the lithosphere to stock in use, plus near-complete recycling of the metals from that point forward.

During the 20th century, the global production of refined lead increased nearly

---

* Mao J S*, Graedel T E. 2009. Lead In-Use Stock: A Dynamic Analysis. Journal of Industrial Ecology, 13 (1): 112-126.

eight-fold (Fig. 3.1), and some of that production is currently retained as in-use stock. The amount, however, has remained uncertain. Several studies have been made for different cities and countries, and the per capita stock estimates that have resulted range between 25-210 kgPb/capita (Table 3.1), a substantial but poorly-quantified fraction being contained in cable sheathing. Because these studies were all done in highly developed regions, it seems quite unlikely that the results reflect accurate pictures of the standing stocks of lead. In the current chapter we take a broader view, characterizing the global in-use stock of lead and then disaggregating that amount to the world's regions and to selected countries.

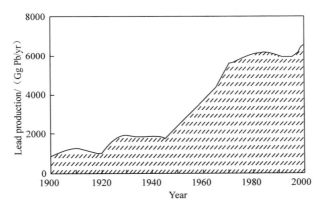

Fig. 3.1 Global lead production in the 20th century
The data are from DiFrancesco and Smith (2003)

Table 3.1 Per capita in-use lead stocks in various regions

Region	Year	In-use Stock[a]/[kg Pb/capita]	TD/BU[b]	Reference
Japan	2004	25	TD	Murakami, 2006
Stockholm	1995	73 (−30)	BU	Sörme et al., 2001
Netherlands	1998	115 (−90)	BU	Elshkaki et al., 2004c
USA	2000	146 (−50)	TD	Sullivan, 2003
Vienna	1991	120-210 (50-80)	BU	Obernosterer and Brunner, 2001

a Values in parentheses indicated estimated stocks without cable sheathing.
b TD = top-down analysis; BU = bottom-up analysis (see text for explanation).

In related work, we have generated a methodology for analyzing the life cycle of lead (Mao et al., 2008a; hereafter "Paper Ⅰ"), characterized the lead cycles for 52 countries, eight regions, and the planet (Mao et al., 2008b; hereafter "Paper Ⅱ"), and discussed the losses to the environment from the regional and global lead cycles

(Mao et al., hereafter "Paper Ⅲ"). Using this information as a basis, we present in this chapter an estimation of the annual lead addition to in-use stock throughout all of the last century, leading to a characterization of the in-use stock as of the year 2000. We then attempt to relate the levels of in-use stock to the development level for countries and regions. Finally, we compare the results with those developed previously, and comment on the accuracy and utility of stock determinations by ourselves and others.

3.2 Methodology

3.2.1 "Top-Down" Computation of In-Use Stocks

One can determine in-use "standing" stocks in either of two ways: by measuring input and output flows over time and computing the difference (this is the top-down approach), or by measuring the contents of metal in-use reservoirs directly (the bottom-up approach). The latter method utilizes inventories of the different service units that contain metal, such as buildings, factories, and vehicles. The content of metal per service unit is obtained from engineering data and combined with information on the number of units in a given geographic area for a determination of the metal stock in use. This method is very labor-intensive and is impractical where the supporting information is not available. In the present instance, for which we wish to quantify lead stocks on national, regional, and global levels, it is thus necessary to utilize the top-down approach. We describe below the methodology by which this assessment is performed.

The framework for our analysis is the anthropogenic cycle of lead shown in Fig. 3.2. Lead mined from the lithosphere is transformed into metal, fashioned into products, used for varying periods of time, and eventually discarded (perhaps to be reused, perhaps not). At all life stages there are potential emissions to the environment, a particular concern in the case of lead because of lead's toxicity (Paper Ⅲ). In the present work we focus on the degree to which lead inputs to the use stage exceed outflows; these net additions over time create the in-use stock of lead.

S_{U_τ}, the additions to lead in-use stock in year τ, are related to the balance of the flows of lead at the use stage:

$$S_{U_\tau} = U_\tau - D_{U_\tau} - E_{U_\tau} \qquad (3.1)$$

where U is the lead entering use, D_U the lead in end-of-life discards, and E_U the lead in emissions to the environment during use of products (mainly through corrosion, wear, or dissipative uses).

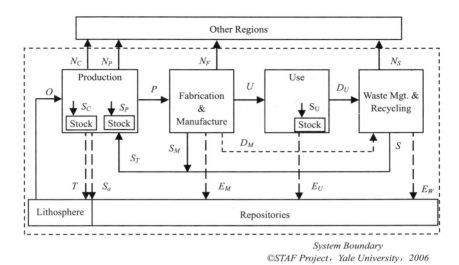

Fig. 3.2 A simplified schematic diagram of the comprehensive anthropogenic lead cycle with successive life stages plotted from left to right

Symbols and definition

O	Lead from ore	S_a	Lead in slag
P	Refined lead	S_P	Lead entering stock of refined lead
U	Lead entering use	S_C	Lead entering concentrate stock
D_U	End-of-life lead to waste management & recycling	S_U	Lead entering in-use stock
D_M	Lead in F&M discards to WMR	S	Lead in scrap from WMR
N	Lead in net export (export minus import)	S_T	Lead in total scrap to production
N_P	Lead in net export of refined lead	S_M	Lead in new scrap from Manufacture
N_C	Lead in net export of concentrate or mine	E	Lead in emissions to the environment
N_S	Lead in net export of scrap	E_M	Lead in F&M emissions
N_F	Lead in net export of semi- or finished products	E_U	Lead in Use emissions
T	Lead in tailings	E_W	Lead in WMR emissions

Lead entering into use is discarded or emitted at the end of its service lifetime $\Delta\tau$ (see details in section 2.3):

$$D_{U_\tau} + E_{U_\tau} = U_{\tau-\Delta\tau} \quad (3.2)$$

Substituting equation (3.2) into equation (3.1) gives

$$S_{U_\tau} = U_\tau - U_{\tau-\Delta\tau} \quad (3.3)$$

Thus, the annual addition to in-use stock is the difference between the lead entering use in the present year and that entering use one life span ago (A similar approach to this type of problem was taken by Kleijn et al., 2000). Over a chosen interval of time, the lead added to in-use stock is given by integrating the annual net additions during the period:

$$\sigma \equiv \int_0^T S_U \mathrm{d}t \equiv \sum_{\tau=0}^T S_{U_\tau} \quad (3.4)$$

where T is the chosen time interval.

3.2.2 Considerations for Specific Lead Product Groups

Lead products can be divided into five groups (Paper Ⅰ), as shown in Fig. 3.3. The "alloys" group includes Pb-Sb-Sn, Pb-Sb-Cu, and others with specific stable uses, but not solders. "Others" excludes dissipative uses (lead tetraethyl for petrol additives is typical example), but includes such lead uses as stabilizers in PVC and in glass crystal. The product groups have quite different in-use lifetimes – short for batteries and long for pipe, for example. This requires that the year-to-year flows into the Use phase be apportioned among the five groups, and that

$$U = \sum_{i=1}^N U_i \quad (3.5)$$

where i is the product group index.

Equation (3.3) and equation (3.5) can be combined to give

$$S_{U_\tau} \equiv \sum_{i=1}^N S_{U_{\tau_i}} = \sum_{i=1}^N \left(U_\tau - U_{\tau-\Delta\tau_i} \right)_i \quad (3.6)$$

The total in-use stock generated over the chosen interval of time is then calculated as

$$\sigma \equiv \sum_{\tau=0}^T \sum_{i=1}^N \left(U_\tau - U_{\tau-\Delta\tau_i} \right)_i \quad (3.7)$$

where N is the number of product groups under consideration.

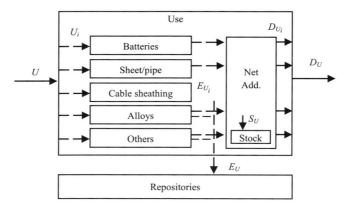

Fig. 3.3 A model of lead flows entering and leaving in-use stock

3.2.3 Types of Stocks and Losses

The actual in-use stock calculation is more complex than suggested in the previous section because of differences in the life span and eventual fate for specific products. It is useful to discuss three types of products: those involving lead in stable use, those involving lead in hibernating use, and those involving lead in dissipative use.

We define lead used in products with predictable life spans and reasonable potential for recycling as the lead in "stable" use. Lead-acid batteries and sheet & pipe are typical examples. For those products, we can calculate the annual lead entering in-use stock with equation (3.3), and their addition to in-use stock for fixed periods with equation (3.4).

Lead in hibernating use refers to lead used in products with a predictable life span but with low commercial value at end of life (Elshkaki et al., 2004a, b). Lead in cable sheathing is the principal example. When obsolete, this lead is almost universally left un-recycled where it was originally installed (e.g., under streets or on the ocean floor). We thus treat the lead in cable sheathing in a special way in our study: not as an addition to in-use stock, but as a distinct and immediate form of lead transfer to the environment, into which it will be dissipated over long time periods (Paper Ⅲ).

The *lead in dissipative uses* is defined as the lead used in products with short lifetimes and low recycling potential. The lead in ammunition, in gasoline additives, and in pigments for paints are typical examples. Because the lead in these products will be emitted to the environment when the products are used, no lead is added to in-use stock for these lead uses.

It follows from these considerations that evaluations of the lead in-use stock need only consider the lead in stable uses, including batteries, sheet & pipe, non-solder alloys, and non-dissipative "other" uses.

3.3 Lead In-Use Stock in the 20th Century

We focus on year 2000 for the estimation of lead in-use stock, and the last century as an example of a period of accumulation of lead into in-use stock.

3.3.1 Additions to Stock in Year 2000

As we concluded above, additions to in-use stock can only occur with lead in some form of stable use. In paper Ⅱ, we estimated the annual lead addition to in-use stock in 2000 according to equation 3. We analyzed both the lead entering use in 2000 and the lead existing one life-span previously, based on data from ILZSG(1992, 2005a, 2005b, and 2007). For example, to estimate the annual lead addition to the in-use stock of batteries, we utilize 4 and 12 years for the average life-span of starting-lighting-ignition (SLI) and stationary batteries(Table 3.2). The lead leaving the in-use stock of batteries is then the lead entering the Use stage in SLI batteries in 1996, and in stationary batteries in 1988.

Table 3.2 Data used in estimation of end-of-life discards

Lead product	Life span, years[a]	Potential recycling percentage
SLI battery	4	100
Stationary battery	12	100
Pipe/sheet	30	100
Alloys	12	40 (excludes solder)
Others	12	50 (excludes dissipative uses)

a) Adapted from work by ILZSG (2005b).

The global lead uses and use pattern for the period 1960–2000 are shown in Fig. 3.4. It is apparent that the lead used in batteries has been increasing, while the lead used in sheet & pipe, cable sheathing, alloy, and other products has been nearly constant or slightly decreasing in the past two decades. This indicates that the lead addition to in-use stock in 2000 is strongly related to increased battery use.

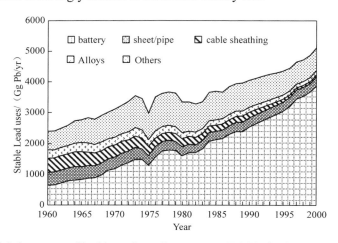

Fig. 3.4 The global spectrum of lead input flows. Data source: ILZSG (1992, 1999, 2003, 2005)

3.3.2 Global In-Use Stock, 1900–2000

1. Historical Lead Use Data

The principal difference between the annual addition of lead to in-use stock, which can be calculated from contemporary data, and the total in-use stock, is that the latter must estimate additions to and subtractions from in-use stock for all years during which any of the current in-use stock was deployed. Given that lead in roofing or vertical cladding may be in place for a hundred years, the contemporary lead in-use stock computation requires integrating the annual addition to in-use stock year by year for essentially the whole 20th century. Because the lead in-use stock is related to increases or decreases of lead used in various applications, we must address historic lead use application patterns.

World lead use data were mainly taken from statistics by ILZSG (2007) for 1960–2000, and the lead use patterns estimated according to the data by ILZSG (1992) for 1960–1990 and ILZSG (1999, 2003) for 1995–2002 respectively. World production (which is used as a surrogate for lead use) is estimated based on the reported data on world smelter production for the years 1900–1954 and on world mine production for the years 1955–1998 (DiFrancesco and Smith, 2003), adjusted for the percentage of secondary lead in the total (Smith, 2000).

2. Lead in Product Groups

We concluded above that the lead addition to in-use stock is only related to "stable" lead uses. The lead in-use stock of specific categories of products can be computed according to equations (3.3) and (3.4). To do that, we need to estimate the lead entering each use group year by year throughout the whole period, as well as the life span of each group of lead products. We assume that the lead entering use each year can be expressed as equation (3.8) for a particular lead product group:

$$U_b = P \cdot f_i \cdot f_{mi} \tag{3.8}$$

in which P is the rate of flow of refined lead into fabrication and manufacturing and f_i and f_{mi} represent the product group fraction and the fabrication efficiency (i.e., the fraction of lead entering manufacturing that leaves as part of finished products) for the product group.

The annual lead addition to in-use stock in different lead product groups takes the following information into account:

The *lead in-use stock of batteries*. Batteries have been the greatest consumer of lead since 1960 (28% in that year, 74% in 1999; Thornton et al., 2001). About 80% of lead-acid batteries are SLI batteries with a 1–4 year life-span; the remainder are stationary batteries with an 8–15 year life-span. The percentages of lead used in batteries in early years were estimated (Table 3.3).

The *lead in-use stock of sheet & pipe*. Lead sheet and pipe have long histories of use, and lead in those uses has a long lifetime. ILZSG (2003) reports that the life spans of lead in sheet for roofing systems and vertical cladding systems are over 100 years.

The *lead in-use stock of alloy*. Lead alloy can be dissipative or stable. The lead used in solder is a typical dissipative application, unlikely to be recycled. On the contrary, lead alloys such as in Pb-Sb-Sn or Pb-Sb-Cu may be readily recycled upon discard. We estimate that 60% of lead alloy use belongs to stable alloys with a 12 year lifespan.

The *lead in-use stock of other uses*. Lead in other uses similarly can be further classified into dissipative and stable uses. The lead used as pigments in paints and gasoline additives constitutes typical dissipative uses in this category. The lead used as a stabilizer in PVC, as shielding in CRTs, and as a constituent of glass crystal, however, may be recycled after it is obsolete. We estimate that approximately 50% of lead in the other uses category belongs to such uses, with a 12-year life span.

The lead use patterns, life spans, and fabrication efficiencies used in our calculations are given in Table 3.3–Table 3.5, respectively. The life spans used increase over time to reflect technological progress.

Table 3.3 Lead use patterns used in determining in-use lead stock in the 20th century

Lead product	1900–1929	1930–1939	1940–1949	1950–1954	1955–1959	1960–1964	1965–1969	1970–1974	1975–1979	1980–1984	1985–1989	1990–1994	1995–1999
Batteries	0	0	0.05	0.1	0.2	0.29	0.33	0.41	0.47	0.52	0.60	0.66	0.73
Sheet/pipe	0.8	0.7	0.5	0.48	0.35	0.17	0.14	0.11	0.09	0.08	0.07	0.07	0.06
Cable sheathing	0	0	0.12	0.2	0.21	0.18	0.15	0.11	0.08	0.07	0.05	0.04	0.02
Alloys	0	0	0.13	0.17	0.14	0.11	0.10	0.08	0.07	0.05	0.04	0.04	0.03
Others	0.2	0.3	0.2	0.05	0.1	0.25	0.28	0.29	0.30	0.27	0.23	0.20	0.16
Total	1	1	1	1	1	1	1	1	1	1	1	1	1

Table 3.4 Life spans (years) used in determining in-use lead stock in the 20th century

Lead product	1901-1940	1941-1960	1961-1970	1971-1980	1981-1990	1991-2000
SLI battery		1	2	3	3	4
Stationary battery		8	9	10	11	12
Pipe	20	20	25	25	30	30
Sheet	20	20	25	30	40	50
Alloys		2	4	6	8	10
Others	1	3	5	7	9	12

Note: Essentially all lead use prior to 1941 was for sheet and pipe.

Table 3.5 Fabrication efficiencies used in determining in-use lead stock in the 20th century

Lead Product	1901-1940	1941-1960	1961-1970	1971-1980	1981-1990	1991-2000
Batteries		0.6	0.7	0.8	0.86	0.91
Sheet/pipe	0.6	0.7	0.75	0.82	0.86	0.89
Alloys		0.6	0.7	0.8	0.85	0.88
Others		0.7	0.78	0.89	0.93	0.96

Note: Essentially all lead use prior to 1941 was for sheet and pipe.

3. Global Lead Stock Results

Our result for the global accumulation of in-use lead stock during the last century is shown in Fig. 3.5. The in-use stock of lead in 2000 we estimate at 25.0 teragrams (Tg) Pb. Given the total population of the 52 countries in our analysis in 2000 (about 4.48 billion), the average per capita in-use lead stock was about 5.6 kg. The lead stock in batteries increased rapidly from 1950 onward, caused by the rapid increase in the number of vehicles. As a result, lead stock in batteries reached about 68% of the total by 2000 (40% in SLI batteries, 28% in stationary batteries). Pipe and sheet composed 20% of the total, alloy 4%, and other uses 8%. The share of lead in-use stock in sheet and pipe is expected to decrease further in the future, because lead in those uses tends not to be replaced. The share of stock in batteries might be expected to increase somewhat in the next decade or so, however, as the number of vehicles with lead batteries increases worldwide.

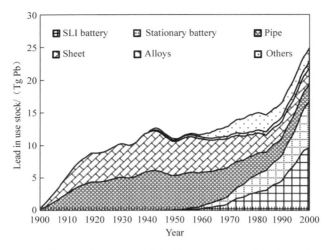

Fig. 3.5 The history of growth of global in-use lead stock in the 20th century

3.4 In-Use Lead Stocks in Regions and Countries

3.4.1 The Disaggregation Methodology

As we have discussed, historical statistics on lead flows in different countries and regions are much less widely available than that for the entire planet. As a consequence, it is not straightforward to determine in-use lead stocks at other than global level. It is nonetheless useful to develop approaches to disaggregating the global data so that the regions can be compared, at least approximately. We choose to do this by defining three different disaggregation methods, computing the results, and taking the numerical average of the three as the best estimate currently achievable.

Method 1. In this approach, the global in-use lead stock is distributed according to the percentage of total lead use in the region or country of interest relative to the world total for the period 1960–2000–that is,

$$Q_{r,c} = Q_w \left(P_{r,c} / P_w \right)_{1960-2000} \qquad (3.9)$$

where $Q_{r,c}$ is the in-use lead stock for the chosen region or country and Q_w is the global in-use lead stock. The value $P_{r,c}/P_w$ is the regional or country percentage of total global lead flows into F&M during 1960 to 2000. This is calculated based on statistics by ILZSG (2007).

The obvious advantage of method 1 (equation 3.9) is that we utilize the percentage of lead uses $(P_{r,c}/P_w)$ to disaggregate the total lead stock; these historical data are of good quality and available at the country level. The disadvantage of this method is that we must adjust lead use in a country by the trade of lead products, the lead use pattern, and fabrication efficiency to derive the amount of lead entering the use stage. This approach may therefore be most appropriate for countries with standard use patterns and relatively low lead trade.

Method 2. Compared to method 1, two main differences are employed in this approach. The first is that the percentages relate to total lead flows into the Use stage rather than F&M; the second is that data are for year 2000. This method is justified on the basis of year 2000 data being more accurate than that for 1960–2000. Thus,

$$Q_{r,c} = Q_w \left(U_{r,c}/U_w \right)_{2000} \quad (3.10)$$

The distinct advantage of method 2 (equation 3.10) is that the percentage of lead entering the use stage in a country or region $(U_{r,c}/U_w)$ is precisely the variable we wish to employ. This method, however, discounts the historical data and thereby historical changes in the percentage of lead entering use.

Method 3. This method is similar to method 1, but directly calculates the specific lead uses in different lead product groups, as shown in equation (3.11).

$$Q_{r,c} = Q_w \left(\sum P_{r,c}^i / \sum P_w^i \right)_{1960-2000} \quad (3.11)$$

where the subscript i refers to specific category lead product.

The data for specific lead uses are available for 28 of our 52 countries (ILZSG, 1992, 1999, 2003, 2005), which account for 57% of the total lead use during 1960–2000. The other 24 countries were estimated as proportions of their average values by methods 1 and 2.

This method focuses on lead uses in specific categories of lead products and is therefore more closely related to the in-use stock than to general lead uses. This method is, however, dependent on historical lead use patterns, which are only estimated in the present study due to data deficiencies.

3.4.2 The Results of Disaggregation

These approaches are clearly rough approximations, necessitated by the absence or scarcity of appropriate data. We defend them largely on the basis that lead use and additions to stock were heavily weighted toward the last few decades of the 20th century

(Fig. 3.5), that a substantial fraction of in-use stock was added within recent years, and thus the recent data are thus a not unreasonable surrogate for the entire century. The results of averaging the three determinations discussed above, and then multiplying by the total global in-use stock of 25.0 Tg Pb are given in Table 3.6.

Table 3.6 In-use lead stock for countries and regions in 2000 (Gg Pb, except kg Pb/capita)

Regions	World	AF	AS	CIS	EU	LAC	ME	NA	OC
Total	24 946	363	5 867	523	8 057	1 018	781	8 009	328
Per capita	5.6	2.0	2.1	3.2	19.7	3.5	5.0	19.3	14.3
Countries	DZ	EG	KE	MA	ZA	TN	CN	IN	ID	JP	MY	PH	SG
Total	65	45	11	46	180	16	1 843	326	162	1 487	455	159	81
Per capita	2.2	0.7	0.0	1.6	4.1	1.6	1.4	0.3	0.8	11.7	19.8	2.1	20.2
Countries	KR	TH	TW	KZ	RU	UR	AT	DK	FI	FR	DE	GR	IE
Total	443	284	628	28	444	51	198	120	50	1 079	1 502	130	150
Per capita	9.5	4.7	0.0	1.8	3.0	0.0	24.5	22.6	9.7	18.2	18.3	12.0	39.4
Countries	IT	NL	NW	PL	PT	ES	SE	GB	BL	SW	AR	BR	CL
Total	1 039	310	82	375	129	594	261	1 543	410	82	256	344	160
Per capita	18.1	19.5	18.4	9.7	12.9	0.0	6.4	26.3	0.0	9.2	6.9	2.0	10.5
Countries	CO	PE	VE	IR	SA	TR	IL	AE	CA	MX	US	AU	NZ
Total	59	79	120	278	71	290	94	48	534	597	6 878	295	32
Per capita	1.4	3.1	0.0	4.2	3.2	4.2	0.0	0.0	17.4	6.0	24.1	15.4	8.6

Note: The regional codes are AF = Africa; AS = Asia; CIS = the Commonwealth of Independent States; EU = Europe; LAC = Latin America and the Caribbean; ME = Middle East; NA = North America; OC = Oceania. Country codes are given in ISO (2007).

On a regional basis, Fig.3.6 shows that the lead in-use stock in 2000 is predominantly in Europe and North America, each of which contains around 1/3 of the total global value. The stock in Asia constitutes around 24% of the total. No other region contains as much as 5% of the total global in-use lead stock. The per capita stocks in each of the regions or countries are given in Table 3.4. These range from 2.0 kg Pb/capita (Africa) to 19.7 kg Pb/capita (Europe).

The country-level in-use lead stock estimates produced by our disaggregation of the global value are also given in Table 3.4. The United States stands out in its in-use stock, with more that one-fourth of the entire global total. In striking contrast, highly-developed countries such as the United Kingdom and Japan are each estimated to contain about 6% of the entire global stock. India, a developing country of significant current interest from a materials perspective, contains only about 1% of global in-use lead stocks.

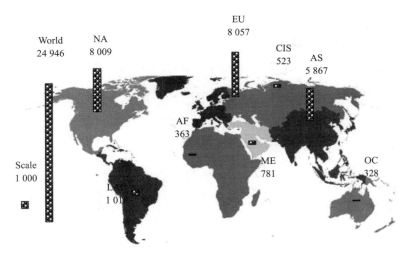

Fig. 3.6 The regional distribution of in-use stock in 2000

The units are Gg Pb. NA = North American; EU = Europe; CIS = Commonwealth of Independent States; LAC = Latin America and the Caribbean; AF = Africa; ME = Middle East; OC = Oceania.

3.5 Discussion

Having characterized the in-use stocks of lead, it is of interest to relate that stock to potential driving variables. We do so in Fig. 3.7 for per capita stock as a function of per capita income, There is a modest dependence of stock on income.

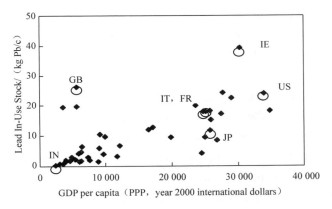

Fig. 3.7 The relationship between per capita in-use lead stock and the per capita gross domestic product for selected countries

A similar result is obtained when per capita stock is plotted as a function of the human development index (UN Development Programme, 2005). A modest relationship is again seen (Fig. 3.8). It appears that countries further along in the development cycle are indeed more likely to have higher standing stocks of lead (abandoned cable sheathing not being included).

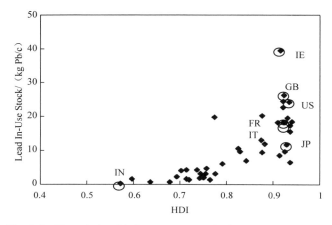

Fig. 3.8 The relationship between per capita in-use lead stock and the human development index for selected countries

It is also of interest to explore the potential long-term sustainability of the lead supply from the perspective of the standing stock, in view of the possible supply limitations identified for some metals (Gordon et al., 2006). According to our study, a total of around 110 Tg of primary lead was produced in the last century, which means that around 200 Tg of lead were extracted from natural repositories given estimated recovery rates of 75% and 70% in concentration and primary refining. There are no extant estimates of the lead resource, but lead and zinc commonly occur together in nature and the zinc resource has been estimated at 1800 Tg (Gordon et al., 2006). The lead resource can be roughly scaled from that of zinc by taking the ratio of their typical concentrations in Earth's crust (15 ppm for lead, 65 ppm for zinc; Mason, 1958), giving 415 Tg Pb. The US Geological Survey (USGS, 2006; Gabby, 2006) is much more expansive, suggesting a resource of about 1500 Tg Pb. For the planet's current population of about 6.5 billion to employ lead so as to have the in-use stock of a typical developed country citizen (about 20 kg Pb/capita) would require about 130 Tg Pb. It thus appears that no constraints on lead supply are likely from an absolute abundance standpoint.

We have discussed in this chapter a characterization of the lead in-use stock, with emphases on linking it to lead use patterns, recycling potential, life span, and the historic change in rates of lead entering use. A quantification of the in-use stock of a metal can never be highly accurate because of limitations in data and knowledge. Nonetheless, the present effort presents an order-of-magnitude estimate that can be improved upon in the future as more information becomes available.

Among the most interesting and significant results are the following:

Lead addition to in-use stock is essentially limited to stable lead uses. Lead releases from in-use stock involve various lead products, but primarily batteries.

In 2000, about 5930 Gg Pb entered the Use stage, of which nearly 660 Gg Pb was added to in-use stock, i.e., 11% of net lead entering stock remained there.

By 2000, the world's total in-use lead stock was about 25.0 Tg Pb, largely distributed throughout Europe, North America, and Asia.

At least 415 Tg Pb exist in ore resources. About 200 Tg Pb has been extracted in the last century. Because around 130 Tg Pb is sufficient for all of Earth's people to live like those of the developed world, the lead resource appears more than sufficient to satisfy demands in the future.

Acknowledgement

This work is part of the Stocks and Flows Project at Yale University's Center for Industrial Ecology. The participation of J.S. Mao was made possible by the Henry Luce Foundation of the USA and the National Basic Research (973) Program (Project 2005CB724204) of the Ministry of Science and Technology of China.

Notes

1. One kilogram (kg, SI) $\approx$ 2.204 pounds (lb).
2. Sb=antimony; Sn=tin; Cu=copper.
3. PVC is polyvinylchloride. CRTs are cathode-ray tubes.
4. One teragram (Tg) = 10^9 kilograms (kg, SI) = 10^6 tonnes (t) $\approx 1.102 \times 10^6$ short tons.
5. PPM denotes parts per million.
6. One gigagram (Gg) = 10^6 kilograms (kg, SI) = 10^3 tonnes (t) $\approx 1.102 \times 10^3$ short tons.

References

Aktina. 2008. Lead replacement for RoHS. www.aktina.co.uk/rohs.php. (Accessed 15 July 2008)

DCWASA (DC Water and Sewer Authority). 2007 third quarter lead service replacement program summary. www.dcwasa.com/news/publications/ProgramSummary.pdf. (Accessed 30 October 2007)

DiFrancesco, C. A. and G. R. Smith. 2000. Lead statistics (1900-2000). http://minerals.usgs.gov/minerals/pubs/of 01-006/lead.pdf. (Accessed 13 December 2003)

Elshkaki A, E Van Der Voet, M Van Holderbeke, et al. 2004a. Development of a dynamic model for substance flow analysis: Part 1. General stock model. Report No 2004/IMS/R/292. www.leidenuniv.nl/cml/ssp/. (Accessed 13 December, 2005)

Elshkaki A, E Van Der Voet, M Van Holderbeke, et al. 2004b. Development of a dynamic model for substance flow analysis: Part 2. Integration of stock and flow model. Report No 2004/IMS/R/293. www.leidenuniv.nl/cml/ssp/. (Accessed December 2005)

Elshkaki A, E Van Der Voet, M van Holderbeke, et al. 2004c. The environmental and economic consequences of the developments of lead stocks in the Dutch economic system. Resources Conservation and Recycling, 42 (2): 133-154.

Gabby, P. Mineral commodity summaries. Lead. 2006. http://minerals.usgs.gov/minerals/pubs/commodity/lead/lead_mcs06.pdf. (Accessed 7 January 2006)

Gordon R B, M Bertram, T E Graedel. 2006. Metal stocks and sustainability. Proceedings of the National Academy of Sciences of the United States of America, 103 (5): 1209-1214.

Hawkins T, C Hendrickson, C Higgins, et al. 2007. A mixed-unit input output model for environmental life-cycle assessment and material flow analysis. Environmental Science and Technology, 41 (3): 1024-1031.

ILZSG (International Lead and Zinc Study Group). 1992. Principal uses of lead and zinc 1960-1990. London: ILZSG.

ILZSG. 1999. Lead and Zinc Statistics: Monthly Bulletin of the International Lead and Zinc Study Group. 39 (11).

ILZSG. 2003. Lead and Zinc Statistics: Monthly Bulletin of the International Lead and Zinc Study Group 43 (11).

ILZSG. 2005a. Principal Uses of Lead and Zinc 2005. July, 21-31. London: ILZSG.

ILZSG. 2005b. Lead and zinc: End use industry statistical supplement, 1994-2005. London: ILZSG.

ILZSG 2007. Lead and zinc in the world. Interactive statistical database. www.ilzsg.org. (Accessed 16 November 2007)

ISO (International Organization for Standardization). Country names and codes, ISO 31661. www.iso.org/iso/english_country_names_and_code_elements. (Accessed 21 December 2007)

Kleijn R, R Huele, E Van Der Voet. 2000. Dynamic substance flow analysis: The delaying mechanism of stocks, with the case of PVC in Sweden. Ecological Economics, 32 (2): 241-254.

Mao J S, Z W Lu. 2003a. A study on causes of low recovery of scrap lead. World Nonferrous Metals 11: 24-32.

Mao J S, Z W Lu. 2003b. On the lead scrap resources for lead industry. World Nonferrous Metals 7: 10-14, 36.

Mao J S, J Dong, T E Graedel. 2008a. The multilevel cycle of anthropogenic lead: I. Methodology. Resources Conservation and Recycling, 52 (52): 1058-1064.

Mao J S, J Dong, T E Graedel. 2008b. The multilevel cycle of anthropogenic lead: II. Results and discussion. Resources Conservation and Recycling 52 (52): 1050-1057.

Mason B. 1958. Principles of geochemistry. 2nd ed. New York: Wiley.

Murakami S. 2006. Material flows and stocks of metals surrounding Japan. In Symposium on advanced material flow analysis for the sustainable society, 187-190. Sendai, Japan. Tokyo: The Society of Non-Traditional Technology.

Nakamura S, S Murakam, K Nakajima, et al. 2008. Hybrid input-output approach to metal production and its application to the introduction of lead-free solders. Environmental Science and Technology 42 (10): 3843-3848.

Obernosterer R, P H Brunner. 2001. Urban metal management: The example of lead. Water, Air, and Soil Pollution: Focus 1 (3-4): 241-253.

Smith G R. Lead (1994−2003). http: //minerals.usgs.gov/minerals/pubs/commodity/lead/380494-380499, 380400.pdf. (Accessed 15 July 2008)

Sörme L, B Bergbäck, U Lohm. 2001. Century perspective of heavy metal use in urban areas. Water, Air, and Soil Pollution Focus 1: 197-211.

Sullivan, D. E. 2003. Productive capacity indicator 5.2: Indicators of stocks-in-use in the United States for aluminium, copper, gold, iron and steel, lead, and zinc. Unpublished report. Washington, DC: U.S. Geological Survey.

Thornton I, R Rautiu, S Brush. 2001. Lead: The facts. www.ila-lead.org/fact_book.html. (Accessed 18 July 2008)

Tukker A, H Buist, L van Oers, et al. 2006. Risks to health and environment of the use of lead in products in the EU. Resources Conservation and Recycling, 49 (2): 89-109.

USGS. 2006. Mineral commodity summaries 2006. Washington, DC: USGS.

VanDer Voet E, J B Guinee, H A Udo de Haes. 2000. Heavy metals: A problem solved? Dordrecht, the Netherlands: Kluwer.

Van Der Voet E, R Kleijn, R Huele, et at. 2002. Predicting future emissions based on characteristics of stocks. Ecological Economics, 41 (2): 223-234.

Verhoef E V, G P J Dijkema, M A Reuter. 2004. Process knowledge, system dynamics, and metal ecology. Journal of Industrial Ecology 8 (1-2): 23-43.

Von Gleich A, R U Ayres, S Gößling-Reisemann, eds. 2006. Sustainable metals management. Frankfurt, Germany: Springer Publishing.

第4章 中国铅元素人为释放动力分析及其环境累积
A Dynamic Analysis of Environmental Losses from Anthropogenic Lead Flow and Their Accumulation in China[*]

4.1 Introduction

In recent decades, lead pollution incidents have happened frequently in China (Qin, 2010). Specially, the children lead poisoning is extremely severe and has attracted much attention (Zhao, 2013). Therefore, the research concerning lead is arousing more and more concerns and thus treatment of lead pollution is highly pressing at present.

Probably as the earliest metal to be used on the earth (Gale, 1981), the manufacturing history of lead is more than several thousand years. Lead is mined as lead ores, and enters the anthropogenic flow by coming into the human society, which is dominated by human activities (Mao, 2009). And then it is released from the flow as emissions. It can even accumulate in the environment, threatening the ecological security. Many studies on lead emissions have been carried out from different perspectives. For example, with the help of monitoring techniques, environmental geosciences mainly focus on recognizing the changes in resources or environment and the impacts imposed by human activities. Undeniably, it is challenging to evaluate pollution sources and quantities in geosciences. In this work, to address this key issue, a quantitative flow model will be established based on the substance flow analysis (SFA). This can greatly help to change the ineffective management pattern, which means remedying the existing pollutions, and therefore remarkably improve the potentials for source controlling. Meanwhile, the implementation of quantitative measures, which obeys the rules of lead life-cycle, can greatly promote the efficiency and scientificity of environmental management.

---

[*] Liang J, Mao J S*. 2014. A dynamic analysis of environmental losses from anthropogenic lead flow and their accumulation in China. Transactions of Nonferrous Metals Society of China, 24 (4): 1125-1133.

Substance flow analysis (SFA) is one of the most widely used techniques in the analysis of material flow which is confined to a specific boundary. Until now, detailed frameworks for cycles such as iron, zinc and lead have already been established (Mao et al., 2008b; Wang et al., 2008. Graedel et al., 2005). SFA is not only an important tool to identify the pollution sources (Spataria, 2005), but can help people to get a better understanding of resources utilization, and even explore the disciplines, which reflect the impacts of natural resources or socio-economic situation on lead emissions. In a word, the application of SFA will help to guide the waste management and evaluate the sustainability of resources in a long time.

Although there have already been a great deal of studies on lead emissions, unfortunately those studies only cover some of all life-cycle stages as to production, manufacture, use or recycling (Zhang et al., 2011; Tukker et al., 2006; Lohmu et al., 1994; Sormel et al., 2001), or just report on a static study at a specific time as a snapshot of the cycle (Mao, 2008b), without a clear idea of the historical characteristics or accumulative effects of lead emissions. It encourages us to wonder how much processed lead has been released into the environment in various ways over the years; this issue can only be satisfactorily addressed from a quantitative and comprehensive flow analysis (Graedel et al., 2004). In this work, lead emissions covering all life stages and the historical accumulation will be studied on the anthropogenic lead flow in China. The situation in 2010 will be analyzed to obtain the characteristics of national emission intensity, and accumulative lead in the environment is to be estimated during the period of 1960-2010.

4.2　Methodology

4.2.1　The Model of Lead Emissions in Anthropogenic Cycle

Anthropogenic lead flow consists of four stages: production, fabrication & manufacture(F&M), use and waste management & recycling(WMR). Lead is extracted from lead ore in the lithosphere, which is the source of lead in anthropogenic cycle, and ends in the environment as landfills or sediments. Therefore, environment acts as both the source and sink of lead cycle. Specifically, lead cycle starts with the mining of ore resources in production, and then lead is transformed into products or semi-products in F&M, after which lead enters use stage and offers services to human. Finally, it is obsolete after use in WMR and some is reclaimed as secondary resources while the other is left in the environment. Especially, the production of lead can be further divided into primary and secondary lead production. The virgin material supplies of primary lead are lead ores after the processes of mining, concentration and smelting while the secondary lead is refined from scrap which contains relatively high amount of lead. Compared with primary lead from ores, the cost of the secondary lead from scraps is less expensive (Zhang and Qiao, 2004), which promotes the recycling of resources and improves environmental protection as well.

The wastes discharged in production include tailings after mining, slag from refining and smelting, clinker and smoke dust containing lead. During manufacturing, dross and lead ash are emitted, together with leftovers or off cuts from fabrication (Guo et al., 2009). At use stage, lead products such as petrol additives, soldering alloys and ammunition are regarded to be permanently lost into the ambient environment and are not recyclable. Finally, many obsolete lead products are recycled or enter the environment as landfills. Lead acid batteries dominate the raw materials of the secondary lead, accounting for over 85% of the wastes, followed by cable sheeting, lead clad and alloys (Cao, 2006).

This research was based on the established framework of lead anthropogenic cycle in 2008 (Mao et al., 2008a; 2008b) (Fig. 4.1, Table 4.1). In this work, we will further study the lead cycle by focusing on the quantitative estimation of lead emissions. We make the following assumptions when carrying out our study: the international market is relatively stable in a long period of time; there is a certain quantitative relationship among lead imports, exports and lead consumption; the recycled lead scraps go through the process of secondary refining.

The historical data for a long period of time are not always available for us, with only some known or easy to get, such as lead consumption P_t. We set up the quantitative relationships between various components such as U, D_U, M and E_M, and lead consumption by quantifying the anthropogenic flow other than simply relying on the scattered data. Obviously, the loss from lead anthropogenic flow is influenced by a variety of factors or parameters. These parameters refer to technology progress or social development, such as λ, which depends on the ability of secondary lead production. We covered all those parameters from the published literatures. Overall, in order to study the loss, we treated other components of the flow as functions of lead consumption with the parameters. The related components are with the unit of t/a, and the subscript t stands for a certain year. Those parameters are shown in Table 4.1.

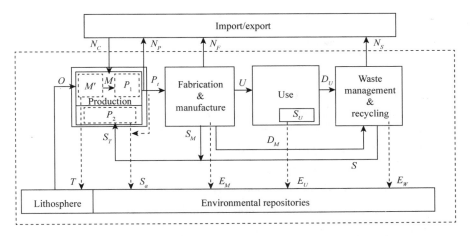

Fig. 4.1　Framework of lead emissions from anthropogenic cycle

O—Lead from ore; M—Lead in total lead concentrate; M'—Lead in domestic lead concentrate; Pt—Refined lead in the year of t; U—Lead entering use; SU—Lead entering in-use stock; D_U—End-of-life lead to waste management and recycling; D_M—Lead in F&M discards to WMR; S_T—Lead in total scrap to production; S_M—Lead in new scrap from manufacture; S—Lead in scrap from WMR; T—Lead in tailings; S_a—Lead in lag; E_M—Lead in F&M emissions; E_U—Lead in use emissions; E_W—Lead in WMR missions; N_C—Lead in net export of concentrate; N_P—Lead in net export of refined lead; N_F—Lead in net export of semi- or finished products; N_S—Lead in net export of scrap.

Table 4.1　Parameters used in estimation of anthropogenic lead emissions

Parameter	Definition	formula
f_m	Fabrication efficiencies, reflecting the technology progress in manufacture and processing	$f_m = \dfrac{U}{P_t}$
α	Accumulation ratio, addition to in-use stock as a fraction of lead entering use	$\alpha = \dfrac{S_U}{U}$
β	Ratio of scrap from manufacturing to total scrap	$\beta = \dfrac{S_M}{S_T}$
μ	Discard-management ratio, reflecting the efficiency of waste management during fabrication	$\mu = \dfrac{D_M}{D_M + E_M}$
λ	Ratio of primary lead to total lead production, depending on the production of secondary lead	$\lambda = \dfrac{P_1}{P_1 + P_2}$
θ	Secondary supply ratio, scrap supplied to production as a fraction of refined lead input to F&M	$\theta = \dfrac{S_T}{P_t}$
ζ	End of life recycling ratio, indicating the applied management and recycling technology	$\zeta = \dfrac{S+N_S}{D_U+D_M}$
ε_{11}	Import ratio of lead concentrate to the domestic demand	$\varepsilon_{11} = \dfrac{N_C}{M}$

Continued

Parameter	Definition	formula
ε_{12}	Net export ratio of refined lead to lead consumption	$\varepsilon_{12} = \dfrac{N_P}{P_t}$
ε_2	Net export ratio of lead in semi- or finished products to lead consumption	$\varepsilon_2 = \dfrac{N_F}{P_t}$
ε_3	Net export ratio of lead in slag to lead consumption	$\varepsilon_3 = \dfrac{N_s}{P_t}$

Although lead state, structure and occurrence change a great through the anthropogenic flow, lead obeys the mass conservation law, namely the balance between inflows and outflows (Kleijn et al., 2000). Based on the characteristics of lead cycle and the balance of mass, the emission model could be successfully established. Using the parameters and lead consumption P_t, we can formulate the components of the lead flow. This complex but rewarding process requires more than precise math deduction, and a deep understanding of the anthropogenic lead flow is also needed. The components formulated by P_t and the parameters can be seen in Table 4.2.

Table 4.2 Symbols formulated by parameters and Pt in lead anthropogenic flow

Symbol	Formula relationship with P_t
N_C	$N_C = \dfrac{\xi_{11}(1-\lambda)(1+\varepsilon_{12})P_t}{\eta_{21}}$
N_F	$N_F = \varepsilon_2 P_t$
S_T	$S_T = \theta P_t$
S	$S = (1-\beta)\theta P_t$
M	$M = \dfrac{(1-\lambda)(1+\varepsilon_{12})P_t}{\eta_{21}}$
P_1	$P_1 = (1-\lambda)(1+\varepsilon_{12})P_t$
U	$U = (f_m - \varepsilon_2)P_t$
D_M	$D_M = \mu(1 - f_m - \beta\theta)P_t$
N_P	$N_P = \varepsilon_{12} P_t$
N_S	$N_S = \varepsilon_3 P_t$
S_M	$S_M = \beta\theta P_t$
S_U	$S_U = \alpha(f_m - \varepsilon_2)P_t$
O	$O = \dfrac{(1-\lambda)(1+\varepsilon_{12})(1-\varepsilon_{11})P_t}{\eta_1 \eta_{21}}$
P_2	$P_2 = \lambda(1+\varepsilon_{12})P_t$
D_U	$D_U = \left[\dfrac{(1-\beta)\theta + \varepsilon_3}{\xi} - \mu(1 - f_m - \beta\theta)\right]P_t$

4.2.2 Intensity and Accumulative Equations of Lead Emissions

1. Intensity equations of lead emissions

Human hazards and ecological risks of lead vary with the lead composition, state and chemical structure (Eckelman and Graedel, 2007; ICMTI, 2013). Because the emissions released from the same stage tend to have similar properties, we classified the different emissions according to stages and calculated the loss at each stage.

With the model built in the article, we can successfully use parameters and P_t to formulate the lead emissions. T, namely the lead emissions in tailings, and S_a, the loss in slag released in lead refining or smelting, are

$$T = \frac{(1-\eta_1)(1-\varepsilon_{11})(1-\lambda)(1+\varepsilon_{12})P_t}{\eta_1\eta_{21}} \quad (4.1)$$

$$S_a = \left[\frac{(1-\eta_{21})(1-\lambda)(1+\varepsilon_{12})}{\eta_{21}} + \frac{(1-\eta_{21})\lambda(1+\varepsilon_{12})}{\eta_{22}}\right]P_t \quad (4.2)$$

E_M, the lead emissions in F&M from anthropogenic flow, is

$$E_M = (1-\mu)(1-f_m-\beta\theta)P_t \quad (4.3)$$

E_U, the lead emissions at the stage of use, is

$$E_U = \left[(f_m-\varepsilon_2)(1-\alpha) - \frac{(1-\beta)\theta+\varepsilon_3}{\xi} + (1-f_m-\beta\theta)\right]P_t \quad (4.4)$$

E_W, the lead emissions during WMR, is

$$E_W = \left(\frac{1}{\xi}-1\right)\left[(1-\beta)\theta+\varepsilon_3\right]P_t \quad (4.5)$$

Because all the emissions in the above equations refer to the refined lead contained in them, the quantity of total environmental emissions can be reached by adding up all the emissions from every stage, namely emissions in Production, F&M, Use and WMR. Thus Q_t, the total emissions at the year of t, is

$$Q_t = \left[1-\theta-\varepsilon_2-\varepsilon_3-(f_m-\varepsilon_2)\alpha + \frac{(1-\eta_1\eta_{21})(1-\varepsilon_{11})(1-\lambda)(1+\varepsilon_{12})}{\eta_1\eta_{21}} + \frac{(1-\eta_{22})\lambda(1+\varepsilon_{12})}{\eta_{22}}\right]P_t \quad (4.6)$$

2. The accumulative equations of lead emissions

There are altogether two kinds of lead flows: the natural flow and the anthropogenic flow. The natural flow includes all the lead transfer and transformation in soil, atmosphere or hydrosphere, under the influence of natural processes, such as weathering of rocks or the volcanic eruptions. Lead goes through human society while offering specific services or satisfying human demands. It is known that the scale of anthropogenic lead flow is 12 times that of its natural flow (The Research Board on Sustainable Development of China Academy of Science, 2000). What's more, due to the increasing lead use, the current lead anthropogenic flow has already exceeded the carrying capacity of the nature. We can find that the loss from the anthropogenic lead flow, which is difficult to be absorbed by natural purification, dominates the sources of lead pollutants in the environment. Additionally, those pollutants gradually accumulate in the environment, impairing the environmental quality and threatening the ecological safety. Under the increasingly strong impacts of anthropogenic lead flow, the accumulation of lead raises a lot with the pass of time. In the specific time interval of T, the accumulative quantity Q_T is

$$Q_T = \int_{t}^{t+T} Q_t \mathrm{d}t = \sum_{t}^{t+T} Q_t \qquad (4.7)$$

where Q_t is the accumulative rate, which only has a specific value at the time t. As Q_t is a function of t, the accumulation quantity is usually a continuous function of time under the ideal condition. However, due to statistics limits, the data available are usually discontinuous in reality. Therefore, when the differential is set as a year, lead accumulation in the environment is the sum of emission rates or intensity of each year. Therefore, Q_t can be expressed as a discrete function. In this study, we selected the last half century (1960-2010) as our study period. We chose this period on the knowledge of lead industry situation, which has a sharp increase in lead consumption. During the period of 1960-2010, the accumulative lead emissions can be formulated as

$$Q_T = \sum_{1960}^{2010} Q_t \qquad (4.8)$$

4.2.3　Values of the Parameters and the Data Sources

1. Values of the parameters

If we intend to get the quantity of lead emissions and the accumulative amount, we must manage to get the values of the parameters.

At the stage of F&M, it was reported that the ratio of F&M scrap to the total scrap is 0.0416 on a global scale (Smith, 2013), and in China this ratio is approximately

0.103 in 2006（Guo et al.，2009）. Therefore，we can estimate the approximate values of β as listed in Table 4.3. Moreover，D_M occupies 30% of the total lead that is not embodied in final products，and E_M takes up 20%（Mao et al.，2009），so it is easy to get the discard-management ratio μ，0.6，and this ratio is 0.659 globally in 2000（Mao et al.，2008b），then the estimation values of μ can be approximately obtained. The fabrication efficiencies，f_m，of various lead products have been determined during the period of 1900-2000 in a relevant research（Mao and Graedel，2009）and f_m of China is about 90%（Mao and Lu，2003）. For the reason that domestic large-scale refineries have already had a comparable smelting ability to those in the foreign countries（Jiang，2004），the values can be derived according to global fabrication efficiencies.

Table 4.3　Estimated values of parameters determining environmental emissions

Year	f_m	α	β	μ	λ	θ	ζ	ε_{11}	ε_{12}	ε_2	η_1	η_{21}	η_{22}
1960-1964	0.701	0.126	0.13	0.57	0.142	0.122	0.60	−0.109	−0.192	−0.09	0.75	0.920	0.909
1965-1969	0.699	0.127	0.12	0.58	0.099	0.092	0.61	−0.091	−0.126	−0.08	0.76	0.922	0.917
1970-1974	0.809	0.128	0.11	0.59	0.051	0.046	0.62	−0.073	−0.142	−0.07	0.77	0.924	0.925
1975-1979	0.816	0.129	0.10	0.60	0.113	0.105	0.63	−0.055	−0.118	−0.06	0.78	0.926	0.934
1980-1984	0.825	0.13	0.09	0.61	0.145	0.133	0.61	−0.036	−0.131	−0.05	0.81	0.928	0.942
1985-1989	0.821	0.14	0.08	0.62	0.126	0.133	0.615	−0.018	0.002	0.10	0.82	0.930	0.950
1990-1994	0.876	0.15	0.07	0.63	0.121	0.172	0.62	0.091	0.37	0.174	0.826	0.944	0.959
1995-1999	0.865	0.16	0.08	0.64	0.179	0.281	0.64	0.201	0.491	0.176	0.851	0.932	0.968
2000-2004	0.910	0.17	0.06	0.65	0.172	0.261	0.65	0.405	0.451	0.178	0.868	0.938	0.976
2005-2010	0.935	0.18	0.04	0.65	0.268	0.320	0.66	0.545	0.154	0.18	0.85	0.946	0.985

At use stage，the accumulation ratio of in-use stock is 0.112 in 2000（Mao et al.，2008b）. Additionally，it is known that this parameter is largely affected by a couple of social and economic factors，such as GDP，population size，consumer preference（Elshkaki et al.，2004）. Based on the above knowledge，the values of α can be estimated.

As to the stage of WMR，the scrap coming back to production takes up 0.662 of the WMR input（Smith，2013）and it is believed that the domestic recycling rate is comparatively modest（Wang et al.，2012）. The late 1980s saw a rapid development in the local lead smelting industry（Cai et al.，1996），while the corresponding management measures appeared to lag behind. Moreover，in 1996，the State Council issued the document of Decision of the State Council on several issues concerning environmental protection，forbidding the development of small-sized lead enterprises and pollution-intensive industrial projects, after which a further progress in waste management has been achieved. Based on all the above, we can estimate the end of life recycling ratios ζ.

For production stage, we can get the data of mining, concentrating and refining efficiencies of primary lead from 1990 to 2010 in China Nonferrous Metals Industry Yearbook. Besides, it was recorded that in 1981, the flotation technology without cyanogen of high-sulphur lead-zinc ores was explored successfully and was applied by more than 40% enterprises in the 1980s (China Nonferrous Metals Industry Association, 2009), so we can see that the mining efficiencies changed a great since the 1980s. Reports show that the refining efficiencies of the secondary lead range from 0.86 to 0.985 (Mao et al., 2008a). Therefore, it is not difficult to estimate the mining and refining efficiencies.

The ratio of secondary lead to refined lead (λ), can be acquired from the total refined lead production divided by the secondary lead production. The supply ratio of secondary lead (θ), can be obtained based on the lead consumption, secondary lead production and the secondary refining efficiency.

The net import and export parameters, namely ε_{11}, ε_{12}, ε_2 can be determined by simple calculation of total lead production, consumption and lead imports or exports from statistics data. Because the imports and exports of scrap are very little (Guo et al., 2009) and are negligible compared with the refined lead (Mao et al., 2007), we assume that the net scrap export ratio is zero. Before the 1980s, it was reported that lead raw materials were exported and lead products were imported in China, and in the latter years the situation changed completely. Since 1996, China began to import lead concentrate (Feng et al., 1996). It has been shown that the imports of lead ore accounted for 1/3 of the total domestic demand in 2000 (Mao et al., 2007), and in the following year, the domestic lead concentrate can only satisfy 49.1% of the refined lead production (Huang, 2004). The international trades are expected to become more frequent since China joined WTO in 2001. China has already become one of dominant countries in lead production and net exports all over the world. In recent years, the exports, especially exports in the manufacturing industry, are the driving force of large resource demands and imports of lead raw material (Roberts and Rushu, 2012). Owing to the export rush before the implement of the policy of canceling export rebates, the exports of refined lead reached a high record while the export of refined lead in 2007 and 2008 decreased heavily, and in 2009 China began to import refined lead (Feng, 2012). So we can see that the export ratio of refined lead during 2005-2010 reduced sharply compared with the previous years. The values of import or export ratio (ε_{11}, ε_{12}, ε_2) can be got on the knowledge with the trade situation in China.

2. The sources and values of the basic data

Lead production, consumption, imports and exports are the basic data of this study. Lead consumption for period 1996-2010 in China can be collected directly from China Nonferrous Metals Industry Yearbook (1998, 2001, 2004, 2007, 2010 and 2011) and lead consumption from 1986 to 1996 can be acquired from reference (Mao and Lu, 2003). The data of lead consumption for every five years from 1960 to 1985 are available in reference (Cai et al., 1996). Due to the consistency in lead consumption in a short time, we can estimate the lead consumption of each year during 1960-1985 based on the proportion that consumption occupies in lead production.

The total production of refined lead, primary lead and secondary lead in China during 1960-2010 can be taken directly from China Nonferrous Metals Industry Yearbook (2011). The exports and imports of lead concentrate, refined lead, lead semi- or finished products and lead scrap can be obtained from the statistics of each year which are organized according to the national customs recordings.

4.3 Results and Discussions

4.3.1 The Intensity of Lead Emissions

The lead emissions released to the environment at every stage in 2010 can be calculated according to equation (4.1) -equation (4.6). It turns out that the emissions in 2010 from anthropogenic flow amounted to 1.89×10^6 t. The lead flow can be referred in Fig. 4.2. It was recorded that the lead consumption was 3.95×10^6 t in 2010. Therefore, we can see that for every 1 kg of lead consumed or put into end use, 0.48 kg lead is lost to the ambient environment. Those emissions or even pollutants will not only deplete the lead resources on the earth, but also inflict severe ecological and health damage (Tukker et al., 2006).

Lead emissions from the use stage were most, which was 740.1×10^3 t, contributing 39.20% to the total amount, followed by 625.4×10^3 t from WMR (33.13%). Additionally, the emissions of production and F&M were 450.2×10^3 t (23.85%) and 72.2×10^3 t (3.82%), respectively (Table 4.4).

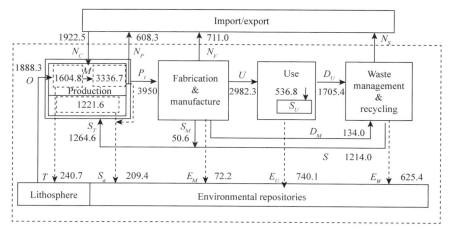

Fig. 4.2 Anthropogenic flow of lead in 2010 (10^3t)

Table 4.4 Environmental emissions from lead anthropogenic flow in 2010

Emission	Lead flow/10^6t	Total loss/%
Tailing	0.241	12.75
Slag	0.209	11.09
F&M emission	0.072	3.82
Use emission	0.740	39.20
WMR emission	0.625	33.13
Total	1.888	100

It was reported that the recyclable lead products such as lead acid batteries, sheet and pipe account for merely 66% in the consumption structure of China (Mao and Lu, 2003) and the dissipative lead (Mao et al., 2009) is lost permanently to the environment during use. Compared with 66% in China, the recyclable lead can be more than 80% internationally (Gao, 2000), so the humble recyclable lead proportion is probably the most important reason for the large emissions in China during use. In addition, the recycling rate in WMR is comparatively low. This can be seen from the fact that the recycling rate of acid lead batteries is only 90% but almost 100% in developed countries across the world (Li, 2000). What's more, the domestic refineries tend to be small in scale and usually apply the less advanced technology. They are quite energy-consuming (Wang et al., 2012) with a low recycling rate, and are responsible for the large amount of wastes and emissions.

4.3.2　The Accumulative Lead Emissions

According to equation (4.8), the accumulative emission from lead anthropogenic flow was 19.54×10⁶ t, respectively from the stages of production (26.67%), F&M (5.01%), use (46.47%) and WMR (21.85%). Fig. 4.3 shows the accumulative quantity of lead emissions during 1990–2010. If the population of China is set to be 1.37 billion, on a per capita basis, the accumulative lead will be 14.26 kg. If the lead emissions were distributed evenly in the land area of China, and there would be 2.04 g/m² lead. With the plough layer assumed to be 20 cm and the soil density 2.65 g/cm³, the pollution concentration would be about 3.84 mg/kg, which means that 3.84 mg/kg was added to the background value. Although this may not seem a sharp increase, lead concentration in some specific area can be extraordinarily high. For example, lead concentration for lightly polluted soil is close to 100 mg/kg while soil seriously polluted by lead-zinc mining can reach 10⁴ mg/kg (Wang and Shao, 2009). The potential harm of lead to human is tremendous.

While lead consumption in the society is increasing and the in-use stock (Mao and Graedel, 2009) is enlarging over time, the corresponding management and recycling technology innovations fall behind. Even though the accumulation rates (lead emissions as a fraction of total lead consumption) reduced from 1.149 to 0.478, the lead emissions of every year intensified during this period.

Besides, the emissions at the WMR stage increase most, in which the emissions increased by 81 times from 1960 to 2010 while the increase of emissions in F&M is far less (Fig. 4.3). The accumulative lead emissions from different stages can be seen in Table 4.5.

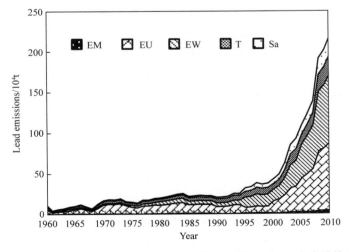

Fig. 4.3 Lead emissions from anthropogenic flow in China for period 1960–2010

Table 4.5 Accumulated environmental emissions from lead anthropogenic flow during period of 1960−2010

Emission	Lead flow/10^6t	Total loss/%
Tailing	3.148	16.11
Slag	2.063	10.56
F&M emission	0.979	5.01
Use emission	9.081	46.47
WMR emission	4.270	21.85
Total	19.542	100

4.3.3　Data Uncertainty and Perspectives

The statistics can only date back to the year of 1990, and for the former years, there are only limited descriptions in literatures and only a few scattered records from lead enterprises or the whole industry. In this study, we referred to the overall development situation of lead industry during 1960−2010 and speculated the possible development level. Although the gap between true values and estimations is unavoidable, we tried our best to make it consistent with reality. Furthermore, the following shortcomings and perspectives are brought out.

(1) The life spans of lead products in the flow are not taken into consideration. Various products have different life spans which greatly influence the quantity of lead wastes in use, and this is what the further study needs to address.

(2) Lead flow is affected by governmental regulations and policies, and studies on relation to policies' impacts need to improve.

(3) Lead emissions on the global scale from anthropogenic flow remain to be studied further because the historical accumulation of lead estimated is the essential groundwork for waste treatment. Studies on the real lead stock in the environment and its environmental risks need to be carried out, considering the purifying capacity of nature.

4.4　Conclusions

In 2010, the lead emissions from anthropogenic flow totaled 2.18×10^6 t which mainly formed at the stages of Use (38.07%) and WMR (38.32%). Given the lead consumption was 3.95×10^6 t, this means that for every 1 kg of lead consumed or put into end use, 0.55 kg is lost to the ambient environment. The high proportion of dissipative lead, together with the less advanced recycling technology is the most important reason for the large emissions.

In 2010, the total lead emissions from anthropogenic flow were 1.89×10^6 t which mainly formed at the stages of use (39.20%) and WMR (33.13%). The lead consumption was 3.95×10^6 t, which means that for every 1 kg of lead consumed or put into end use, 0.48 kg of lead is lost to the ambient environment. The high proportion of dissipative lead, together with the less advanced recycling technology is the most important reason for the large emissions.

During the period of 1960–2010, the accumulative lead emissions amounted to 19.54×10^6 t. If the lead emissions were evenly distributed on the polluted farmland, and there would be 2.04 g/m^2. At the present population, it is 14.26 kg on per capital base. Moreover, the pollution concentration would be about 3.84 mg/kg, which means that 3.84 mg/kg was added to the background value. The corresponding management and recycling technology innovations fall behind the increase in lead consumption, which leads to the increasing emissions released to the environment.

Therefore, it is suggested that in order to reduce the lead emissions, China should: ①optimize the lead consumption structure and reduce the use of dissipative lead while searching for the substitutions of lead positively; ②make full use of the lead scrap to increase the proportion of secondary lead, establish an effective recycling system, improve the application of advanced technology and encourage the centralized processing ability of wastes; ③make effective laws and regulations to regulate the emissions from lead enterprises and conduct necessary macro-control; and ④reduce the exports of lead products reasonably and prolong the life spans of products in society.

References

Cai X D, Feng J C, Jiang D, et al. 1996. The lead consumption and recycling in China. World Nonferrous Metal, 6: 37-40. (in Chinese)

Cao Y S. 2006. The status and perspectives of recycling nonferrous metals resources. China Metal Bulletin, (16): 7-12. (in Chinese)

CNMIA (China Nonferrous Metals Industry Association). 2009. China nonferrous metals industry 60 years. Changsha: Central South University Press, 166. (in Chinese)

Eckelman M J, Graedel T E. 2007. Silver emissions and their environmental impacts: A multilevel assessment. Environmental Science and Technology, 41 (17): 6283-6289.

Elshkaki A, van der Ester V, van HOLDERBEKE M, et al. 2004. The environmental and economic consequences of the developments of lead stocks in the Dutch economic system. Resources, Conservation and Recycling, 42 (2): 133-154.

Fan L. 2012. Analysis of factors which influence imports and exports of lead concentrate in China. Modern Mining, (4): 1-6. (in Chinese)

Feng J C, Zhang F, Huang G P. 1996. The trends of world market and future of lead/zinc industry in China. World Nonferrous Metal, (12): 40-45. (in Chinese)

Gale N H, Stos-gale Z. 1982. Lead and silver in the ancient Aegean. Scientific American, 244 (6): 176-191.

Gao S. 2000. The world lead consuming trends. World Nonferrous Metal, 3: 14-16. (in Chinese)

Graedel T E, Vanbeer S D, Bertram M, et al. 2004. Multilevel cycle of anthropogenic copper. Environmental Science and Technology, 38: 1242-1252.

Graedel T E, Vanbeer S D, Bertram M, et al. 2005. The multilevel cycle of anthropogenic zinc. Industry Ecology, 9 (3): 67-90.

Guo X Y, Zhong J Y, Song Y, et al. 2009. Substance flow analysis of lead in China. Journal of Beijing University of Technology, 35 (11): 1554-1561. (in Chinese)

Huang Z Q. 2004. The present lead and zinc industry development and suggestions in China. World Nonferrous Metal, (8): 4-6, 63. (in Chinese)

Indiana Clean Manufacturing Technology Institute. 2013. Indiana relative chemical hazard ranking system. http://cobweb.ecn.purdue.edu/CMTI/IRCHS/

Jiang J M. 2004. The status and sustainable development of lead and zinc smelting industry in China. The Chinese Journal of Nonferrous Metals, 14 (1): 52-62. (in Chinese)

Kleijn R, Huele R, van der Voet E. 2000. Dynamic substance flow analysis: The delaying mechanism of stocks, with the case of PVC in Sweden. Ecological Economics, 32: 241-254.

Li F Y. 2000. The status and countermeasures of the recovery of obsolete lead-acid batteries. China Resources Comprehensive Utilization, 4: 3-7. in Chinese)

Lohm U, Anderber G S, Bergback B. 1994. Industrial metabolism at the national level: A case-study on chromium and lead pollution in Sweden, 1880-1980//AYRES R U, SEMONIS U E. Tokyo: United Nations University Press, 1994.

Mao J S, Lu Z W. 2003. A study on causes of low recovery of scrap lead. World Nonferrous Metal, (11): 24-28, 32. (in Chinese)

Mao J S, Yang Z F, Lu Z W. 2007. Industrial flow of lead in China. Transactions of Nonferrous Metals Society of China, 17 (2): 400-411.

Mao J S, Dong J, Graedel T E. 2008a. The multilevel cycle of anthropogenic lead: I. Methodology. Resource, Conservation and Recycling, 52: 1058-1064.

Mao J S, Dong J, Graedel T E. 2008b. The multilevel cycle of anthropogenic lead: II. Results and discussion. Resource, Conservation and Recycling, 52: 1050-1057.

Mao J S, Cao J, Graedel T E. 2009. Losses to the environment from the multilevel cycle of anthropogenic lead. Environmental Pollution, 157: 2670-2677.

Mao J S, Graedel T E. 2009. Lead in-use stock. Journal of Industrial Ecology, 13 (1): 112-126.

Qin J F. 2010. Status, sources of pollution and control measures of Chinese children lead poisoning. Guangdong Trace Elements Science, 17 (1): 1-13. in Chinese)

Roberts I, Rushu A. 2012. Understanding China's demand for resource imports. China Economic Review, 23 (3): 566-579.

Smith G R. 2013. Lead [EB/OD]. http: //minerals.usgs.gov/minerals/pubs/commodity/lead/380400. pdf. [2013-12-03].

Sörme L, Bergback B, Lohm U. 2001. Goods in the anthroposphere as a metal emission source a case study of Stockholm, Sweden. Water, Air, and Soil Pollution, 1 (3-4): 213-227.

Spataria S, Bertram M, Gordon R B, et al. 2005. Twentieth century copper stocks and flows in North America: a dynamic analysis, Ecological Economics, 54 (1): 37-51.

The Research Board on Sustainable Development of China Academy of Science. 2000. The strategic report on sustainable development of China in 2000. Beijing: Science Press: 207.

Tukker A, Buist H, van Oers L, et al. 2006. Risks to health and environment of the use of lead in products in the EU. Resources, Conservation and Recycling, 49: 89-109.

Wang T, Muller D B, Graedel T E. 2007. Forging the anthropogenic iron cycle. Environmental Science and Technology, 41 (14): 5120-5129.

Wang Z, Shao Z Q. 2009. Soil pollution of lead and treatment. Agriculture Technology and Equipment, (158): 6-8.

Wang H M, Liu Q, Wang F F, et al. 2012. The status of lead-acid battery recycling and study on the layout of its management. Environmental Science and Management, (6): 51-54.

Zhang S Y, Qiao F H. 2004. Practice and exploration of lead recycling method. Storage Battery, 41 (2): 84-88.

Zhang X W, Yang L S, Li Y H, et al. 2011. Estimation of lead and zinc emissions from mineral exploitation based on characteristics of lead/zinc deposits in China. Transactions of Nonferrous Metals Society of China, 21 (11): 2513-2519.

Zhao T T, Chen B, Wang H P, et al. 2013. Evaluation of toxic and essential elements in whole blood from 0- to 6-year-old children from Ji'nan, China. Clinical Biochemistry, 46: 612-616.

第 5 章　铅酸电池系统的铅流分析
The Lead Flow Analysis for Lead-Acid Battery Systems[*]

近年来，伴随中国铅产量和铅制品产量的快速增长，铅矿资源明显短缺，不得不依靠进口铅矿石来满足中国的需求，2000 年铅精矿进口量达国内消费量的 1/3（中国有色金属工业年鉴编辑委员会，2001），与此同时，铅的人为排放量高达铅的自然循环量的 12.9 倍（中国科学院可持续发展研究组，2000），严重威胁到人体的健康。为此，急需寻找措施，改善中国铅业与外部环境之间的关系，降低中国铅的环境影响水平。

据统计，中国有 66%～70%的铅用于生产铅酸电池，并且该比例逐年上升（李福元等，1999）。因此，研究铅酸电池系统中铅的流动，可望获得保护铅矿资源、改善铅污染现状，有利于可持续发展的战略思路。

本研究在分析铅的工业流动过程的基础上，构建铅酸电池生命周期铅流图，建立铅酸电池系统与外部环境的定量关系，获得铅的工业流动规律，并提出评价铅的工业流动状况的指标。此外，以中国铅酸电池系统为例，分析铅的流动现状，通过与瑞典某铅酸电池系统的对比，找出差距，提出改善对策。

5.1　研　究　方　法

5.1.1　铅酸电池系统及其铅的流动

铅酸电池系统是指以自然资源为起点，从工业原料生产，到铅酸电池制造、使用，以及报废后的回收与物质再生等各个过程的组合（Graedel and Allenby，1995）。为研究其中铅的流动规律，把铅元素作为铅酸电池的代表物质。这种情况下，铅酸电池系统简化为原生铅的生产（含采/选、冶炼、电解等）、铅酸电池制造、使用，以及使用寿命终了后的废铅回收与再生几个阶段。

从铅矿资源中的铅，到形成铅酸电池的组成元素，再作为废铅资源，返回铅的冶炼部门，铅元素将顺次经过每一生命周期阶段，从而形成铅酸电池系统中铅

---

[*] 毛建素，陆钟武，杨志峰. 2006. 铅酸电池系统的铅流分析. 环境科学，27（3）：43-48.

的流动。与此同时，在铅元素流经每一阶段时，都不可避免有一部分以含铅废物、污染物的形式，排放到外部环境之中。整个铅酸电池系统与外部环境的关系表现为：①向社会提供铅酸电池，满足储存并转移电能的需求；②从自然资源中索取铅矿石，形成铅矿资源负荷；③向外部环境排放含铅废物、污染物，形成铅排放负荷。铅矿资源负荷和铅排放负荷统称环境负荷。

5.1.2　评价指标

为定量估计铅酸电池系统与外部环境之间的铅流关系，引入生态效率（山本良一，2003）的概念。

定义单位环境负荷提供的铅酸电池数量为铅酸电池系统的生态效率。其中，与铅矿资源负荷有关的生态效率称为资源效率，用符号 r 表示，单位是（t/t）；与铅排放负荷有关的生态效率称为环境效率，用符号 q 表示，单位也是（t/t）。

不难看出，生态效率越高，意味着获得相同数量的铅酸电池，消耗的铅矿石越少，或者向环境排放的含铅废物、污染物越少，越有利于资源保护和环境改善。

研究中，把铅的资源效率和环境效率作为铅酸电池系统的铅流状态的评价指标。

5.1.3　铅酸电池生命周期铅流图

考虑到铅酸电池通常可使用几年甚至十几年，相比之下，各生产过程所经历的时间十分短暂，因此假设：①不计各生产过程所经历的时间；②铅酸电池的使用寿命为 $\Delta\tau$ 年，所有铅酸电池在其生产年份 $\Delta\tau$ 年以后全部报废，形成折旧废铅，并且，在报废当年返回铅业，进行再生处理。这种条件下，根据各生命周期阶段的物质流入量等于流出量（Kleijin，2000），可以绘制出各生命周期阶段铅的流动数量和方向，如图5.1所示，称为铅酸电池生命周期铅流图。考虑原生铅的生产和废铅再生同属于铅的生产过程，因此，图中将其合并到一起，并用符号 I 表示。

图5.1中，P_τ、$P_{\tau-\Delta\tau}$ 分别为第 τ 年和第 $\tau-\Delta\tau$ 年铅酸电池的产量，以含铅量计算，t；α 为铅的循环率，是铅酸电池产量中，报废后返回到铅的提炼阶段的铅所占的比例（t/t）；显然，$\alpha_{\tau+\Delta\tau}P_\tau$ 是第 τ 年生产的铅酸电池在第 $\tau+\Delta\tau$ 年报废时返回铅业的铅量（t）；$\alpha P_{\tau-\Delta\tau}$ 是产自第 $\tau-\Delta\tau$ 年的铅酸电池在第 τ 年报废时返回铅业的铅量（t）；β 为加工废铅实得率，是在铅酸电池制造阶段，从加工切屑废料中回收的铅量占该年铅酸电池含铅量的比例（t/t）；γ_1、γ_2 分别为Ⅰ、Ⅱ生命周期阶段

的铅排放率,是相应生命周期阶段排放的铅量与该年铅酸电池产量的比值（t/t）；为便于应用,定义两者之和为生产阶段的总铅排放率,简称铅排放率,用符号 γ 表示,即 $\gamma = \gamma_1 + \gamma_2$ (t/t)。

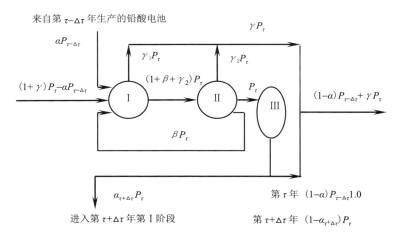

图 5.1 铅酸电池生命周期铅流图

Fig. 5.1 The lead flow diagram of lead-acid battery in its life cycle

第Ⅰ阶段——铅的采选、冶炼；第Ⅱ阶段——铅酸电池加工；第Ⅲ阶段——铅酸电池使用

这种通过追踪产品中的某一元素,同时考虑从产品生产到报废之间的"时间差",来分析产品系统与资源、环境之间的关系的方法,可以称作"元素流分析法"。"元素流分析法"是众多物流分析方法（Ayres, 1994; Joosten et al., 1999）中的一种,以考虑从产品生产到报废之间的时间差异为显著特征,由陆钟武院士于 2000 年提出。曾经多次成功运用到钢铁工业废钢资源（陆钟武, 2000）、铁排放量的源头指标（陆钟武, 2002）、铅在铅酸电池中的资源服务效率（毛建素和陆钟武, 2003）等问题的研究中。

5.2 铅流基本规律

5.2.1 资源效率

由图 5.1 可知,在第 τ 年,投入 $(1+\gamma)P_\tau - \alpha P_{\tau-\Delta\tau}$ 铅矿石,可获得含铅量为 P_τ 的铅酸电池。根据前文中铅的资源效率的定义,可整理得到

$$r = \frac{1}{1+\gamma-\alpha p} \tag{5.1}$$

式中，$p = P_{\tau-\Delta\tau}/P_\tau$，是铅酸电池生命周期产量变化比[(t/t)]。

可见，铅在铅酸电池系统中的资源效率（r）是铅的循环率（α）、排放率（γ）、和铅酸电池产量变化比（p）的函数。进一步分析发现，提高铅的循环率、降低铅的排放率、提高铅酸电池的产量变化比（即产量下降），均有利于提高铅的资源效率。

5.2.2 环境效率

同样的思路，由图 5.1 可知，在第 τ 年为生产含铅量为 P_τ 的铅酸电池，铅酸电池系统将向环境排放 $\gamma P_\tau + (1-\alpha)P_{\tau-\Delta\tau}$ 含铅废物、污染物，由此可整理得到

$$q = \frac{1}{\gamma + (1-\alpha)p} \tag{5.2}$$

式中各符号意义同式（5.1）。

可见，铅在铅酸电池系统中的环境效率（q）也是铅的循环率（α）、排放率（γ）、和铅酸电池产量变化比（p）的函数。进一步分析发现，提高铅的循环率，或者降低铅的排放率，有利于提高铅的环境效率；降低铅酸电池产量变化比（即产量增长），将有利于提高铅的环境效率，这是与产量变化比对资源效率的影响截然不同的。

5.2.3 环境效率与资源效率的关系

将式（5.1）与式（5.2）联立，解得

$$\frac{1}{q} - \frac{1}{r} = p - 1 \tag{5.3}$$

可见，资源效率与环境效率之间的关系，与铅酸电池生命周期产量变化比密切相关。进一步分析发现：①在产量不变的情况下，资源效率恒等于环境效率；②在产量增长的情况下，资源效率恒小于环境效率；③在产量下降的情况下，资源效率恒大于环境效率。这主要是由于铅酸电池产量的变化引起了产品系统的涨缩：产量增长，系统扩张，资源索取量大而排放量较小，因而，资源效率小于环境效率；反之，产量下降，系统收缩，资源索取量小而排放量较大，使得资源效率大于环境效率。

从本节的分析发现：铅的循环率、排放率、铅酸电池生命周期的产量变化比是影响铅酸电池系统中铅的生态效率的内在因素，反映系统内部各组分之间铅的流动关系，应用中称为生态效率的驱动因子。可作为评价铅酸电池系统的铅流状态的内部指标。

5.3　铅酸电池系统中铅的流动（实例分析）

5.3.1　中国铅酸电池系统的铅流

以中国 1999 年铅酸电池系统为研究背景，该年精炼铅的国内消费量为 525.00kt（中国有色金属工业年鉴编辑委员会，2001）。据估计，其中 66.8%（李富元等，1999），相当于 350.70kt，用于生产铅酸电池。

在铅酸电池生产中，平均每投入 1t 铅金属，将有 0.92t 进入铅酸电池中，另有 0.0356t 以加工废铅的方式得到回收利用，其余 0.0444t 以含铅废物、污染物的形式散失到环境中。铅酸电池的平均寿命估算为 3 年（兰兴华和殷建华，2000）。

根据 1999 年回收的废铅数量和废铅的构成（李富元等，1999），估计有 90.90kt 废铅酸电池和 12.48kt 加工废铅投入铅的再生。铅的再生收率估计为 80%～88%（马永刚，2001），本文按 86.37%计算，共可获得 89.29kt 再生铅，其余 14.09kt 铅以含铅废物、污染物的形式散失到环境之中。用于铅酸电池生产的其余部分（261.41kt），按原生铅计算。

在原生铅的生产中，涉及选矿、冶炼等过程，据中国有色金属工业统计，1999 年选矿收率为 83.8%，冶炼综合收率为 92.78%，因此，为获得 261.41kt 原生铅，需要投入含铅量为 336.22kt 的铅矿石，同时有 54.47kt、20.34kt 的铅分别损失在选矿和冶炼过程中。

另外，由于铅酸电池的平均寿命估计为 3 年，因此，1999 年回收的废铅酸电池是 1996 年生产的。表 5.1 中列出了中国 1990～2000 年的铅酸电池产量。考虑到表 5.1 中铅酸电池产量占全国总产量的 75%～85%，若 1996 年、1999 年分别按 77%、78%计算，则可估算出 1996 年铅酸电池的产量为 291.67kt。由于 1999 年回收废铅酸电池 90.90kt，因此，有 200.77kt 废铅酸电池未能得以回收。或者说，没有能够进入统计数据。

表 5.1　中国近 11 年精炼铅和铅酸电池产量[1]

Table 5.1　The annual outputs of refined lead and lead-acid battery in the last 11 years of China

年份	1990	1991	1992	1993	1994	1995	1996	1997	1998	1999	2000
铅酸电池产量/GWh	6.980	5.146	6.837	7.773	—	7.080	9.487	—	—	10.394	11.881

1）数据来源：中国机械工业年鉴。

根据以上数据，整理得到中国 1999 年铅酸电池生命周期铅流图，如图 5.2 所示。图中单位为 kt。

| 第 5 章 | 铅酸电池系统的铅流分析

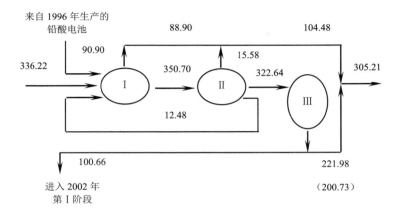

图 5.2 中国铅酸电池生命周期铅流图（1999）

Fig. 5.2 The lead flow diagram of lead-acid battery in its life cycle for China（1999）

5.3.2 结果与讨论

1. 中国铅酸电池系统的铅流评价指标

根据前文中的铅流分析，可整理得到 1999 年中国铅酸电池系统的铅流评价指标，计算结果如表 5.2 所示。可见，该系统铅的资源效率和环境效率分别为 0.960（t/t）、1.057（t/t）。

表 5.2 中国与瑞典铅的工业流动的对比

Table 5.2 The comparison of the lead industrial flow in China to those in Sweden

国家	资源效率/（t/t）	环境效率/（t/t）	循环率/（t/t）	铅排放率/（t/t）	产量比/（t/t）
中国	0.96	1.057	0.312	0.324	0.904
瑞典	79.02	79.02	0.99	0.002 655	1

2. 与瑞典铅酸电池系统的对比

按照同样的方法，根据文献（Karlsson，1999）中的数据，可分析瑞典铅酸电池系统的铅流状况。为便于对比，将瑞典的有关计算结果也汇总到表 5.2 中。

可见，在铅酸电池系统中，瑞典铅的资源效率和环境效率都高达 79.02（t/t），分别是中国的 82.31 倍和 74.76 倍。其原因在于：在铅的循环方面，瑞典铅的循环率高达 0.99（t/t），几乎全部循环，而中国仅为 0.312（t/t）；在铅排放方面，瑞典铅的排放率仅为 0.002 655（t/t），而中国却高达 0.324（t/t）；在产量变化方面，瑞典至少稳定了一个铅酸电池的生命周期——5 年,而中国仍在持续增长。

3. 原因分析

(1) 循环率低下的原因:不难设想,铅的循环率主要与铅酸电池的国内消费与回收、废铅贸易,以及统计数据完整情况等因素有关。

表 5.3 中列出了 1986~1996 年铅酸电池的生产与国内消费数量。由此推测,由于铅酸电池的出口,大约有 7.62%的废铅酸电池不能返回中国铅酸电池系统。

表 5.3 中国铅酸电池产量与国内消费量[1]

Table 5.3 The output and domestic consumption of lead-acid batteries in China

年份	1986	1987	1988	1989	1990	1991	1992	1993	1994	1995	1996	平均
铅酸电池产量/GWh	3.220	5.072	4.550	—	6.980	5.146	6.837	7.773	—	7.080	9.487	—
铅酸电池出口量/GWh	0.158	0.477	0.007	—	0	1.968	0.020	0.014	0.022	0.011	0	—
出口量∶产量/%	4.91	9.40	0.15	—	0	38.24	0.29	0.18	—	0.16	0	7.62

1) 数据来源:产量数据来源于中国机械工业年鉴;出口量数据来源于中国对外经济贸易年鉴。

表 5.4 中列出了 1990~2000 年废铅的贸易额。可见,在废铅的贸易方面,虽然经历了从净出口到净进口的过渡,但其数量相对于铅酸电池产量而言极小,特别是近几年,废铅的进出口基本平衡,因此可以忽略废铅贸易对铅的循环率的影响。

表 5.4 中国废铅贸易数量[1]

Table 5.4 The trade of lead scraps in China

年份	1990	1991	1992	1993	1994	1995	1996	1997	1998	1999	2000	平均
出口量/kt	6.132	5.250	1.500	0.621	1.510	0.589	0.152	0.061	1.751	1.060	0.037	—
进口量/kt	1.690	0.140	5.712	7.322	5.793	5.690	0.820	0.204	0.007	—	0.050	—
净出口量/kt	4.442	5.110	−4.212	−6.701	−4.283	−5.101	−0.668	−0.143	1.744	—	−0.013	−0.983

1) 数据来源:进、出口量数据来源于中国对外经济贸易年鉴。

在中国,废铅酸电池的回收与再生铅厂购买废铅的渠道紧密相关。据记载,1999 年中国曾有 300 家再生铅厂,其中只有 3 家规模较大,生产能力在 100kt/a 以上,总生产能力 500kt 左右,实际再生量占中国再生铅总量的一半左右(杨春明和马永刚,2000;马永刚,2000)。由此估算,另有一半的废铅,相当于循环率为 0.312,被私家再生铅厂收购并再生,而未列入统计数据。

由此推算,1999 年约有 29.98%,相当于 87.44kt 的废铅酸电池没有得到回收,散失到环境中。

(2) 排放率偏高的原因:根据前文中的铅流分析,将 1999 年铅酸电池生命周期生产阶段的铅损失数量列入表 5.5。由此可见,在与生产过程有关的铅损失中,

52.13%发生在选矿阶段，19.47%、13.49%分别发生在原生铅的冶炼和废铅再生阶段，其余 14.91%发生在铅酸电池的制造阶段。

表 5.5 中国铅酸电池系统生产过程中的铅损失构成
Table 5.5 The profile of the lead losses in the production processes of lead-acid battery system in China

生产过程	选矿	原生铅冶炼	废铅再生	铅酸电池制造	总计
铅损失量/kt	54.47	20.34	14.09	15.58	104.48
百分比/%	52.13	19.47	13.49	14.91	100

根据北京矿冶研究院的调查报告，中国铅的选矿收率在 81%～86%，较国外低 5%～15%。这主要是由于中国的铅矿多属于以锌为主的铅锌共生、多金属伴生、金属嵌布关系复杂的矿石。铅锌比一般为 1∶2.6，铅矿中伴生铜、银、金、锡、锑、镉等 50 余种有用金属，并且，金属矿物之间共生密切，粒度不均，多以星点状、细脉状或树枝状等分布于矿石中。造成选矿难度较大。

在原生铅的冶炼中，存在着企业多、规模小、集中度低、冶炼水平落后等问题。据估计，目前全国约有几百家冶炼厂，其中仅 23 家的产量在 10kt 以上，10 家最大企业的产量，仅占国内总产量的 55.27%。在冶炼工艺方面，仍以传统的烧结-鼓风炉为主。整体看来，相当于国外 20 世纪 80 年代以前的水平。

在废铅再生中，存在着企业数量多、规模小、工艺技术落后等问题。据估计，目前，全国约有 300 家废铅再生厂，生产能力从几千吨到上万吨不等，万吨以上的厂家屈指可数；工艺上，约有一半的厂家仍使用传统的小反射炉、鼓风炉和冲天炉等熔炼工艺，回收率一般为 80%～85%，最高不过 90%。较先进工艺低 10%左右（杨春明和马永刚，2000）。

在铅酸电池的制造阶段，铅的加工利用率通常为 0.85～0.95（姜松，2000）。根据本研究中获得的调研数据，约有 28.06kt 铅未得到有效利用，其中 12.48kt，占 44.48%，以加工废铅的形式得到回收，其余部分，或散失到环境中，或未列入统计数据。据称，有很大一部分废铅，是经过私人转送到废铅再生企业的（马永刚，2000），这部分数据很难收集得到。

4. 改善对策

在铅的资源效率和环境效率的 3 个驱动因子中，产量变化比取决于社会需求与经济发展，近期内很难降下来。因此，改善中国铅酸电池系统中铅的资源效率和环境效率，将主要通过提高铅的循环率，降低铅的排放率来实现。

为提高铅的循环率，建议：①借鉴国外先进的管理经验，建立废铅回收法规，完善回收机制，制约和引导废铅电池的回收工作步入正常轨道；②针对铅酸电池试行"出租产品"的消费形式，促进社会从"产品消费型"向"服务消费型"转变(Stahel, 1994; Stahel, 1997);③尝试"以旧换新"的销售形式,保障废铅回收工作的顺利进行。

为降低铅的排放率,建议进一步加强矿冶联合企业的管理,对铅的生产企业实行特别许可证管理制度;开发或引进采矿、选矿、冶炼新工艺;推行"清洁生产",淘汰落后的工艺、技术和设备,从而提高铅的综合收率,降低废铅损失。

5. 改善前景

若假定 20 年后,中国铅酸电池系统中铅的资源效率和环境效率都能达到 60(t/t),则意味着获得与1999年相同数量的铅酸电池,只需要投入5.38kt的铅矿石,铅排放负荷也将降低到5.38kt,分别是1999年铅矿资源负荷的1.6%和铅排放负荷的 1.76%。届时,铅矿石消耗水平和铅排放水平都将低于环境承载力,资源与环境状况将大为改善。

5.4 结　　论

(1)应用"元素流分析法"分析了铅在铅酸电池系统中的流动。构建了铅酸电池生命周期铅流图。这是分析铅酸电池系统与外部环境之间定量关系的基础。

(2)确立了铅流评价指标,其中资源效率和环境效率用于评价铅酸电池系统和外部资源与环境之间的关系;铅的循环率、排放率、产量变化比用于评价系统内部铅的流动状态。

(3)获得了铅酸电池系统的铅流规律。结果表明,提高铅的循环率、降低铅的排放率有助于提高铅的资源效率和环境效率。

(4)分析了中国铅酸电池系统的铅流状况。结果表明,与瑞典铅酸电池系统相比,中国铅的资源效率和环境效率十分低下。其原因主要在于中国铅的循环率较低,而排放率偏高。

(5)挖掘了中国铅的循环率低下和排放率偏高的深层原因,提出了改善对策,并预测,20年后,中国的铅矿资源和环境状况将明显好转。

参 考 文 献

姜松. 2000. 中国再生有色金属资源的开发利用. 中国资源综合利用,(1):18-21.
兰兴华, 殷建华. 2000. 发展中的再生铅工业. 中国资源综合利用,(8):19-21.
李富元, 李世双, 王进. 1999. 国内外再生铅生产现状及发展趋势. 世界有色金属,(5):26-30.
陆钟武. 2000. 关于钢铁工业废钢资源的基础研究. 金属学报, 36(7):728-734.
陆钟武. 2002. 钢铁产品生命周期的铁流分析——关于铁排放量源头指标等问题的基础研究. 金属学报, 38(1),

58-68.

马永刚. 2000. 关于改变中国再生铅现状和制定有关政策法规的建议. 资源节约和综合利用, 3（1），15-19.

马永刚. 2001. 铅污染现状、原因及对策. 中国资源综合利用，(2)，26-27.

毛建素, 陆钟武. 2003. 铅在铅酸电池中的资源服务效率. 东北大学学报（自然科学版），24（12）：1173-1176.

山本良一. 2003. 战略环境经营生态设计——范例100. 北京：化学工业出版社，177-179.

杨春明, 马永刚. 2000. 中国废铅蓄电池回收和再生铅生产. 见(In)：中国电工技术学会铅酸蓄电池专业委员会. 第七届全国铅酸蓄电池学术年会论文全集. 广东南海，198-202.

中国科学院可持续发展研究组. 2000. 年中国可持续发展战略报告. 北京：科学出版社，2000.207.

中国有色金属工业年鉴编辑委员会. 2001. 中国有色金属工业年鉴2001. 北京：中国有色金属工业年鉴出版社，323-344.

Ayres R U. 1994. Industrial metabolism: theory and policy. The greening of industrial eco-systems. Washington D.C: National Academy Press，23-37.

Graedel T E，Allenby B R. 1995. Industrial ecology. New Jersey：Prentice Hall，110.

Joosten L A J，Hekkert M P，Worrell E，et al. 1999. STREAMS: a new method for analysis material flows through society. Resources Conservation and Recycling，27（3）：249-266.

Karlsson S. 1999. Closing the technospheric flow of toxic metals，modeling lead losses from a lead-acid battery system for Sweden. Journal of Industrial Ecology，3（1）：23-40.

Kleijn R. 2000. In=out，the trivial central paradigm of MFA?. Journal of Industrial Ecology，3（2-3）：8-10.

Stahel W.R. 1994. The utilization-focused service economy: resource efficiency and product life extension. The Greening of Industrial Ecosystems. Washington DC：National Academy Press，178-190.

Stahel W.R. 1997. The functional economy: cultural and organizational change. The industrial green game. Washington DC：National Academy Press，91-100.

第6章 中国铅流分析
The Industrial Flow of Lead in China[*]

6.1 Introduction

Lead is naturally deposit in the racks of the lithosphere, and is transferred and cycled through the soils, atmosphere, hydrosphere, and biosphere by means of weathering of rocks, volcanic emissions, atmospheric sedimentation, water and wind erosion, and biological ingestion. These lead-flow processes are collectively termed the 'natural' lead flow (Annegrete and Bruvoll, 1998). In recent centuries, lead has been widely used in many industrial fields such as mechanical, electronic, and chemical engineering with a deeper understanding of its properties and a big progress in industrial technology (Dong, 2000). To meet the human demand, it is needed to mine lead ore, then produce various lead products through series of processes, such as lead concentration, smelting and machining. An anthropogenic lead flow is thus been formed, which can be named the 'industrial flow of lead' (IFL) because it is tightly related to industrial processes.

With rising lead consumption in the world, the present scale of IFL has far surpassed the environmental carrying capacity, which may result in unsustainable use of lead ore resource and worse environment quality due to an accumulation of lead pollutants. This situation has become particularly serious in China in recent years. Statistical data (Editorial Staff of the Yearbook of the Nonferrous Metals Industry, 1990-2001; Editorial Staff of the Yearbook of China Machinery Industry, 1986-2001) indicate that the annual production and consumption of metallic lead have been increased rapidly, especially for the production of lead-acid batteries (LABs), which is the main lead product in China (Table 6.1). Meanwhile, the lead ore reserves of China seem to be exhausted, many lead ores had to be imported to meet the demand,

---

[*] Mao J S., Yang Z F., Lu Z W. 2007. Industrial flow of lead in China. Transaction of Nonferrous Metals Society of China, 17 (2): 400-411.

and the imports of lead ore accounted for 1/3 of the total domestic demand in 2000. On the other hand, the anthropogenic lead flow has reached about 12 times higher than its estimated natural flow (The Research Board on Sustainable Development of China Academy of Science, 2000). Thus, it is necessary to find ways to improve the relationship between China's lead industry and its environment and to reduce the environmental impacts of the lead system.

Table 6.1 Annual production of lead and lead-acid batteries, and domestic lead consumption in China

Year	1990	1991	1992	1993	1994	1995	1996	1997	1998	2000	2001
Lead production/kt	296.5	319.7	366.0	411.9	467.9	607.9	706.2	707.5	756.9	918.4	1 099.9
Lead consumption/kt	244.0	249.9	259.3	299.6	298	447.7	464.3	529.9	530.2	525.0	663.0
Battery production/ (GWh)	6.980	5.146	6.837	7.773	—	7.080	9.487	—	—	10.394	11.881

LABs are the main lead product and account for around 70% of the total domestic lead consumption in China in 1996 (Li et al., 1999). Mao and Lu (2003a) thus studied the impacts of China's LAB system on the lead ore resource, and Mao et al (2006; 2008) analyzed the lead flow and the eco-efficiency of lead in lead-acid battery system. However, this study seemed insufficient to capture the system for the whole lead industry because more than one kind of lead product is produced throughout the overall system, and different product systems may impact the environment in different ways. For this reason, we believe that the impacts of the overall lead industry system on the lead ore resource may be much more complex than is indicated by an examination of just the LAB system. Moreover, lead wastes or pollutants (called "lead emissions" hereafter) are inevitably produced in the IFL, and the quantity of lead emissions is tightly related to the quality of the environment. Therefore, a theoretical study of the relationships between the systems for various lead products, the system for the lead industry as a whole (the "lead industry system" henceforth), and their respective environments will have significant implications for the environment.

As one of the largest country of the world in lead production and consumption, China accounted for 16.35% and 10.62% of the totals, respectively (Editorial Staff of the Yearbook of the Nonferrous Metals Industry, 1990–2001). Thus, the status of the IFL in China will greatly influence the global lead ore resource and lead-related environmental quality. Therefore, the present study will have important and practical significance for lead ore conservation and environmental improvement.

In this chapter, we studied the lead industry system in China, with an emphasis on the flow of lead within the system. Lead mining, concentration, smelting, and refining, as well as the manufacturing, use, disposal, and recovery of lead products are the main components of the system.

The study was composed of two parts: a theoretical study and a case study. In the theoretical study, we developed a model of the IFL based on the analysis of the industrial flow of lead and derived a quantitative relationship between a product's system and its environment, allowing us to formulate the fundamental rules for IFL. Based on these rules, we proposed several indices to evaluate IFL and the relationship between the lead industry system and its environment. In the case study, we analyzed China's IFL in 1999, evaluated its status, and identified the problems existed in lead industrial system by comparing the IFL for China's lead industry system with that of Sweden's lead-acid battery system. Moreover, we analyzed the main factors supporting these findings and proposed countermeasures for improvement of the IFL in China.

Since we specially emphasized the flow of lead in our study, the lead content in a material we used represented the quantity of the corresponding material.

6.2 Theoretical Study

6.2.1 The IFL Model

1. The IFL model of a lead product's system

In general, a lead product system is a series of processes in the product life cycle that starts from the natural resource (lead ore) and consists of several stages such as the production of primary materials, manufacturing of products, use of products and its recovery (Graedel and Allenby, 1995). In order to obtain the rules for the behavior of lead flow in a system, we let lead itself to represent the product. Based on this assumption, the components of the product system can be simplified into the following phases: primary lead production (including mining, concentration, smelting and refining), manufacturing of the product, the use and recovery of the product.

Lead flows through every stage in a product life cycle, not only from the lead ore resource to the lead product, but also from lead scrap (as secondary resource for lead refining) to lead metals. While the lead passes through these stages, some of the lead will also flow into the environment as wastes or pollutants (emissions). Thus, the relationship between a product's system and its environment will appear as follows:

①to provide products to society; ②to consume lead ore (thus forming a load on the lead ore resource); ③to emit lead wastes or pollutants into the environment (thereby causing an emission load). Both the load on the lead ore resource and the lead emission load are collectively considered the environmental impacts.

In order to study the relationship between a product system and its environment, we assumed as follows:

(1) The life span of a product is $\Delta \tau$ years.

(2) The time expense in various production processes can be ignored, since it is relatively very short compared to the product life span.

(3) Each product becomes obsolete $\Delta \tau$ years after its production, and some of the obsolete products become scrap lead (termed "old scrap lead") through a collection process.

(4) All the scrap lead is recycled and refined as secondary lead in the year when the scrap is formed.

Based on these assumptions, we may illustrate the lead-flow diagram for a product life cycle in Fig. 6.1, which reflects the directions and distribution of the lead flow during every stage of the product life cycle. This flow obeys the "conservation law", in which inputs equal outputs (Kleijn, 2000) for every stage. In addition, the production of primary and secondary lead is combined into a single stage since they both belong to lead production, and are represented by stage I in Fig. 6.1.

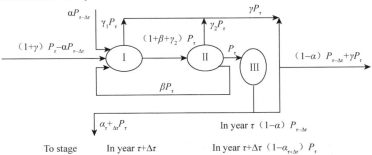

Fig. 6.1 Lead-flow diagram for lead product life cycle

Stage I Lead mining, concentration, smelting, and refining;
Stage II Lead product manufacturing; Stage III Use of lead product

In Fig. 6.1, we assume that the annual production of products changes yearly, and the productions in years τ and $\tau - \Delta \tau$ are P_τ and $P_{\tau - \Delta \tau}$, respectively, with units of t/a. Similarly, it is clear that if the life-span of a product is $\Delta\tau$, then the products manufactured in year τ will become obsolete and form its old scrap in year $\tau+\Delta\tau$, and the scrap that becomes production inputs in year τ will come from the lead products produced in year $\tau-\Delta\tau$. Some of the obsolete products are collected and returned to the lead production stage through recycling processes. In order to simplify the formula, we may define the ratio of the old scrap lead that is recycled in year τ to the total production of products $\Delta\tau$ years τ ago as the recycling rate, and can represent this rate as α. Under these conditions, $\alpha_{\tau+\Delta\tau}P_\tau$ of old scrap lead will become inputs for lead production in year $\tau+\Delta\tau$, and $\alpha_\tau P_{\tau-\Delta\tau}$ will become the inputs in year τ. The subscript τ for the recycling rate in year τ has been omitted in Fig. 6.1 for simplicity.

Some other indices involved in Fig. 6.1 are explained as follows:

β, the manufacturing recycling rate (with units of (t/t)), is defined as the ratio of the scrap lead produced in the manufacturing of products (termed as "prompt scrap lead") that is recycled to the total production of lead products in the same year.

γ_1 and γ_2 are the lead emission rates in stages Ⅰ and Ⅱ (with units of (t/t)), respectively, are defined as the ratio of the lead emissions in the corresponding stage to the production of lead products in the same year. The sum of the two ratios is defined as the "overall lead emission ratio" and is represented by γ (i.e., $\gamma=\gamma_1+\gamma_2$).

The method we have used to describe IFL is called "element flow analysis" (EFA), which is one of many kinds of "material flow analysis" (MFA) (Kleijn, 2000; Ayres, 1994; Hansen and Lassen, 2003; Joosten et al., 1999). Some basic characteristics of EFA are as follows:

(1) Only one element in the product studied is traced. (In the present study, that element is lead.)

(2) The time interval between manufacturing and disposal of the products is considered.

(3) Changes in the annual production of the products are considered.

Lu (2000) first proposed this method in 2000 and successfully used it to study scrap steel, iron emissions (Lu, 2002) and energy (Liu et al., 2003). Mao and Lu (2003a) improved this method by focusing on the final products and taking the fiscal year as the statistical period to permit a more direct link with the social environment and to facilitate data gathering.

2. The IFL model of a complex lead product system

It is easy to understand that to meet various social demands on lead products, a lead industry system should include more than one product system. Thus, a series of product systems must be considered together to represent the overall lead industry system, and the system for each of these products acts as a subsystem of the overall system. If we assume that each subsystem concerns only one kind of product, and the annual production of that product is expressed by P_i, the load on the lead ore resource and the lead emission load can be expressed by R_i and Q_i, respectively. The overall lead industry system will thus have a total annual production of $P = \sum P_i$, a total load on the lead ore resource of $R = \sum R_i$, and a total lead emission load of $Q = \sum Q_i$. This model is illustrated in Fig. 6.2, which reflects the relationship between the lead industry system and its environment.

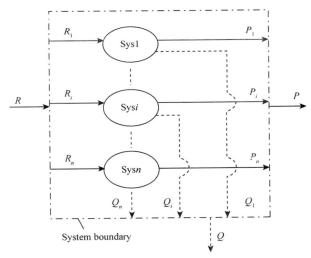

Fig. 6.2 Conceptual model of relationship between lead industry system and its environment

6.2.2 Evaluation Indices: External Indices

In order to estimate the quantitative relationship between a product system and its environment, we have introduced the concept of eco-efficiency (OECD, 1998). This concept defines the output of final products (the social benefit) per unit of environmental impact as the eco-efficiency of the product system. In this study, we focused on lead, and the eco-efficiency has two components: one is related to the consumption of lead ore, and is named the resource efficiency (RE) and represented by r, the other is related to lead emissions, and is termed the environmental efficiency (EE) and represented by q. RE and EE can be expressed as follows, respectively:

$$r = \frac{P}{R} \quad (6.1)$$

$$q = \frac{P}{Q} \quad (6.2)$$

Where R and Q represent the load on the lead ore resource and the lead emission load per year, respectively. Both have the units of tons of lead content per year (t/a).

From equation (6.1) and equation (6.2), we can see that a higher eco-efficiency means reduced consumption of lead ore, reduced lead emissions, or both simultaneously for a given level of output provided by a product system. Thus, increased eco-efficiency means better resource conservation and environmental protection. The above analysis shows that eco-efficiency forms the bridge between a product's system and its environment, and can therefore be treated as an evaluation index for the IFL within a system.

6.2.3 Primary Regulation of the System

1. Resource efficiency

Fig. 6.1 shows that for a single product's system, a lead ore input of $(1+\gamma)P_\tau - \alpha P_{\tau-\Delta\tau}$ will produce P_τ of product. Based on equation (6.1), we can derive the following equation for resource efficiency:

$$r = \frac{1}{1+\gamma-\alpha p} \quad (6.3)$$

where p represents the production ratio (t/t) in a product's life cycle and is defined as $p = P_{\tau-\Delta\tau}/P_\tau$. This value, which represents the ratio of the quantity of a product in year $\tau-\Delta\tau$ to that in year τ, is always positive.

Equation (6.3) shows that the RE of lead in a product's system is a function of the lead recycling rate (α), the lead emission rate (γ), and the production ratio (p) in the system. Further analysis shows that a higher recycling rate, a reduced lead emission rate, or a decreased production of the product will improve the RE of a product's system.

For the lead industry system, the RE equals the total lead products produced by the system divided by the total lead ore consumption. Under the model illustrated in Fig. 6.2, we obtain the following equation:

$$r' = \left[\sum f_{Pi} \cdot \frac{1}{r_i}\right]^{-1} \quad (6.4)$$

where f_{Pi} represents a fraction equal to the production of lead product i divided by total production of all lead products, and is expressed as $f_{Pi} = P_i/P$ with units of (t/t). The sum of all f_{Pi} in the system equals 1 (i.e., $\sum f_{Pi} = 1$).

Equation (6.4) shows that the RE of lead in the lead industry system is tightly related to both the composition of the range of products and the individual RE in each product's system. We thus conclude that in order to improve the RE of lead in the lead industry system, we must improve the RE of lead in each product's system and optimize the composition of the various lead products in the system by increasing the f_{Pi} of the products with higher values of RE.

In practice, the RE of the lead industry system can also be expressed as Equation (6.3), and the corresponding parameters (such as the lead recycling rate, lead emission rate, and the production ratio) can be treated as the nominal parameters of the lead industry system.

2. Environmental efficiency

Fig. 6.1 also shows that a product's system will produce $\gamma P_\tau + (1-\alpha)P_{\tau-\Delta\tau}$ of lead emissions into the environment while producing P_τ of products. Based on Equation (6.2), we can derive the EE of lead in a single product's system as follows:

$$q = \frac{1}{\gamma + (1-\alpha)p} \quad (6.5)$$

Equation (6.5) shows that the EE of lead in a product's system is also a function of the lead recycling rate (α), the lead emission rate (γ), and the production ratio (p). Further analysis shows that a higher recycling rate, a reduced lead emission rate, an increase in the total production of lead products, or all three changes together will improve the EE of lead in a product's system. Note that in this case, the influence of the production ratio (p) on EE is very different from that for RE.

For the lead industry system, EE equals the total products produced by the system divided by the total lead emissions into the environment. Based on the model illustrated in Fig. 6.2, we obtain the following equation:

$$q' = \left[\sum f_{Pi} \cdot \frac{1}{q_i}\right]^{-1} \quad (6.6)$$

where q_i represent the EE of lead in the system for product i and f_{Pi} has the same meaning as in previous equations.

Equation (6.6) shows that the EE of the lead industry system is tightly related to the composition of the range of lead products in the overall system and the individual

EE of lead in each product's system. We thus conclude that in order to improve the EE of lead in the lead industry system, we must improve the EE of lead in each subsystem and optimize the composition of products in the system by increasing the fraction (f_{Pi}) of products with higher values of EE.

In practice, the EE of lead industry system can be expressed as equation (6.5), with the corresponding recycling rate, lead emission rate, and production ratio used as the same nominal parameters described in the previous section.

3. The relationship between EE and RE

By combining equation (6.3) with equation (6.5), we can describe the relationship between EE and RE:

$$\frac{1}{q} - \frac{1}{r} = p - 1 \qquad (6.7)$$

Equation (6.7) shows that the relationship between EE and RE is tightly related to the production ratio (i.e., to the ratio of the production in year $\tau - \Delta\tau$ to that in year τ). Further analysis shows that EE will equal RE when the production of lead products remains constant, whereas the RE will be less than and greater than EE, respectively, with increasing and decreasing production of lead products. These results are mainly due to the expansion or shrinkage of a product's system under different situations; that is, the system expands with increasing production, and consumes more lead ore while emitting less lead into the environment, thus the value of RE is less than that of EE. Conversely, the system shrinks with decreasing production, and consumes less lead ore while emitting more lead into the environment, thus the value of RE is greater than that of EE.

6.2.4 The Evaluation Indices: Internal Indices

The above analysis of equation (6.3) and equation (6.5) shows that the lead recycling rate, lead emission rate, and production ratio in the product's life cycle are the internal factors that affect the eco-efficiency of lead in a specific product's system. Because they reflect the links among the internal components of the product's system, they can thus be termed the "driving factors" of eco-efficiency. The three factors can thus be treated as internal indices for evaluating the lead flow within a product's system.

For the lead industry system, equation (6.4) and equation (6.6) reveal that the composition of the products (or the structure of lead consumption) and the individual eco-efficiency in each product system are the internal factors that drive the eco-efficiency of the system, and can thus be treated as the internal evaluation indices for lead flow in the lead industry system as well.

Because the production ratio in the product life cycle is related to the life span of lead product, and different life spans result in different production ratios for the same annual growth in production. For instance, if we assume that the annual growth rate of production is 0.10, the production ratio will be 0.7 when the life span is 3 a, whereas the ratio will be 0.6 when the life span is 4 a. The average life span of lead products or the individual product life span can be treated as internal evaluation indices for the lead flow in the lead industry system as well.

6.3　Case Study: the Industrial Flow of Lead in China

6.3.1　Brief Description of Lead Flow in China

The case study described in this section is based on the statistical data for all of Chinese lead-related industries in 1999. It is reported that the domestic consumption of refined lead in 1999 is about 525 kt (Editorial Staff of the Yearbook of the Nonferrous Metals Industry, 1990-2001). Of this amount, 66.8% was used in the manufacturing of LABs, 11.6% was used in construction materials and cables, and the remaining 21.6% was used in chemical engineering (Ma, 2000).

During the process of manufacturing LABs, every 1 t of lead inputs produces an average of 0.9200 kt of LABs, 0.0356 t of scrap lead that would be promptly recycled, and 0.0444 t of lead emissions into the environment (Mao and Lu, 2003). The average life-span of Chinese LABs has been estimated at 3 a (Lan and Yin, 2000).

During the manufacturing of lead-related construction materials and cables, the lead utilization rate ranges from 0.85 to 0.95 (Jiang, 2000). In the present case study, we chose a value of 0.87 (87%). An estimated 11.26% of the lead input for these processes is transformed into scrap lead that is promptly recycled, and the remaining 1.74% percent is dissipated into the environment as lead emissions. The average lifespan of these lead products was estimated as 15 a (Jiang, 2000).

During the production of lead products used in chemical engineering, the lead utilization rate is estimated at 0.9524 (95.24%), and the remaining 4.76% is dissipated into the environment as lead emissions. These lead products cannot be recycled after their uses, and most of the lead will be permanently dissipated into the environment;

thus, the average life span of these products is considered to be 0 year.

It has been estimated that 110.84 kt scraps were recycled in 1999, of which 82.01% represented obsolete LABs, 0.54% represented obsolete lead-related construction materials and cables, and the remaining 17.45% represented scrap lead that was promptly recovered for reuse during the manufacturing of lead products (Li et al., 1999). The overall recovery rate in secondary lead smelting and refining changed from 80% to 88% (Ma and Yang, 2001). We used a value of 82.31% for the calculations in the present case study. Thus, 91.23 kt of secondary refined lead was recovered in total. The remaining domestic lead consumption would be primary refined lead, at an estimated total of 433.77 kt.

During the production of primary lead, many processes are involved, including mining, concentration, smelting, and refining. The recovery rate in lead mining and concentration averaged 83.82% in 1999, versus 93.49% for lead smelting and refining (Editorial Staff of the Yearbook of the Nonferrous Metals Industry, 1990-2001). Therefore, the production of 433.77 kt of primary refined lead would consume 553.54 kt of lead ore.

Based on the above analysis, the industrial flow of lead in China in 1999 is illustrated in Fig. 6.3.

With an assumed LAB lifespan of 3 years, obsolete LABs recycled in 1999 would have been manufactured in 1996. Based on the data in Table 6.1, the production of LABs in 1996 accounted for an estimated 291.67 kt of lead content based on the assumptions that the data in Table 6.1 represent 77% and 78% of the total national production of LABs in 1996 and 1999, respectively, and that both the LAB life-span and the LAB specific energy (a coefficient used to convert the production of LABs from energy units into a lead content) remained constant. Because 90.90 kt of obsolete LABs were recovered in 1999, the remaining 200.77 kt of obsolete LABs had not been recycled (or had not been included in the statistical data) and were thus treated as lead losses into the environment. For the product system related to construction materials and cables, with an estimated production in 1984 of 55.3 kt, the old scrap lead from these products recovered in 1999 was estimated at 0.6 kt, and we thus estimated that about 54.70 kt of obsolete lead-related construction materials and cables had not been recycled (or had not been included in the statistical data) and was thus treated as lead losses into the environment. All lead using in chemical engineering products was considered to be lead losses into the environment.

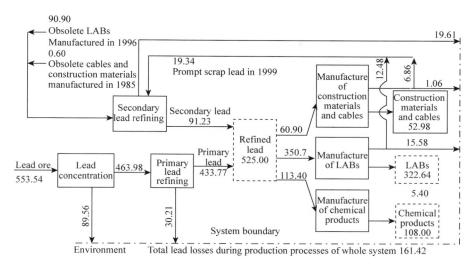

Fig. 6.3 Lead flow diagram for China in 1999 (kt)

6.3.2 Data Sources

The sources of the data related to China's IFL in 1999 are listed in Table 6.2.

Table 6.2 Sources of data for the case study

Data type or name	Data source	Agency responsible for compiling the data
Recovery rate in lead mining	China Investigation Report on the Exploitation and Utilization of Lead-zinc Mineral Resource 2000	Beijing General Research Institute of Mining and Metallurgy
Recovery rate in lead concentration, smelting, and refining	China Nonferrous Metals Industry Yearbook (1990-2001)	Editorial staff of the Yearbook of the nonferrous metals industry
Data related to scrap lead and lead recycling	Published literature or actual manufacturing data	Partly provided by the China Association for Metals Recycling
Data related to battery manufacturing	Report on the environmental impacts for some lead-acid battery companies	Research Institute of Environment Science
Battery performance and profiles	China Statistic Report on Lead-acid Batteries	Shenyang Research Institute of Storage Battery
Annual production of lead-acid batteries	China Machinery Industries Yearbook China Power and Electrical Equipment Yearbook	Editorial staff of the Yearbook of machinery industries Editorial staff of the Yearbook of nonferrous metals industry, power and electrical equipment
Export of lead-acid batteries and scrap lead	China Foreign Trade Yearbook	Editorial staff of China Foreign Trade Yearbook

6.3.3 Evaluation of the IFL in China

1. The evaluation indices

(1) Eco-efficiency.

For the lead industry system in 1999, Fig. 6.3 shows that the total consumption of lead ore was 553.54 kt, and the total production of lead products was 483.62 kt; a total of 524.89 kt of lead were lost into the environment, which includes 200.77 kt of obsolete LABs, 54.70 kt of discarded lead-related construction materials and cables, 108.00 kt of chemical engineering products, and 161.42 kt of lead emissions in production-related processes. Using equation (6.1) and equation (6.2), we calculated the RE and EE of lead in the lead industry system of China as follows:

$r' = 0.874$ and $q' = 0.921$

These results could also be obtained by estimating the eco-efficiency of lead for each product system and summing up the values using equation (6.4) and equation (6.6). The results are the same, which indicates that the calculation is correct.

(2) The composition of the lead products.

Fig. 6.3 shows that the production of lead products in 1999 totaled 483.62 kt, of which the production of LABs, construction materials and cables, and chemical engineering products totaled 322.64, 52.98, and 108.00 kt, respectively. We thus can estimate the contributions of the three kinds of lead product to the total production of lead products as 66.71%, 10.95%, and 22.34%, respectively.

(3) The production ratio.

Substituting the values of r' and q' into equation (6.7) produces a result of $p' = 0.942$. Because this value is less than 1, the production of lead products in China is increasing yearly.

(4) The lead emission rate.

Fig.6.3 shows that the total lead emissions into the environment during production processes in 1999 equaled 161.42 kt. Based on the definition of the lead emission rate in the first part of this chapter, $\gamma' = 0.334(t/t)$, which indicates that 0.334 t of lead emissions will result during production processes for every 1 t of lead products that is produced.

(5) The lead recycling rate.

Substituting the values of r', p', and γ' that we obtained earlier in this section into equation (6.3), $\alpha' = 0.202(t/t)$, which indicates that only 20.2% of the total lead products were recycled in 1999. If we substitute the values of q', p', and γ' into equation (6.5), we also derive $\alpha' = 0.202(\text{t}/\text{t})$, which confirms that the above calculation is correct.

In the same way, we can estimate the IFL in each lead product's system by distinguishing its data from that of the other products. For instance, we could analyze the data for collected obsolete LABs, the scrap lead created during the manufacture of LABs, and the refining of this scrap lead for reuse in the LAB system. Such work is always very complex. For the present study, we have summarized the evaluation indices for the LAB system, the construction materials and cables related system, the chemical engineering products system, and the lead industry system as a whole in Table 6.3.

Table 6.3 Summary of the evaluation indices for several lead product systems in China

Item	Product profile /%	Resource efficiency	Environmental efficiency	Lead recycling rate	Lead emission rate	Production ratio	Life span /a
Lead-using industry	100.00	0.874	0.921	0.202	0.334	0.942	4
Lead-acid batteries	66.71	0.960	1.057	0.312	0.324	0.904	3
Construction materials and cables	10.95	0.751	0.727	0.011	0.342	1.044	15
Chemical engineering products	22.34	0.736	0.736	0.000	0.359	1.000	0

2. Discussion

Table 6.3 shows that the different product systems had different eco-efficiency values. The eco-efficiency of the LAB system was highest, with both the RE and the EE greater than the average for the lead industry system and the values for the other two product systems. In contrast, the eco-efficiency of the construction materials and cables system was intermediate, and was lower than the average for the lead industry system (the lower EE in this system was mainly due to decreasing production), whereas the eco-efficiency of the chemical engineering system was the lowest and was lower than the average for the lead industry. Further analysis shows that the different eco-efficiencies resulted mainly from different abilities to recycle the lead from the various lead products. Ayres sorted the use of materials into three classes (Ayres, 1994).

(1) Uses that are economically and technological compatible with recycling under present prices and regulations, such as the use of lead in LABs in the present study, where the lead recycling rate in China reaches 0.312.

(2) Uses that are not economically compatible with recycling but where recycling is technically feasible, such as the use of lead in construction materials and cables in the

present study, where the lead recycling rate in China is only 0.011.

(3) Uses for which recycling is inherently not feasible, such as the use of lead in chemical engineering products, where the lead recycling rate is zero.

Of the three lead product systems in the present study, the LAB system is the system that is potentially most able to harmonize with its environment. The relatively small difference between the eco-efficiencies of the three lead product systems is mainly due to the fact that each had similar lead emission rates (Table 6.3). The production ratio in the LAB system was the smallest of the three systems, which indicates that the production of LABs is increasing fastest compared with that of the other products, and that the contribution of LABs to the total production of lead products has been increasing in recent years.

To improve the eco-efficiency of lead in the lead industry system, we have two main options. First, we can increase the contribution of LABs to the total production of all types of products and gradually reduce the use of lead in the other systems. Second, we can improve the eco-efficiency of the LAB system.

In order to assess the potential for improving China's lead industry system, we compared the evaluation indices for China's LAB system with those for Sweden (Table 6.4), where the data for Sweden has been estimated based on the data provided by Karlsson (1999).

Table 6.4 Comparison of lead flow in the lead-acid battery systems of China and Sweden

Country	Resource efficiency	Environmental efficiency	Lead recycling rate	Lead emission rate	Production ratio	Life span/a
China	0.960	1.057	0.312	0.324 000	0.904	3
Sweden	79.020	79.020	0.990	0.002 655	1.000	5

Table 6.4 shows that both the RE and the EE of lead in Sweden's LAB system had reached 79.02 (t/t), which represents 82.31 and 74.76 times the corresponding values for China. The main reasons for this difference are as follows:

(1) The lead recycling rate for Sweden has reached 0.99 (t/t), which means that nearly all of the obsolete LABs are recycled. In contrast, the corresponding rate for China is only 0.312, which means that nearly 70% of the obsolete LABs are not recycled.

(2) The lead emission rate for Sweden's LAB system is only 0.002 655 (t/t), which means that there are almost no lead emissions from the system. In contrast, China's emission rate is 0.324 which means that nearly 33% of the lead inputs used in the LAB system are lost into the environment.

(3) The production of LABs in Sweden has remained constant for at least 5 years (a period equal to one LAB life span in Sweden), whereas production in China has increased rapidly during the same period.

6.3.4 Analysis of Causes and Proposed Improvements

The previous analysis indicates that to improve the eco-efficiency of lead in China's lead industry system, attention should focus on increasing the lead recycling rate and reducing the lead emission rate. Thus, additional discussion of the causes of China's low recycling rate and high emission rate is necessary to permit us to propose improvements in the system.

1. Reasons for the low recycling rate

It is easy to see from Fig. 6.1 and Fig. 6.2 that the recycling rate of the lead industry system relates mainly to the composition of the various lead products in the system, the domestic consumption of these products, the collection of obsolete lead products for recycling, and the trade in scrap lead, as well as on the integrity of the data gathered and other factors (Mao and Lu, 2003b).

Fig. 6.3 shows that in 1999 the contributions of LABs, construction materials and cables, and chemical engineering products to the total production of lead products in China amounted to 66.71%, 10.95%, and 22.34% of the total, respectively. Because the lead recycling rate in the construction materials and cables system was only 0.011 (Table 6.3) and the lead used in chemical engineering is not recyclable, both can be effectively ignored in the present case study. Thus, we can deduce that 33.29% of the total lead in the lead products system was lost into the environment in 1999.

The annual production and domestic consumption of LABs for a recent 10-year period are summarized in Table 6.5, which indicates that about 7.62% of the obsolete LABs cannot be recycled back into China's LAB system because of the export of LABs.

Table 6.5 The production, domestic consumption, and export of lead-acid batteries in China

Year	1986	1987	1988	1989	1990	1991	1992	1993	1994	1995	1996	Average
Production/ (GWh)	3.220	5.072	4.550	—	6.980	5.146	6.837	7.773	—	7.080	9.487	
Export/ (GWh)	0.158	0.477	0.007	—	0.000	1.968	0.020	0.014	0.022	0.011	0.000	
Export/production/%	4.91	9.40	0.15	—	0.00	38.24	0.29	0.18	—	0.16	0.00	7.62

The trade in scrap lead from 1990 to 2000 is summarized in Table 6.6, which shows that the trade in scrap lead is very small compared with the total production of lead products (for instance, an average of 0.983 kt vs. the 483.62 kt of lead products produced in 1999), and has been nearly balanced in recent years. Thus, we can ignore the influence of the trade in scrap lead on the lead recycling rate in the present study.

Table 6.6 The trade in scrap lead in China

Year	1990	1991	1992	1993	1994	1995	1996	1997	1998	1999	2000	Average
Exports/ (kt)	6.132	5.250	1.500	0.621	1.510	0.589	0.152	0.061	1.751	1.060	0.037	
Imports/ (kt)	1.690	0.140	5.712	7.322	5.793	5.690	0.820	0.204	0.007	—	0.050	
Net imports/ (kt)	−4.442	−5.110	4.212	6.701	4.283	5.101	0.668	0.143	−1.744	—	0.013	0.983

Ma (2000) and Yang and Ma (2000) have reported that there were about 300 secondary lead refineries in China in 1999, among which only three could be considered large-scale operations that together produced nearly half of the secondary lead. They also reported that most of these facilities were privately owned and were operated on a small scale. About half of the obsolete LABs are collected and recovered by these private refineries, and this data may not be fully accounted for in the recorded statistics. Many small units are engaged in the collection of obsolete LABs in China in 1999, such as the supply and marketing systems of business, individuals or small groups that collect obsolete LABs for secondary lead refineries, of which the individuals played a dominant position in the collection and gathering of obsolete LABs. In other words, no national or regional network exists for the collection of obsolete LABs, and this work is still being done by individual, uncoordinated operations (Ma, 2000). This situation potentially decreases the recycling rate by 0.202. Consequently, we can estimate that about 18.69% (i.e., the result of 1-0.3329-0.0762-0.202-0.202) of the obsolete lead products in 1999 were not recovered and were thus lost into the environment.

From the above discussion, we conclude that the main reasons for the low lead recycling rate in China are the use of non-recyclable lead, which still contributes strongly to the total, and the inefficient collection of obsolete lead products.

2. for the high emission rate

Fig. 6.3 shows that 161.42 kt of lead was lost in various production-related processes in China's lead industry system in 1999. Further analysis of this total is summarized in Table 6.7, which shows that most of the lead losses occurred during lead concentration, followed by refining, then manufacturing.

Table 6.7 Lead losses during the production processes in China's lead industry system

Item	Lead concentration	Primary lead refining	Secondary lead refining	Manufacturing of lead products	Total
Lead loss/kt	89.56	30.21	19.61	22.04	161.42
Percentage of total/%	55.48	18.72	12.15	13.65	100.00

The recovery rate during lead concentration was only 81% to 86%, which is 5% to 15% lower than the recovery levels in other countries (Beijing General Research Institute of Mining and Metallurgy, 2001). This is mainly because of the poor quality of Chinese lead ore resources, with lead-zinc para-generated, many metals concomitant, and a complex distribution of metals in the minerals. Chinese lead ore usually has a low lead to zinc ratio (1∶2.6) and contains more than 50 kinds of useful metals (e.g., copper, silver, gold) in a complex distribution that makes the concentration process unusually difficult (Beijing General Research Institute of Mining and Metallurgy, 2001).

In the smelting and refining processes, Ma and Yang (2001) and Yang and Ma (2000) reported that about half of the lead (including both the primary and the secondary) was smelted and refined by small-scale enterprises through outdated technology. Most of these enterprises still utilized traditional sintering and blast-furnace, which means the technology during the 1980s in the advanced countries has been replaced by more advanced processes.

In the manufacturing of lead products, Jiang (2000) reported that the lead utilization rate is usually about 0.85 to 0.95. Based on the data obtained in this study, it is estimated that about 41.38 kt of lead was not utilized efficiently in these processes, of which 46.74% (Lea, 19.34 kt of lead) was promptly recycled as lead scrap, while the remained one was dissipated into the environment as emissions, or was omitted from the statistical data. Jiang (2000) also reported that much of the lead scrap was collected and transferred to lead refineries by individuals, and thus accurate data were very difficult to obtain.

3. Possible countermeasure to improve eco-efficiency

To improve the lead recycling rate, several countermeasures might be adopted (Mao and Lu, 2003a; Ma, 2000; Yang and Ma, 2000; Mao and Lu, 2003c).

(1) To improve the types of lead products produced by the overall system (or the lead consumption) so as to increase the contribution of LABS, which has the highest eco-efficiency, to the total production of lead products, while gradually eliminating the use of lead in non-recyclable products.

(2) To take advanced management method and treat lead wastes as resources (David and Nasrw, 1994). To implement laws, regulations, and mechanisms for lead recycling so as to lead recycling of lead scrap to more effective paths (Sakuragi, 2002). Thus, we could gradually enclose the system in terms of the flow of lead (Frosch, 1994).

(3) To extend the responsibility of LAB companies so that they can sell services instead of only products (Stahel, 1994, 1997), or levy a tax upon the consumers of lead products (Bruvoll, 1998) and charge them for lead emissions into the environment (Turner et al., 1998), thereby encouraging the recycling of obsolete lead products.

To reduce lead emissions, we suggest that China should develop or introduce new technology for lead mining, concentration, smelting, and refining; improve the management of companies involved in lead production and implement a special license for these companies that strictly stipulates the production scale, technology used, and measures required for environmental protection; eliminate the use of outdated technology by small-scale companies; and promote the spread of clean production technology for lead. Thus, we could improve the overall rate of lead utilization and reduce lead emissions.

4. Forecast for China's LAB system

Ma (2004) reported that a technical policy for the prevention of pollution resulted from obsolete lead-acid batteries has been implemented. The situation for Chinese lead industry system is thus expected to be improved in the coming years.

If Chinese lead industry system can be improved, the following targets will be realized within 20 years:

(1) LABs will represent 95% of the total lead products; other products contribute only 5% of the total.

(2) Both the RE and EE of lead in the LAB system will reach 60, whereas those for other systems will reach 1.

Then, the eco-efficiency of lead in China would rise to 15.19 by around 2020, which means that for the same total production of lead products as in 1999 (i.e., 483.62 kt), the following changes would occur.

(1) The production of LABs would increase to 459.44 kt (i.e., 1.499 times the present level), which means an annual average growth rate of 2.495%. Simultaneously, the production of other lead products would decrease to 24.18 kt (i.e., only 15% of their present level), which means an annual average growth rate of -4.25%.

(2) The consumption of lead ore will decrease to 1/15.19 of the present value (i.e., to 36.44 kt of lead ore per year). This will alleviate the growing scarcity of lead ore.

(3) Lead emissions will decrease to 1/15.19 of the present level, which is lower than the environmental carrying capacity (i.e., the estimated natural flow; RBSDCAS 2000). Thus, the current overload of lead pollutants in the environment will be eliminated and the environmental quality will gradually improve.

6.4　Conclusions

(1) A model of the industrial flow of lead in the lead product system was developed based on the application of element flow analysis. This model served as the foundation of a quantitative case study on the relationships of lead flow between the lead product system and its natural and socio-economical environments.

(2) We developed evaluation indices for the industrial flow of lead. We used the eco-efficiency (including RE and EE) as external indices to reflect the relationship between the lead product system and its environment, and used the types of lead products, the lead recycling rate, the lead emission rate, and the production ratio as the internal factors that drive the lead flows and thus are treated as internal indices.

(3) Three countermeasures can improve the eco-efficiency of an individual lead product system or the lead industry system as a whole, i.e. to increase the contribution of LABS to the total lead products; to increase the lead recycling rate and to reduce the lead emission rate.

(4) The current state of Chinese IFL in 1999 was studied. The results show that the RE and the EE in both LAB system and lead industry system as a whole are only around 1, which indicates a level of around 1/80 of that achieved by Swedish LAB system. The main reasons for this difference were Chinese low lead recycling rate and high lead emission rate.

(5) Additional reasons for Chinese low eco-efficiency of lead (i.e., the reasons for the low lead recycling rate and higher lead emissions) were also studied. The main reasons for low lead recycling rate are the low contribution of LABS to the total lead products and inefficient management of the recycling of lead scraps. The main reasons for the high lead emission rate are the poor quality of lead ore resource, and an abundance of small-scale lead-related plants using outdated technologies. Several countermeasures were proposed to improve this situation and the future status of lead ore resource and environment quality was forecasted to be improved substantially within 20 years by

implementation of these countermeasures.

Acknowledgement

The authors thank several people for explaining and contributing data on specific industrial processes: SUN Chuan-yao, YANG Ling, FENG Jun-cong, and YAO Hong-fei.

References

Annegrete, Bruvoll. 1998. Taxing virgin materials: an approach to waste problems. Resources Conservation and Recycling, 22: 15-19.

Ayres R U. 1994. Industrial Metabolism: Theory and Policy, the Greening of Industrial Eco-Systems. Washington DC: National Academy Press, 23-37.

Beijing General Research Institute of Mining and Metallurgy. 2001. China Investigation Report on the Exploitation and Utilization of Lead-Zinc Mineral Resource. Beijing: Beijing General Research Institute of Mining and Metallurgy.

Bruvoll A. 1998. Bruvoll taxing virgin materials: An approach to waste problems. Resources Conservation and Recycling, 22: 15-19.

Chen J S. 1990. Environmental Geochemistry. Beijing: Ocean Press, 196-198.

David T A, Nasrw B. 1994. Wastes as Raw Materials, the Greening of Industrial Ecosystems. Washington DC: National Academy Press, 69-89.

Dong J G. 2000. The Manual of Application Materials. Beijing: China Machine Industry Press, 1122-1128.

Editorial Staff of the Yearbook of China Machinery Industry. 1986-2001. China Machinery Industries Yearbook. Beijing: China Machinery Industry Yearbook Press.

Editorial Staff of the Yearbook of the Nonferrous Metals Industry. 1990-2001. China Nonferrous Metals Industry Yearbook. Beijing: China Nonferrous Metals Industry Yearbook Press.

Frosch R A. 1994. Closing the Loop on Waste Materials, the Greening of Industrial Ecosystems. Washington DC: National Academy Press, 37-47.

Graedel T E, Allenby B R. 1995. Industrial Ecology. New Jersey: Prentice Hall, 110.

Hansen E, Lassen C. 2003. Experience with the use of substance flow analysis in Denmark. Journal of Industrial Ecology, 6 (3/4): 201-219.

Jiang S. 2000. The development and utilization of secondary nonferrous resource in China. China Resources Comprehensive Utilization, 1: 18-21.

Joosten L A J, Hekkert M P, Worrell E, et al. 1999. STREAMS: A new method for analysis material flows through society. Resources, Conservation and Recycling, 27 (3): 249-266.

Karlsson S. 1999. Closing the technospheric flows of toxic metals-Modeling lead losses from a LAB system for Sweden. Journal of Industrial Ecology, 3 (1): 23-40.

Kleijn R. 2000. In-out. The trivial central paradigm of MFA?. Journal of Industrial Ecology，3（2/3）：8-10.

Lan X H，Yin J H. 2000. The lead-recycling industry in developing. China Resources Comprehensive Utilization，8：19-21.

Li F Y，Li S S，Wang J. 1999. The present production status of recycled lead and its future in domestic and oversea. World Nonferrous Metals，5：26-30.

Liu L R，Lu Z W，Yu Q B，et al. 2003. Influence of material flow in alumina manufacturing process with Bayer method on its energy intensity. The Chinese Journal of Nonferrous Metals，13（1）：265-270.（in Chinese）

Lu Z W. 2000. A study on the steel scrap resources for steel industry. Acta Metallurgica Sinica，36（7）：728-734.

Lu Z W. 2002. Iron-flow analysis for the life cycle of steel products：A study on the source index for iron emission. Acta Metallurgica Sinica，38（1）：58-68.

Ma Y G. 2000. Some suggestions on changing present state of secondary lead refining in China and its related policy and regulations. Resources Saving and Comprehensive Utilization，3（1）：15-19.

Ma Y G. 2000. The present state of lead emissions，reasons and countermeasures. China Resources Comprehensive Utilization，(2)：26-27.

Ma Y G. 2004. The promulgation of the technical policy for the prevention of pollution by obsolete lead-acid batteries. Chinese Journal of power Sources，28（2）：100.

Ma Y G，Yang H Y. 2001. The recovery of the obsolete lead-acid battery and the secondary lead refining. Environmental Herald，1：52-53.

Mao J S，Lu Z W. 2003a. Resource-service efficiency of lead in lead-acid battery. Journal of Northeastern University，24（12）：1173-1176.

Mao J S，Lu Z W. 2003b. A study on causes of low recovery of scrap lead. World Nonferrous Metals，11：4-32.

Mao J S，Lu Z W. 2003c. Study on resource efficiency of lead for China. Research of Environmental Science，17（3）：78-80.

Mao J S，Lu Z W，Yang Z F. 2006. Lead flow analysis for lead-acid battery system. Environmental Science，27（3）：442-447.

Mao J S，Lu Z W，Yang Z F. 2008. The eco-efficiency of lead in China's lead-acid battery system. Journal of Industrial Ecology，10（1-2）：185-197.

OECD（Organization for Economic Cooperation and Development）. 1998. Eco-Efficiency. Paris：OECD.

Sakuragi Y. 2002. A new partnership model for Japan：Promoting a circular flow society. Corporate Environmental Strategy，9（3）：292-296.

Stahel W R. 1994. The Utilization-Focused Service Economy：Resource Efficiency and Product Life Extension，the Greening of Industrial Ecosystems. Washington DC：National Academy Press，178-190.

Stahel W R. 1997. The Functional Economy：Cultural and Organizational Change，the Industrial Green Game. Washington DC：National Academy Press，91-100.

The Research Board on Sustainable Development of China Academy of Science. 2000. The Strategic Report on Sustainable

Development of China in 2000. Beijing: Science Press, 207.

Turner R K, Salmons R, Powell J, et al. 1998. Green taxes, waste management and political economy. Journal of Environmental Management, 53: 121-136.

Yang C M, Ma Y G. 2000. The recovery and recycling of discarded lead-acid batteries in China. China Electrician Technique Association LAB Council, Thesis Collection for the 7th National Annual Meeting on Lead-Acid Batteries. Nanhai: Shenyang Storage Battery Institute, 198-202.

第 7 章 2005 年北京市铅的使用蓄积研究
Lead in Use Stock of Beijing in 2005[*]

物质的使用蓄积是指人类对物质的使用使得大量物质从自然资源中转移并积累到人类经济系统中,从而形成了物质的使用蓄积库(in-use-stock)(Gordon et al., 2006)。研究物质在社会经济系统中的使用蓄积量有助于定量评估物质的使用所带来的环境影响,为资源战略、产业政策和环境政策的制定提供科学基础和合理建议。因此,物质的使用蓄积研究成为热点。在国外,以美国耶鲁大学为代表的科研小组相继开展了对钢铁、铜、镍等金属的蓄积研究(Drakonakis et al., 2007; Spatari et al., 2005; Rostkowski et al., 2007); 在国内,楼俞等对 2005 年邯郸市的钢铁和铝的使用蓄积量进行了调查研究(楼俞和石磊,2008),这表明,物质蓄积的研究已经成为产业生态学领域的研究热点。

目前,铅矿资源短缺,据美国地质勘探局(U.S. Geological Survey,USGS)数据(USGS,2005)测算,全球铅矿资源仅可保证使用约 20 年。因此废铅的再生成为促进铅矿资源持续利用的热点(毛建素和陆钟武,2003)。此外,加强对废铅的回收利用还可以减少废铅带来的环境污染(Mao et al., 2007; 吴纯德等, 2009),而铅的使用蓄积量,正是影响可回收利用的废铅量的重要因素。耶鲁大学曾研究了 2000 年全球 8 个地区 52 个国家的铅元素的人为循环状况,初步估算了 20 世纪全球铅的使用蓄积量(Mao and Graedel,2009)。

由于城市中的资源随着城市的发展在不断地积累着,在某种意义上城市可称为"未来资源的富矿"(Muller et al., 2006)。因此,对于城市中的物质使用蓄积开展研究也十分必要(Gerst and Graedel,2008)。国外也有对若干发达国家城市的蓄积量的研究报道(Drakonakis et al., 2007; Rostkowski et al., 2007),但是尚未见到针对非发达国家城市尺度上铅的使用蓄积进行的研究。

7.1 研 究 方 法

7.1.1 研究对象的确定

在铅的终端消费结构中,铅酸蓄电池一直是其最主要的组成部分。据安泰科技有限公司统计,2007 年铅酸蓄电池行业铅消费量占中国铅消费总量的 75%(唐

---

[*] 曾润,毛建素. 2010. 2005 年北京市铅的使用蓄积研究. 环境科学与技术,33(8): 49-52.

武军，2008）。中国再生铅行业的原料85%来自废铅酸蓄电池，生产铅酸蓄电池的铅50%为再生铅（姜妮，2007）。由此可见，人类活动所使用的铅很大部分集中在铅酸蓄电池的应用上。因此，本研究选择铅酸蓄电池作为铅制品的代表产品。

研究表明，铅的消费量主要与城市的机动车有关（王金良和马扣祥，2005），北京是中国的首都，政治经济文化中心，正处在高速发展时期，据统计，2005年民用机动车达246.1万辆（北京市统计局，2006），并且近年来以10.4%的速度高速增长。因此，本研究选择北京作为案例城市，并以2005年作为参考年份。

7.1.2 自下而上（bottom-up）法

目前对物质的使用蓄积的分析方法，主要有自上而下（top-down）和自下而上（bottom-up）两种方法。其中自上而下的方法是通过分析特定系统在某一时期内的物质净蓄积量的累计结果来计算出系统内物质的使用蓄积量的方法，例如Spatari等对北美洲铜的蓄积的研究。而自下而上的方法则是通过分析含有所研究物质的各种不同的服务单元、系统边界内的服务单元的数量，结合每类服务单元中所研究物质的含量，计算出系统内物质的使用蓄积量的方法。例如Katherine等人对美国纽黑文市镍的使用蓄积进行的研究（2007）。与自上而下的方法相比，城市层面上的使用蓄积研究更适合采用自下而上的方法，这是因为进出城市系统边界的物质输入及输出量难以统计和确定，而产品清单等数据相对容易调查获得。因此本研究中将采用自下而上的方法。

在本研究中，为了应用自下而上的方法对北京市铅酸蓄电池中铅的使用蓄积量进行计算，需要一一清算和调查各类铅酸蓄电池的数目和铅含量，故应用自下而上法估算使用蓄积量的公式为

$$S = \sum_{i=1}^{n}(b_i \cdot m_i) \tag{7.1}$$

式中，S是铅酸蓄电池中铅的使用蓄积量，单位为kt；b_i是第i种铅酸蓄电池产品的数量，单位为unit；m_i是第i种铅酸蓄电池单位产品中所含的铅的质量，单位为kg/unit；n则是铅酸蓄电池的种类数量。

7.1.3 铅酸蓄电池的产品种类

铅酸蓄电池按用途可分为启动型、固定型及牵引型蓄电池，按产品的应用场所则可分为机动车用蓄电池、电动自行车用蓄电池、铁道用蓄电池、通讯用蓄电池、工业电力用蓄电池和不间断电源系统用蓄电池等。为进行本研究，首先要确定系统边界内铅酸蓄电池的使用情况。根据统计资料（王金良和马扣祥，2005），可将北京城市系统中应用的主要铅酸蓄电池分为以下几类产品，如表7.1所示。

表 7.1　北京市铅酸蓄电池产品分类
Table 7.1　Classification of lead-acid products in Beijing

蓄电池类别	产品应用类别	产品应用场所
启动型蓄电池	载客汽车	机动车
	载货卡车	机动车
	摩托车	机动车
	拖拉机	机动车
	铁路机车	铁道
固定型蓄电池	通信用程控交换机	通信
	工业变电站变压器	工业电力
	计算机不间断电源系统（UPS）	UPS
牵引型蓄电池	叉车、铲车等专项作业车	机动车
	电动自行车	电动自行车
	蓄电池车	机动车

7.1.4　铅酸电池使用量及其数据来源

从式（7.1）中可看出，使用自下而上法计算北京市铅酸蓄电池中铅的使用蓄积量，就需要确定各类应用产品的单元数 b_i 以及每个单元产品的含铅量 m_i。因此本研究的关键在于确定 b_i 和 m_i。其中所需数据，部分来源于各类统计年鉴，部分来自于向中国计算机学会 UPS 分会等单位的调研，还有部分通过走访行业专家以及参考国内外相关文献获得。具体数据如表 7.2 所示。

表 7.2　2005 年北京市铅酸蓄电池各类应用产品的使用数量
Table 7.2　The number of various lead-acid battery products in Beijing in 2005

产品应用类别 i	产品数量 b_i /unit	数据来源
载客汽车	1 883 063	2006 北京公安交通管理年鉴
载货卡车[1]	248 977	2006 北京公安交通管理年鉴
摩托车	264 912	2006 北京公安交通管理年鉴
拖拉机	92 457	2006 北京公安交通管理年鉴
铁路机车	670	北京铁路局年鉴 2006
通信用程控交换机[2]	—	北京统计年鉴 2006
工业变电站变压器	640	北京电力公司年鉴 2006
计算机不间断电源系统（UPS）[3]	200 000	中国计算机学会 UPS 分会
叉车、铲车等专项作业车[4]	44 811	2006 北京公安交通管理年鉴
电动自行车	63 017	2006 北京公安交通管理年鉴
蓄电池车[5]	49 003	2006 北京公安交通管理年鉴

1）包括统计资料中军车；
2）数据不可得；
3）咨询自中国计算机学会 UPS 分会；
4）包括统计资料中挂车、牵引车、轮式自行机械以及专项作业车；
5）包括统计资料中低速汽车、电车以及三轮汽车。

7.1.5　单元铅酸电池含铅量

在本研究过程中调查发现,在我国各类铅蓄电池产品对应的铅含量很难获得,因此本文含铅量的计算采用的是来自国际铅酸蓄电池委员会的数据,汇总如表 7.3 所示。

表 7.3　2005 年北京市铅酸蓄电池各类应用产品含铅量
Table 7.3　The content of lead in various lead-acid battery products in Beijing in 2005

产品应用类别 i	产品含铅量 m_i / (kg/unit) [1)]
载客汽车	9.70
载货卡车	17.50
摩托车	2.86
拖拉机	15.00
铁路机车	97.00
工业变电站变压器	816.00
计算机不间断电源系统（UPS）	3.80
叉车、铲车等专项作业车	11.30
电动自行车	15.40
蓄电池车	18.60

1）数据来自国际铅酸蓄电池委员会（Battery Council International）（USGS，2006）。

7.2　结果与讨论

7.2.1　蓄积量及其构成

综上所述,本研究根据自下而上的方法,在调查获得北京市铅酸蓄电池产品的种类、数量及含铅量的基础上,根据式（7.1）,得到各类产品中铅的使用蓄积量,如表 7.4 所示。

表 7.4　2005 年北京市铅酸蓄电池各类应用产品中铅的使用蓄积量
Table 7.4　The in-use stock of lead in various lead-acid battery products in Beijing in 2005

产品应用类别 i	铅的使用蓄积量 S_i/kg	所占百分比/%
载客汽车	18 265 711.1	64.08
载货卡车	4 357 097.5	15.29
摩托车	757 648.3	2.66

续表

产品应用类别 i	铅的使用蓄积量 S_i/kg	所占百分比/%
拖拉机	1 386 855.0	4.86
铁路机车	64 990.0	0.23
工业变电站变压器	522 240.0	1.83
计算机不间断电源系统（UPS）	760 000.0	2.67
叉车、铲车等专项作业车	506 364.3	1.78
电动自行车	970 461.8	3.40
蓄电池车	911 455.8	3.20
合计	28 502 823.8	100.00

由表 7.4 可见，2005 年北京市铅酸蓄电池中铅的使用蓄积量 S 约为 28.5kt。在北京市各类铅酸蓄电池产品中，载客汽车占全部铅使用蓄积量的 64.08%，是影响北京市铅的使用蓄积量的最大因素。其次为载货汽车，占 15%。可以看出，铅的使用蓄积量主要与城市机动车的保有量相关。

若按照电池的分类来分析铅的蓄积量的构成，如图 7.1 所示，可见，北京市的铅酸蓄电池中的铅近 90% 蓄积于启动型蓄电池中。

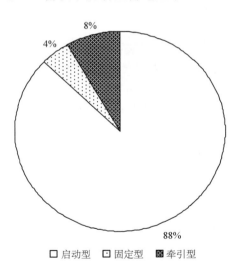

图 7.1 各类蓄电池中铅的使用蓄积构成

Fig. 7.1 The composing of in-use stocks of lead in various lead-acid batteries

若从电池的应用场所来分析铅的蓄积量的构成，如图 7.2 所示，可见，机动车中蓄积的铅含量超过北京市使用铅酸蓄电池蓄积量的 90%。

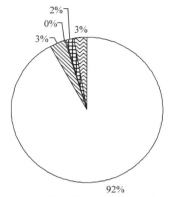

图 7.2 各类应用场所中铅的使用蓄积构成

Fig. 7.2 The composing of in-use stocks of lead in various applications

考虑目前中国铅酸蓄电池行业约占铅消费总量的 75%,可以推测,北京市铅的蓄积总量约为本研究计算结果的 1.3 倍,即约 38.0kt。

7.2.2 与国外部分城市对比

由表 7.4 可知,2005 年北京市铅酸蓄电池中铅的使用蓄积量 S 约为 28.5kt,考虑到当年北京市人口为 1538 万人,可计算得到 2005 年铅的人均使用蓄积量为 1.85 kg/人。将该结果与国外其他研究结果对比,可以看出国外发达国家或城市铅酸蓄电池中铅的人均蓄积量约为北京市的 2 到 6 倍,具体结果见表 7.5。

表 7.5 部分城市和国家铅酸蓄电池中铅的使用蓄积量

Table 7.5 The in-use stock of lead in lead-acid batteries in some cities and countries

地区	时间	蓄积总量/kt	人均蓄积量/(kg/人)	人均蓄积量之比[1]
维也纳(Obernosterer and Brunner,2001)	1991	11.7	7.8	4.2
斯德哥尔摩(Sörme et al.,2001)	1995	7.8	11.0	5.9
荷兰(Elshkaki et al.,2004)	1998	45.6	3.0	1.6
北京	2005	28.5	1.9	1.0

1)人均蓄积量之比为其他城市人居蓄积量/北京人均蓄积量。

由表 7.5 可以看出,国外发达国家及城市,其蓄电池中铅的使用蓄积量的人均值均高出北京市数倍,但在总量上并没有太大优势,由此可见,北京市铅酸蓄电池中铅的人均蓄积量较低主要是由北京市人口基数较大,并且人均机动车占有量较低引起的。

7.2.3 结果的不确定性

自下而上的方法要求本研究清算出北京市全部处于使用中的铅酸蓄电池。而本研究中的数据主要来自北京交通管理年鉴、北京统计年鉴等，根据统计专家会谈，该数据至少涵盖北京市在使用的铅酸蓄电池产品（主要为机动车、电动车等交通工具中的铅酸蓄电池）的 90%。

此外，在实际使用中，蓄电池的铅含量会因电池的类型、规格的不同而异。本研究针对各类型电池，采用主要规格的电池产品的平均含铅量进行计算，并忽略铅酸蓄电池在使用过程中铅的损耗。

7.3 结　　论

本文采用自下而上的方法，在对北京市铅酸电池使用种类、场所、数量等深入调研的基础上，估算 2005 年北京市铅酸蓄电池中铅的使用蓄积量。结果表明，铅的使用蓄积总量约为 28.5kt，如果按蓄电池种类分类，则启动型蓄电池占总量的 88%，为最多；如果按产品应用场所来分类，则机动车用电池占总量的 92%，为最多，因此在今后的进一步研究中，可以将机动车作为研究的重点目标。

人均使用蓄积量为 1.85 kg/人，与国外研究结果对比表明，北京市铅的人均蓄积量约为国外发达国家和城市的 1/6 到 1/2，这主要是由北京市人口基数大以及机动车人均拥有量较低引起的。由于个别产品的清单数据不可得，并且国内缺乏各类单元含铅量数据，本研究计算结果难免存在一定偏差。但本文研究结果仍可为后续研究提供依据，也为国内外铅的蓄积研究提供一个发展中国家的研究案例。

参 考 文 献

北京市统计局. 2006. 北京统计年鉴 2006. 北京：中国统计出版社.

姜妮. 2007. 循环路上难处多——对铅酸蓄电池再生利用的思考. 环境经济，Z1：65-68.

楼俞，石磊. 2008. 城市尺度的金属存量分析——以邯郸市 2005 年钢铁和铝存量为例. 资源科学，30（1）：147-152.

毛建素，陆钟武. 2003. 论铅业的废铅资源. 世界有色金属，7：10-14.

唐武军. 2008. 中国铅业发展路在何方?. 中国有色金属，13：44-45.

王金良，马扣祥. 2005. 铅和铅蓄电池工业现状与发展趋势. 电池工业，10（6）：364-368.

吴纯德，刘燕，朱能武. 2009. 广东省某市城区废电池污染防治建议. 环境科学与技术，3：201-205.

Drakonakis K，Rostkowski K，Rauch J，et al.2007. Metal capital sustaining a North American city：iron and copper in New Haven，CT. Resources，Conservation and Recycling，49（4）：406-420.

Elshkaki A，Van Der Voet E，Van Holderbeke M，et al. 2004. The environmental and economic consequences of the developments of lead stocks in the Dutch economic system. Resources, Conservation and Recycling, 42: 133-154.

Gerst M D, Graedel T E. 2008. In-use stocks of metals: status and implications. Environmental Science and Technology, 42 (19): 7038-7045.

Gordon R B, Bertram M, Graedel T E. 2006. Metal stocks and sustainability. Proceedings National Academy of Sciences, 103 (5): 1209-1214.

Katherine R, Jason R, Konstantine D, et al. 2007."Bottom-up"study of in-use nickel stocks in New Haven, CT. Resources, Conservation and Recycling, 50 (1): 58-70.

Mao J S, Graedel T E. 2009. Lead in-use stock: a dynamic analysis. Journal of Industrial Ecology, 13 (1): 112-126.

Mao J S, Yang Z F, Lu Z W. 2007. Industrial flow of lead in China. Transactions of Nonferrous Metals Society of China, 17 (2): 400-411.

Muller D B, Wang T, Duval B, et al. 2006. Exploring the engine of anthropogenic iron cycles. Proceedings National Academy of Sciences, 103 (44): 16111-16116.

Obernosterer R, Brunner P H. 2001. Urban metal management: the example of lead. Water, Air, and Soil Pollution: Focus, 1 (3-4): 241-253.

Sörme L, Bergbäck B, Lohm U. 2001. Century perspective of heavy metal use in urban areas. Water, Air, and Soil Pollution: Focus, 1 (3-4): 197-211.

Spatari S, Bertram M, Gordon R B, et al. 2005. Twentieth century copper stocks and flows in North America: a dynamic analysis. Ecological Economics, 54 (1): 37-51.

USGS. 2005. Lead: U.S. Geological Survey mineral commodity summaries 2005[EB/OL]. http: //minerals.usgs.gov/minerals/pubs/commodity/lead/lead_mcs05.pdf[2008-09-29].

USGS. 2006. Apparent consumption vs. total consumption—a lead-acid battery case study. http: //pubs.usgs.gov/sir/2006/5155/sir20065155.pdf.

第 8 章 我国耗散型铅的使用现状及趋势分析
Tread Analysis on Dissipative Uses of Pb in China*

8.1 概 述

金属铅是现代化工业的重要原料，被广泛应用于铅酸蓄电池、颜料、汽油等领（熊亚，2003）。研究表明，铅污染会影响人体健康，可导致成人血压升高，引发心脏病、心肌梗死等疾病（Gilbert and Weiss，2006），造成胎儿发育不全和早产死亡，特别是可能导致儿童的智商水平（intelligence quotient，IQ）下降（Lamphear et al.，2005）。因此，近几十年控制铅污染颇受关注。在众多的铅制品中，耗散型铅使用，即那些在使用过程中或者使用完后不可能对其中的铅进行回收的铅制品（Ayers，1994），其用量在铅消费总量中的份额相对较少（Mao et al.，2008a；Mao et al.，2008b），因而时常被忽略。但这类产品中所含的铅无法被回收而散逸入环境中，并且在日常使用过程中容易接近甚至接触人体，对人体健康直接造成危害，因此有必要给予充分重视。此外，从提高我国铅的回收利用情况而言，减少耗散型铅消费也是十分必要和有效的（Mao et al.，2007）。本章将针对几类典型的耗散型铅制品，列举和分析近年来其使用状况，以及一些相关政策和标准，并与美国、欧盟的相应政策和最新标准进行比较和分析，以获得我国关于铅使用政策方面的发展趋势。

8.2 我国典型耗散型铅制品的现状及分析

8.2.1 使用状况

在我国，过去铅的耗散型使用主要集中在汽油、涂料等产品上，如汽油中人为添加的抗爆剂四乙基铅，涂料和油墨作为颜料加入的各种铅化合物等。近年来人们对铅污染的认识不断提高，许多应用上的用量都在随着法律法规的限制而逐渐降低（高山，2000）。含铅汽油已被禁止生产和使用，涂料和油墨开始选用有机

---

* 曾润，毛建素. 2010. 我国耗散型铅使用的变化及趋势分析. 环境科学与技术，33（2）：192-195.

颜料来代替铅化合物颜料，我国也正在计划修改法律来禁止使用含铅物品制造玩具。近年来，这四类物品总是与生活中铅污染的话题有关，因此本文将以这四类物品为例，对我国近年来有关政策和标准进行具体分析。

8.2.2 相关政策

20世纪90年代，我国开始实施车用汽油无铅化，汽油铅含量逐年降低。1998年，国务院办公厅发布通知，要求在全国范围内于2000年1月1日起停止生产车用含铅汽油，2000年7月1日起停止销售和使用含铅汽油。研究表明，推广使用无铅汽油后，环境铅污染在一定程度上得到了控制，儿童血铅水平下降（Luo et al., 2003；颜崇淮等，2002）。据世界卫生组织（WHO）统计，汽油无铅化后，人体血铅以7.8%的年速率递减（WTO，2003）。

室内装修装饰、家具表面等涂有的含铅涂料是形成室内铅污染的一个重要原因。我国于2001年制定了两部限制涂料有害物质含量的标准，对象囊括传统的溶剂型涂料和新兴的水性涂料，对铅等重金属含量的控制制定了非常严格的要求。

在过去，我国有关油墨的标准一般都只规定了颜色、黏度等物理指标，对可能损害人体健康及造成环境污染的成分却没有严格要求。2007年，为了鼓励低污染油墨的研发，引导油墨行业向绿色化发展，我国环境保护总局组织专人制定了两部有关油墨的环境标志产品技术要求标准。标准规定不允许人为添加铅等重金属，并规定了其中可迁移铅含量上限。2008年我国发改委又通过了油墨标准化委员会制定的油墨有害物质限定标准，这条强制性的标准对保护环境、减少油墨带来的铅污染具有深远意义。

儿童玩具一直是我国向北美、欧洲等国家大量出口的商品。欧盟和美国对于儿童玩具铅含量早有相应限制，而我国缺乏此类标准法规，导致出口玩具屡屡被退。2003年我国颁布了国家玩具安全技术规范，参照发达国家标准对可迁移铅含量提出了要求。此外，我国政府还于2007年同美国政府发表了中美联合声明，禁止在出口美国的玩具上使用含铅油漆。

8.2.3 相关标准

经过文献调研以及从各部门获得的相关资料，本文整理出有关车用汽油、涂料、油墨以及儿童玩具的含铅量限制的各类国家标准及行业标准，具体信息如表8.1所示。

表 8.1 我国典型耗散型铅制品的相关标准

Table 8.1 The relevant standards for the representative lead products in dissipative in China

产品类别	铅含量上限	标准名称（标号）	颁布年限	颁布单位
车用汽油	0.35 g/L（90#） 0.45 g/L（93#、97#）	车用汽油（GB 484—1993）	1993	国家技术监督局
	0.013 g/L	车用汽油有害物质控制标准（GWKB 1—1999）	1999	国家环境保护总局
	0.005 g/L	车用无铅汽油（GB 17930—1999）	1999	国家质量技术监督局
	0.005 g/L	车用汽油（GB 17930—2006）	2006	国家质量监督检验检疫总局
水性涂料	无要求	环境标志产品技术要求水性涂料（HJBZ 4—1999）	1999	国家环境保护总局
	可迁移铅 90 mg/kg	室内装饰装修材料内墙涂料中有害物质限量（GB 18582—2001）	2001	国家质量监督检验检疫总局
	可迁移铅 90 mg/kg	环境标志产品技术要求水性涂料（HBC 12—2002）	2002	国家环境保护总局
	可迁移铅 90 mg/kg	环境标志产品技术要求水性涂料（HJ/T 201—2005）	2005	国家环境保护总局
	可迁移铅 90 mg/kg	室内装饰装修材料内墙涂料中有害物质限量（GB 18582—2008）	2008	国家质量监督检验检疫总局
溶剂型涂料	可迁移铅 90 mg/kg	室内装饰装修材料溶剂型木器涂料中有害物质限量（GB 18581—2001）	2001	国家质量监督检验检疫总局
	可迁移铅 90 mg/kg	环境标志产品技术要求室内装饰装修用溶剂型木器涂料（HJ/T 414—2007）	2007	国家环境保护总局
油墨	可迁移铅 90 mg/kg	环境标志产品技术要求凹印油墨和柔印油墨（HJ/T 371—2007）	2007	国家环境保护总局
	可迁移铅 90 mg/kg	环境标志产品技术要求胶印油墨（HJ/T 370—2007）	2007	国家环境保护总局
	可迁移铅 90 mg/kg	油墨中某些有害元素的限量及其测定方法（QB 2930.1—2008）	2008	国家发展和改革委员会
儿童玩具	总铅 2500 mg/kg 可迁移铅 250 mg/kg	玩具安全（GB 6675—1986）	1986	国家标准局
	可迁移铅 90 mg/kg	国家玩具安全技术规范（GB 6675—2003）	2003	国家质量监督检验检疫总局

从表 8.1 可知，我国对于耗散型铅使用的含铅量要求从无到有，从宽松到严格，旨在配合国家政策，保护环境及人体健康。

8.3 国外同类产品的现状及分析

8.3.1 国外使用状况

美国是较早意识到耗散型铅使用的危害的国家之一，早在 20 世纪 70 年代就开始限制涂料和汽油中的铅（James et al.，1998）。美国对儿童玩具的铅污染特别关注，近年来多次因为检验出玩具表面的涂料或油墨含铅量超标而将其退回出口国。美国以前由含铅产品制成的饮用水管也都被取代。由于这些法规和其他方面的努力，美国的铅暴露和血铅水平近年来显著下降（EPA，2000）。

欧盟对于铅的使用限制也非常严格。过去用于下水道管道、汽油添加剂、电缆铅包皮的铅的用量都在逐年递减。欧盟在 15 个成员国内禁止了含铅汽油的生产和使用，并严格限制电子电器设备和儿童玩具中使用含铅材料，禁止含铅油墨在食品包装印刷中使用（Amold et al.，2006）。

8.3.2　国外相关政策

1970 年颁布的美国《清洁空气法》第一次间接约束了铅在汽油中的使用，美联邦政府从 20 世纪 70 年代开始逐步限制含铅汽油，1986 年开始限制四乙基铅的使用，1996 年实现了完全禁止含铅汽油的生产和使用。美国消费者产品安全委员会（Consumer Product Safety Commission，CPSC）早在 1978 年就开始在涂料和其他一些消费品中限制铅的使用。美国华盛顿州最近通过的一项法案还将玩具总铅含量上限标准从 600ppm 降低到了 90ppm，美国环境保护署（U.S. Environmental Protection Agency，EPA）也对铅污染非常重视，一直不遗余力地与相关机构合作来制定各种限制铅使用的法规，推行相关政策如对油漆工进行严格培训以减少房屋的铅污染，并且在其官方网站上辟出有关铅的专栏，提供铅污染防治的各类信息（EPA，2008）。

德国是欧洲第一个对汽油含铅量进行限制的国家，于 1972 年开始将含铅量上限从 0.6 g/L 降低到 0.4 g/L，其后，欧盟在 1987 年通过法规开始禁止含铅汽油的生产和使用（Hans et al.，2003）。欧盟许多国家在彻底实现汽油无铅化以前，则是采取针对含铅汽油和无铅汽油分别征收不同税率的环境税的措施来限制含铅汽油的使用（Robert，2003）。欧盟还相继通过了 RoHS 指令以及 Reach 法案，前者用于限制在各类电子电器设备中使用含铅材料，后者对铅高度关注，极有可能将其列入化学用品的限制使用名单。欧洲标准化委员会制定的 EN71 系列标准也对儿童玩具及其表面涂料、油墨中的含铅量进行严格控制。

8.3.3　国外相关标准

经过文献调研获得的资料，本文整理出近年来美国、欧盟有关车用汽油、涂料、儿童玩具以及油墨的含铅量限制的最新标准，具体信息如表 8.2 所示。

表 8.2　美国及欧盟典型耗散型铅制品的相关标准

Table 8.2　The relevant standards for the representative lead products in dissipative in U.S. and EU

	产品类别	铅含量上限	标准名称（标号）	颁布年限	颁布单位
美国	汽油	13 mg/L	ASTM D4814—08	2008	美国实验与材料协会
	涂料	总铅 600 mg/kg 可迁移铅 90 mg/kg	16 CFR 1303	2006	消费者产品安全委员会
	油墨	总铅 600 mg/kg 可迁移铅 90 mg/kg	ASTM F963—08	2008	美国实验与材料协会
	儿童玩具	总铅 600 mg/kg 可迁移铅 90 mg/kg	ASTM F963—08	2008	美国实验与材料协会
欧盟	汽油	5 mg/L	EN 228—2004	2004	欧洲标准化委员会
	涂料	总铅 1000 mg/kg 可迁移铅 90 mg/kg	EU RoHS EN 71-3	2003 2002	欧洲议会和理事会 欧洲标准化委员会
	油墨	可迁移铅 90 mg/kg	EU RoHS EN 71-3	2003 2002	欧洲议会和理事会 欧洲标准化委员会
	儿童玩具	总铅 1000 mg/kg 可迁移铅 90 mg/kg	EN 71-3	2002	欧洲标准化委员会

8.4　我国相关政策标准趋势分析

对比上述资料发现：我国过去在车用汽油、涂料、儿童玩具以及油墨等铅制品的相关标准中都对其铅含量没有进行限制或是限制比较宽松。但近年来人们开始重视这些铅制品带来的铅污染，相关部门陆续发布了相应的新标准、制定了新政策来严格控制耗散型铅使用。目前看来，我国这几类铅制品含铅量的上限数值都达到了发达国家现今的水准，但总体而言仍然存在以下几点不足。

（1）美国、欧盟对于玩具、文具、儿童房间涂料等都专门制定了严格而全面的标准，我国缺乏针对儿童用品的特别保护的政策法规，这是我国今后需要补充的一个重要方向。

（2）美国、欧盟对涂料、油墨等物品的总铅含量和可迁移铅含量均进行了限制，而同类物品我国标准仅仅限制了可迁移铅含量，因此，制定含铅物品中的总铅上限及其测定方法的标准，是我国未来需要考虑的一项举措。

美国的 ASTM F963、欧盟的 EN 71 系列标准近年来每年都根据最新研究资讯进行修订，因此保证了其科学性和连续性，而我国除了环保部门的环境标志产品技术要求系列的标准是 2~3 年更新一次外，其他多数国标和行业标准的标龄都很长，容易落后于时代，这也是我国需要学习和改进的方面。

8.5 改善建议与展望

综上所述，本文对我国现在铅使用的标准政策提出以下改进建议。

（1）在《室内环境空气标准》中增加对铅含量的要求，并根据相关研究降低大气、水和土壤环境质量标准中允许的铅含量限值。铅从使用物品中进入人体大多通过环境介质，室内空气更是一个重要途径。因此需要根据生理学研究的结论来确认并制定对人体安全的环境铅浓度限值。

（2）增加对含铅物品的含铅总量限制。不仅需要制定总铅含量上限，也要制定对应的且为国际所接受的检测方法。

（3）制定与儿童有关物品的有害物质限制的专项标准。儿童铅中毒是医学研究的重要课题，对于儿童玩具、图书、文具、学校餐具、房间涂料等都需要制定专门的标准，不允许在这些物品中人为加入任何含铅材料，并根据医学研究的结论严格要求这些物品的铅含量。

（4）积极扶持企业研发新技术，同时学习国外先进技术，保障制定的各项标准能切实执行。近年来我国几类铅使用物的标准均是以发达国家为参考，有的标准文献中甚至直接引用发达国家相应标准为附录。这样的标准要求虽然达到了国际先进水平，但检验技术、产品生产工艺能不能达到国际先进水平却没有保障，因此急需这方面相应政策的保证。

（5）增加对废铅酸蓄电池回收进行规范的政策和标准。在目前铅的消费产品中，用量仍在继续增长的，只有被视为可以稳定回收铅的铅酸蓄电池，因此需要相关的政策和标准，规范废铅酸蓄电池的回收市场和回收工艺，确保其含有的大量的铅能被充分回收而不是散逸到周围环境中。

参 考 文 献

高山. 2000. 世界铅的消费趋势. 世界有色金属，（03）：14-16.

熊亚. 2003. 环境铅接触对健康的影响. 微量元素与健康研究，20（1）：48-50.

颜崇淮，吴胜虎，沈晓明，等. 2002. 上海市推广使用无铅汽油对儿童血铅水平影响的追踪调查. 中华流行病学杂志，23（3）：172-174.

Arnold T, Harrie B, Lauran van O, et al. 2006. Risks to health and environment of the use of lead in products in the EU. Resources, Conservation and Recycling, 49（2）：89-109.

Ayres R U. 1994. Industrial Metabolism: Theory and Policy. B.R. Allenby, D.J. Richards. The Greening of Industrial Ecosystems. WashingtonD.C.: National Academy Press, 23-37.

EPA. 2000. Economic analysis of Toxic Substances Control Act Section 403. http://www.epa.gov/lead/pubs/403_ea_d21.pdf [2008-8-16].

EPA. 2008. Lead in paint, dust, and soil. http://www.epa.gov/opptintr/lead/index.html[2008-8-16].

Gilbert S G, Weiss B. 2006. A rationale for lowering the blood lead action level from 10 to 2 μg/dL. Neurotoxicology. 27 (5): 693-701.

Hans von S, Mariza C C, Charlotte H, et al. 2003. Four decades of gasoline lead emissions and control policies in Europe: a retrospective assessment. The Science of The Total Environment, 311 (1-3): 151-176.

James L P, Rachel B K, Debra J B, et al. 1998. Exposure of the U.S. population to lead, 1991-1994. Environmental Health Perspectives, 106 (11): 745-750.

Lanphear B P, Hornung R, Khoury J, et al. 2005. Low-level environmental lead exposure and children's intellectual function: an international pooled analysis. Environmental Health Perspectives, 113 (7): 894-899.

Luo W H, Zhang Y, Li H. 2003. Children's blood lead levels after the phasing out of leaded gasoline in Shantou, China. Archives Of Environmental Health, 58 (3): 184-187.

Mao J S, Dong J, Graedel T. E. 2008a. The multilevel cycle of anthropogenic lead: I. methodology. Resources, Conservation and Recycling, 52 (8-9): 1058-1064.

Mao J S, Dong J, Graedel T. E. 2008b. The multilevel cycle of anthropogenic lead: II. results and discussion. Resources, Conservation and Recycling, 52 (8-9): 1050-1057.

Mao J S, Yang Z F, Lu Z W. 2007. Industrial flow of lead in China. Transactions of Nonferrous Metals Society of China, 17 (2): 400-411.

Robert N S. 2003. Experience with Market-Based Environmental Policy Instruments. K.G. Maler, J.R. Vincent. Handbook of Environmental Economics. Amsterdam: North-Holland, 355-435.

WHO. 2003. Lead: assessing the environmental burden of disease at national and local levels. http://www.who.int/quantifying_ehimpacts/publications/en/leadebd2.pdf[2008-7-16].

第二篇

人为迁移与转化

第 9 章 矿物资源服务归趋：概念、内涵与议题
The Flows of Mineral Resources to Provide Human Service: Concepts, Connotation and Contents[*]

近年来，为满足日益增长的人类需求，大量矿物资源被加工转变成具有特定服务功能的产品，并向人类提供特定服务（ISO，2006），形成了矿产资源服务归趋流动。这一过程有效促进了社会与经济发展，但由于需要消耗大量物质资源和能源，并向环境排放废物，长此以往，造成了诸多资源与环境问题（Schmidt，2010；Elshkaki et al.，2004；Elshkaki et al.，2009）。

由于矿产资源服务归趋是以满足人类需求服务为核心的，因此该服务归趋过程交织着人类与环境间的复杂交互作用关系。其中，服务归趋过程中的物质加工转变是借助设计、生产等多种人类活动实现的（冯桂林和何蔼平，2000；刘楚明等，2010），而在传统设计、生产等各环节中，通常较多地关注人类有怎样的需求，要求物质发生怎样的转变，如何实现这些转变等技术问题（沈春林，2008；刘广林，2011），而相对轻视发生这些转变将可能对外部环境造成的影响。而关注环境变化议题的地学领域，又较多地从地表角度采用 GIS、环境监测等方法来推演环境质量的时空分异、变化程度及其可能导致的生态环境风险（Agustin et al.，2011；He et al.，2009；Li et al.，2010），从而难以弄清诸多环境问题的内在形成机制、污染物总量及其进入环境时的状态等问题，削弱了污染物及环境问题的源头管理潜力。近年来工程技术领域开展了大量面向环境改善的技术革新和环境性能评估方法的研究（Femandes et al.，2012；Maruthamuthu et al.，2011；Zackrisson，2010），为系统了解和全面评估产品、服务或特定工艺的环境影响提供了方法基础与途径，但现有研究较多针对特定产品、服务或特定工艺进行生命周期技术现状评估，而较少关注服务归趋过程中各变化要素间、各影响因素间的内在协同关系，也未关注这一服务归趋过程的时变性、时间累积效应等重要议题，为开展矿产资源服务归趋相关研究提供了重要契机。

本章将从基本概念着手，重点阐述矿物资源服务归趋基本概念和研究框架，基于所关注议题与各学科间的基本关系，阐释矿物资源服务归趋的科学内涵。以

---

[*] 毛建素，梁静，徐琳瑜. 2013. 矿物资源服务归趋：概念、内涵与议题. 环境科学与技术，36（12）：195-201.

期为深入探求人类发展与环境变化间内在关系，以及补充污染物源头定量分析提供新的方法基础。

9.1　概　　念

9.1.1　服务归趋的概念

"归趋"一词主要用于地学领域，用于描述污染物在环境中的迁移转化过程，研究其来源、归趋路径、环境归宿，并关注特定物质的空间转移和赋存形态转化等问题（Mannhan，2005）。如大气中镉、铅等重金属物质经大气输运、沉降等过程进入土壤（Steinnes et al.，2011；Niisoe et al.，2010），而土壤中的镉、铅等重金属物质又可经植物吸收过程转移并蓄积到植物果粒中（严莎等，2008），或经渗流从土壤转移到水体，并沉积到水体淤泥中（严莎等，2008；王碧玲和谢正苗，2008）。不难看出，这些污染物的归趋过程发生在人类以外的环境系统中，归趋动力是自然力。

矿物资源的服务归趋是指矿产资源中金属物质（如铁、铝）、能源物质（如煤、石油、天然气）等从岩石圈到向人类提供某特定最终服务的实现过程，简称服务归趋过程。该过程包括资源获取、材料加工、产品制造和使用等基本环节，如图9.1所示。其中产品的使用过程是物质向人类提供最终"服务（service）"的过程，而其他过程则是物质最终服务的形成过程。从物理属性看，服务归趋过程既包括物质的空间变化过程，如大庆油田将地下原油开采出来并冶炼成为汽油，再经输运管道送往哈尔滨各加油站；也包括物质的技术转变过程，如通过裂解将原油制成汽油、天然气等工业产品。本文中着重分析从矿产资源到"服务"的技术归趋过程。这一服务归趋过程发生在人类技术圈，归趋动力是人类需求。

从系统论观点看，对于同一种物质，其服务归趋过程和环境归趋过程是物质整个生物地球化学循环过程的两个不同阶段，物质服务归趋过程释放的废物恰是环境中污染物的来源（Mannhan，2005；Von Bertalanffy，1968）。这两种归趋过程的"同源性"为基于相似论将分析污染物环境归趋的相关方法用于物质服务归趋研究提供了从概念到研究思路、方法等的科学依据（Szücs，1980）。物质环境归趋研究中关注物质的形态转变、物质与自然环境介质间的结合关系，物质服务归趋中也将关注物质的形态转变。只是由于这些转变是以满足人类特定服务需求为目标的，因此，还将关注形态转变后物质与人类的关系，表现为物质所具有的使用性能，或称服务功能。矿物资源的若干常用服务及其对应的物质功能、形态见表9.1。

第 9 章 | 矿物资源服务归趋：概念、内涵与议题

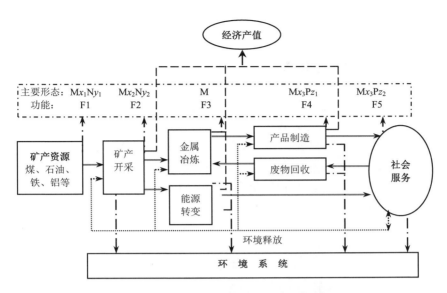

图 9.1 矿物资源服务归趋过程概念框架

Fig. 9.1 Conceptual framework for the flows of mineral resource to providing human service

—— 复合流； ········ 能量流； — — — 价值流； —·—·— 废物流； —··—··— 辅助线

M 为特定金属元素；N、P 为其他物质；x, y, z 为组成变量；F 为物质功能

表 9.1 矿产资源若干常用服务及其对应的物质功能、形态示例

Table 9.1 Several examples for the usual human services, functions and forms of mineral resources

服务	产品	功能	主要物质形态	
			化学结构或物质组分	结构组成
擦干手臂	干手器	向湿手吹风，借助空气对流完成干燥手臂	$OCr_{18}Ni_9$，$[CH_2-CHCl]_n$	电机、风机、叶片、过滤器、开关
物体表面着色	油漆	借助粉刷形成物体表面彩色涂层	$PbCrO_4$，$CaCO_3$	树脂、溶剂、填充剂、特定添加剂
电能储存与转移	铅酸蓄电池	发生电化学反应，借助充电和放电过程实现电能储存和转移	$PbSO_4$, Pb, PbO_2, Pb_xSb_y, Pb_xCa_y	极板、电解液、隔板、外壳

在矿产资源服务归趋过程中，为实现发生物质功能、形态等基本转变，需要提供发生这些变化的基本条件，即需要投入相应的能源动力、人力，以及相应的生产工艺、设备等基础设施，使得能源流动、价值流动融入了物质的服务归趋流动过程。由于物质既是能量载体，也是价值载体，而各股物质流动间遵守物质守恒定律，使得"物质"成为了连接各股物流、能流和价值流间内在关系的纽带。从而形成了矿产资源服务归趋过程中不同属性、品质的物流、能流和价值流间的

特定耦合、协同关系。

9.1.2　服务归趋研究框架

矿产资源服务归趋过程不仅涉及多种物质的功能、形态、数量的变化，还涉及完成这些转变的条件，需要投入大量能源、人力和物力，由于代谢活动，相应地，将涉及废热、余热的排放，以及社会、经济的产出，是一个多要素、多层次、多成分、多尺度的复杂变化过程。因此，为进行矿物资源服务归趋研究，构建了"一点、两线、三层面、四品种"研究框架，如图9.2所示。

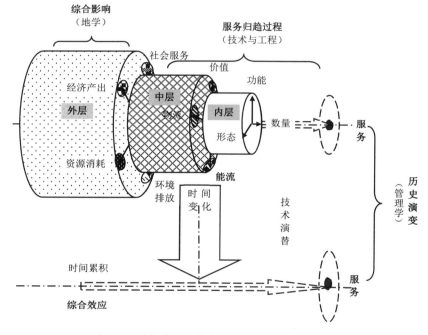

图 9.2　矿物资源服务归趋过程研究框架示意图

Fig. 9.2　Schematic framework of the research on the flows of mineral resources to providing

静态研究中，以人类服务作为矿产资源服务归趋的目标点，以矿产资源到实现最终服务的技术归趋过程为技术维轴线，针对特定断面，又进一步分为内、中、外三个层面，其中内层关注物质形态、功能及其数量分配关系，用以回答这一服务归趋过程中物质发生了怎样的变化，数量如何分配等问题；中层关注物流、能流、价值流及其间耦合关系，用以回答如何实现诸多变化，需要怎样的生产条件等问题；外层关注实现这一过程的外部条件和必然结果，用以回答该过程对外部造成了怎样的影响，影响水平如何等问题。动态研究中，充分考虑需求的时变性，沿时间维轴线，估算该归趋过程及其综合影响的历史演变过程和所造成的结果，用以探求该过程中技术圈诸要素间内在作用关系及外部影响规律。在案例分析和实证研究中，选择矿产资源的若干典型人类服务，结合特定区域矿产资源利用、产品设计与服务模式及其技术历史演变，实证矿产资源服务归趋过程变化规律，形成定量清单，提出改善建议。

9.1.3 物质服务归趋与环境归趋的对比

矿产资源服务归趋过程是为满足人类特种需要形成的，是以自然系统中矿产资源为起点，以实现物质的人类服务功能为终点的物质服务形成与实现过程中所发生的物质在其服务性能、形态间转变、数量分配，以及与社会、经济、环境间的交互作用。与传统地学中的污染物环境归趋概念相比，具有以下不同。

（1）发生的圈层不同。矿产资源服务归趋从资源开采到实现最终服务的各种变化都发生在人类社会经济复合系统，或称技术圈、人类活动圈（Mannhan，2005）。而污染物环境归趋关注物质在人类以外环境系统中的变化。

（2）归趋动力与归趋机制不同。人类需求既是矿物资源服务归趋的原始驱动力，也是其服务归趋的核心和目的，整个归趋过程呈现为满足人类特种需求所进行的一系列设计、生产加工、贸易运输、产品使用等人类行为（Xia and Li，2007；Zhang et al.，2010；Mao et al.，2012）。这与环境中物质依靠其与环境介质间的自然力，如浓度差异、不同物质间的化学力等作为归趋动力并完成环境归趋过程截然不同。

（3）关注的内容与表征参数不同。矿产资源服务归趋以满足人类特种需要为服务目标，关注的是物质对人类的可用性（或称为物质的使用性能、物质的功能）和服务的数量。而物质的功能决定于物质的结构、组合方式等赋存形态（高丕英和李江波，2007），因此物质的功能、形态、和数量成为表征矿产资源服务归趋的基本参数。

（4）物质来源、归趋路径和归趋终点不同。矿产资源服务归趋过程中的物质的初始来源是自然系统中的矿产资源，废物资源（主要来自报废产品，这类资源归根结底来自自然系统中的矿产资源）是其二次来源（曾润和毛建素，2010），而人类的使用蓄积库（in-use stock）（Elshkaki et al.，2004）是其服务归趋终点，并相应地拥有两条归趋路径，分别是从矿产资源开采和以废物回收为起点到物质终产品投入使用、实现其服务功能为终点的归趋过程（Mao et al.，2012）。在物质环境归趋中，环境中污染物的来源主要是人类圈，归趋路径多为大气、水体等环境介质，其环境归趋终点是自然圈层，如土壤圈、生态系统、水圈底泥等（Mannhan，2005）。

9.2 科学意义

作为物质在地球表面系统中运动、迁移、转化的重要组成，矿产资源服务归趋过程中不仅在系统内部发生一系列物质性能、结构的变化，而且还将与自然环境之间发生动态、交互作用关系。因此，具有重要科学内涵和学科意义。

9.2.1 科学内涵

1. 反映自然系统对人类的支撑关系

从系统论角度看，人类技术圈是地表系统的重要组成，矿产资源服务归趋是构成整个循环的有机组成，并与物质自然生物地球化学循环存在着多种内在必然联系。表现在自然圈层既承担了矿产资源服务归趋中物质的"源"，又承担了该过程中代谢废物、污染物的"汇"，对矿产资源服务归趋起到基础支撑作用。

2. 反映人类消费偏好与发展模式

矿产资源服务归趋过程以满足人类特定需求为目的，而这些需求往往借助具有特定服务功能的多种产品来实现，这使得人类获得服务时可以从多种产品中做出选择，反映着人类的消费偏好。如为了擦干手臂，可使用纸巾，也可选择干手器。由于不同产品具有不同的环境性能，使得选择不同的服务将意味着产生不同的环境影响。人类发展推动着人类服务需求数量、服务模式、品种等的变化，也势必带来人类技术圈的技术演替，影响到人类的社会与技术进步。

3. 反映人类对地表环境的干扰方式

矿产资源服务归趋过程是借助人类一系列设计、物质生产和产品使用活动来实现的，其中蕴含、交织着人类对地表环境产生作用关系的方式。如金属矿产开采冶炼活动，促使铁、铜等金属物质发生从地壳内部"岩石圈"到地壳外部"人类技术圈"的转移，化学结构也由 Fe_3O_4、$CuFeS_2$ 等转变为 Fe、Cu 等金属。同时，这一生产活动还需要消耗大量能源、电力，并会向环境释放富含多种不同物质和形态的矿渣、废气、废热等环境污染物，从而对地表生态环境系统产生某种影响。除此之外，物质进入环境时的状态是物质进入环境系统的"起点"，反映出人类活动对环境释放物之间的内在驱动关系。

4. 反映人类发展与自然要素间的动态耦合与时空差异

矿产资源服务归趋过程是人类与自然环境间交互作用的动态过程，意味着该过程中物质的变化既与人类活动因素密切相关，又受到矿产存储地点和禀赋状况的影响。其中，人类活动因素反映在服务模式、生产工艺、需求数量、人口分布等方面，并借助设计、生产、物资运输、消费等技术环节体现出来，整个过程交织着物质在服务类型、功能、形态、时间、空间、数量等多参数间的内在动态耦合关系，也交织着物质在地表各区域、各圈层间的空间再分配。

5. 反映人类对地表环境的干扰结果

人类社会经济系统作为地球表面系统中最具活力的生态组分，其中所发生的矿产资源服务归趋过程将造成物质元素在地球表面的再分配和物质服务属性的改变，表现为，人类对矿产资源的长期使用，将不断削减岩石圈中的矿产资源量，却增加处于人类使用状态（in-use stock）的物质存量（Elshkaki et al.，2004），以及增加环境中低资源品位、污染物的数量；同时，伴随物质服务对象由自然转向人类，其服务功能也由自然系统支撑功能转向人类社会服务功能。这一归趋过程的时间累积结果将可能造成物质从岩石圈矿产资源转向人类技术圈的社会存量，资源品位降低为环境污染物的积累，并产生巨大的生态环境风险。如一些水域富营养化（王碧玲和谢正苗，2008）、食品中重金属超标（莫文莲等，2010）等，深刻反映着矿产资源归趋过程对地表环境的干扰后果。

9.2.2 学科意义

由于矿产资源服务归趋过程是一个以特定人类服务为核心，以多种工程技术手段为技术途径，又以生态系统服务为承载基础的复杂过程，呈现人类社会经济复合系统与外部环境系统间多种物质与能量交换与作用关系，因此该过程将涉及人类社会学、工程技术、环境地学、经济管理学等多种学科，如图9.3所示，具有显著的学科交叉、共融、综合作用。具体表现在以下几方面。

（1）以社会服务为导向：矿产资源服务归趋以向社会提供服务为工作目标，而这一目标与人口数量、分布、需求模式、消费偏好等社会因素密切相关，使得社会学在矿产资源服务归趋过程中起导向作用。

（2）以工程技术为手段：物质特定功能是以物质的微观形态转变为基础的，与物理化学等基础学科密切相关，并借助工程技术手段完成（Mao et al.，2012；左孝青，2000；宋兴诚，2011）。而完成这些变化又需要投入大量能源、物力、人力，凝聚着技术活动中物质、能量、价值等的内在协同。

（3）以人地交互规律为支撑：矿产资源服务归趋过程中蕴含着人类与环境之间的交互作用关系，并客观地反映在地表环境系统和人类活动圈的物质迁移、转化和时空分异等方面，而这些内容正是地学所长期关注的。物质的这些变化关系反映出人类与环境间的关系规律，是协调人类与环境间关系的理论依据。

（4）以经济发展为动力：矿产资源服务归趋过程中涉及多种生产环节，并客观地表现为生产不同产品、半成品的诸多生产企业。作为经济活动基本单元，企业在创造"服务"的同时，还将借助市场交易或贸易，获得经济收益。特别是对于大多数企业而言，经济收益已成为企业存活、发展及运营的主要动力。这使得矿产资源服务归趋过程对整个国民经济起到推动作用。

（5）以宏观管理为保障：实现人类与自然系统的持续发展是开展本项研究的最终目标，而做到这一点，需要在充分认识人类与环境间的内在关系规律的基础上，从微观、中观、宏观不同尺度上，对整个矿产资源服务归趋过程进行科学、有效管理，从而保证人类与自然环境协调、共融、同步演进。

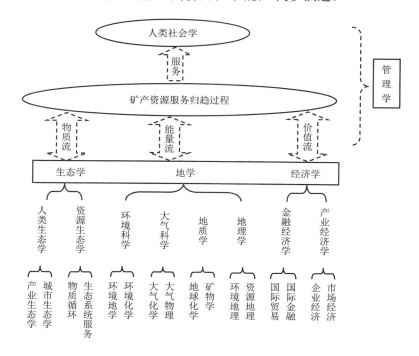

图 9.3 矿物资源服务归趋过程与相关学科间关系

Fig.9.3 The relationship between the flows of mineral resources to providing services and its related disciplines

9.3 核心议题

9.3.1 关键问题

矿产资源服务归趋过程是一个十分复杂的过程，是以人类需求为源头动力，通过设计、一系列加工、转变、产品使用等活动，来实现其人类服务目标的。在这一过程中，不仅每一阶段物质都要完成特定形态、功能转变，而且实现这些转变又需具备特定条件，蕴含着工程技术与物质自然属性间内在微观作用关系，并受到社会经济发展、技术应用、管理手段及政策等人为因素，以及资源与环境状况等环境因素的外在宏观影响，反过来还影响到社会经济发展与资源环境状况。因此，进行矿产资源服务归趋研究，将需要解决以下关键问题。

（1）为满足人类特定需求，物质的功能、形态发生了怎样的转变？数量分配关系如何？

矿产资源从岩石圈到满足人类某特定需求要经历一系列设计、生产、使用等环节，交织着技术圈与物质自然属性的复杂作用关系，物质的组成、形态、功能、数量等将发生一系列变化。综合考虑各环节技术因素对物质功能、形态、数量的影响，剖析这个归趋过程中物质的形态、功能、数量等方面的变化是弄清各要素间内在关系以及外部综合影响关系的重要基础。

（2）在矿产资源服务归趋过程中，物质流、能量流、价值流间存在怎样的耦合关系？

能量投入是驱动物质形态变化的重要条件，而物质价值的提升又是生产部门的运营动力，矿产资源人类服务归趋过程交织着物质、能量和价值流动间的内在耦合关系。然而，物质既是能量在载体，也是价值的载体，抓住"载体"这一关键联系，探究物质流、能量流、价值流间协同关系，是获得复合流动间协同耦合关系的重要基础，也是分析这一过程对社会经济资源环境影响综合效应的基础。

（3）矿产资源人类服务归趋过程将产生怎样的外部影响？

矿产资源服务归趋过程中物质元素将发生一系列重要转变，而完成这些转变不仅需要投入多种物力、人力等，而且还会代谢多种环境废物，从而不可避免地造成对社会、经济、资源、环境等诸多影响。采用科学方法合理构建服务归趋过程与外部要素间内在联系，既是弄清这个过程综合影响的前提，也是进一步分析其时间累积影响结果的基础，还是服务归趋基本规律的重要内容。

9.3.2 工作内容

为能获得以上问题的答案，将设置以下工作内容。

(1) 解析矿产资源服务归趋过程，辨识物质功能、形态转变及数量分配。

分析矿产资源服务归趋基本过程，界定服务归趋基本概念和内涵。针对典型蓄电池实地考察、调研、分析，辨识该过程各阶段中物质使用性能、化学结构、物质组合形式的变化，确定各阶段物质功能、形态类型，定量估算物质数量在各功能、数量间分配。

(2) 构建矿产资源服务归趋复合流动模型，探求服务归趋各流动间静态协同关系。

分析服务归趋过程各股流动中物质与能量、价值三种流动间内在联系，界定载能体、价值载体概念，构建矿物资源人类服务归趋静态模型；针对各种典型蓄电池进行物质流、能量流、价值流分析，辨识物质流特征、能量流特征、价值流特征；辨识物质、能量和价值三种流动间协同耦合关系；凝练不同种类蓄电池间存在共性，探求物质、能量和价值间静态协同规律。

(3) 构建矿产资源服务归趋过程综合影响评估框架，评估该过程综合影响水平。

构建服务归趋过程综合影响评估框架，择定外部影响类型及其表征指标，包括社会、经济、资源和环境几个方面，辨识外部影响指标与服务归趋过程内部各物质、能量、价值相关参数变化的内在定量联系，评估该过程的社会、经济、资源和环境四方面综合影响水平。

(4) 构建矿产资源人类服务归趋动态模型，探求服务归趋基本规律。

研究人类需求的时变性，以及这一变化对服务归趋过程及其外部综合影响的驱动关系，构建归趋动态模型，探求服务模式、数量等需求变化下服务归趋过程的动态响应关系，以及这些变化对社会、经济、资源、环境的综合影响关系；分析这些变化的时间累积效应；对比四种蓄电池间差异与共性，探求矿产资源服务归趋基本规律。

(5) 实证特定区域矿产资源服务归趋过程及其综合效应，提出改善建议。

选择典型区域，静态分析特定年份矿产资源服务归趋过程，评估其综合影响水平，构建定量清单；估算某特定时段矿产资源服务归趋过程的历史演变及其外部综合影响的时间累积结果，实证服务归趋变化规律；辨识实体服务与经济效应间偏差；辨识改善环节，提出改善措施和管理保障建议。

除此之外，还将对所用数据、所做假设、所建模型进行不确定性分析和敏感性分析。

9.4　展望与结语

矿产资源是为了满足人类需求而形成其服务归趋过程的，因此，研究矿产资源服务归趋有望弄清以下诸多问题：①人类活动如何干扰了物质的状态？②在人类活动干扰下，物质发生了怎样的改变？改变的过程如何？③人类活动改变物质的后果怎样？与此同时，伴随相关基本规律的获得，还有望弄清人类应该如何调整自己的活动来协调人类与环境间的关系等问题。

目前，人们已认识到人类活动对地表环境影响的重要性，但苦于传统中社会经济各学科与自然科学等各学科间的割裂，以及受限于对人类社会经济发展与自然环境系统之间的认识水平，难以弄清诸多环境问题发生的根本原因、发生机制和内在规律。为此，迫切需要借助学科交叉、综合来诞生新概念、新方法。希望通过本文对矿产资源服务归趋概念的介绍，帮助人们深入认识人类社会经济系统与自然环境系统的依存关系，充分理解由于人类不合理需求、人类与自然之间不合理作用模式造成了资源耗竭和环境质量下降。同时，也帮助相关科研人员了解矿物资源服务归趋基本概念和内涵，以便根据已有地理学、环境科学、工程技术、管理学等基础知识和可获得数据，对矿产资源的使用和环境污染物累积等典型议题进行科学的评估，进而为可持续发展与管理提供科学依据。

参 考 文 献

冯桂林，何蔼平. 2000. 有色金属矿产资源的开发及加工技术. 昆明：云南科技出版社，45-67.

高丕英，李江波. 2007. 物理化学. 北京：科学出版社，78-86.

刘楚明，林高用，邓运来，等. 2010. 有色金属材料加工. 长沙：中南大学出版社，154-160.

刘广林. 2011. 铅酸蓄电池工艺学概论. 北京：机械工业出版社，38-54.

莫文莲，陈沸清，孙禧华. 2010. 石墨炉原子吸收光谱法测定稻米各加工阶段铅含量研究. 粮食与油脂，8：34-35.

沈春林. 2008. 化学建材原材料手册. 北京：中国标准出版社，268-300.

宋兴诚. 2011. 重有色金属冶炼. 北京：冶金工业出版社，86-72.

王碧玲，谢正苗. 2008. 磷对铅、锌和镉在土壤固相-液相-植物系统中迁移转化的影响. 环境科学，29（11）：3225-3229.

严莎，凌其聪，严森，等. 2008. 城市工业区周边土壤-水稻系统中重金属的迁移累积特征. 环境化学，27（2）：226-230.

曾润，毛建素. 2010. 2005 年北京市铅的使用蓄积研究. 环境科学与技术，33（8）：49-52.

左孝青. 2000. 有色金属矿产资源的开发及加工技术. 昆明：云南科技出版社，112-135.

Agustín C A, Jesús L V G, Diana M F, et al. 2011. Impact of mining activities on sediments in a semi-arid environment: San Pedro River, Sonora, Mexico. Applied Geochemistry, 26（12）：2101-2112.

Elshkaki A, van der Voet E, Van Holderbeke M, et al. 2004. The environmental and economic consequences of the developments of lead stocks in the Dutch economic system. Resources, Conservation and Recycling, 42：133-154.

Elshkaki A, van der Voet E, Van Holderbeke M, et al. 2009. Long-term consequences of non-intentional flows of substances: Modelling non-intentional flows of lead in the Dutch economic system and evaluating their environmental consequences. Waste Management. 29（6）: 1916-1928.

Fernandes A, Afonso J K, Dutra A J B. 2012. Hydrometallurgical route to recover nickel, cobalt and cadmium from spent Ni-Cd batteries. Journal of Power Sources, 220: 286-291.

He K M, Wang S Q, Zhang J L, et al. 2009. Blood lead levels of children and its trend in China. Science of the Total Environment, 407（13）: 3986-3993.

International Organization for Standardization Technical Committee TC-207 on Environmental Management Systems. 2006. ISO 14044-2006. Geneva: Environmental management-Life cycle assessment-Requirements and guidelines. International Organization for Standardization.

Li Y L, Liu J L, Cao Z G, et al. 2010. Spatial distribution and health risk of heavy metals and polycyclic aromatic hydrocarbons （PAHs） in the water of the Luanhe River Basin, China. Environmental Monitoring and Assessment, 163: 1-13.

Mannhan S E. 2005. Environmental Chemistry（8th）. Boca Raton: CRC Press LLC.

Mao J S, Ma L, Niu J P. 2012. Anthropogenic transfer & transformation of heavy metals in anthrosphere: concepts, connotations and contents. International Journal of Earth Sciences and Engineering, 5（5）: 1129-1137.

Maruthamuthu S, Dhanibabu T, Veluchamy A, et al. 2011. Elecrokinetic separation of sulphate and lead from sludge of spent lead acid battery. Journal of Hazardous Materials, 193（15）: 188-193.

Niisoe T, Nakamura E, Harada K, et al. 2010. A global transport model of lead in the atmosphere. Atmospheric Environment, 44（14）: 1806-1814.

Schmidt-Bleek F. 2010. International factor 10 club's 1997 Carnoules' statement to government and business leaders: a ten-fold leap in energy and resource efficiency in one generation. http: // www. springtijfestival. nl/ Portals/ 0/ docs/ Factor10. pdf[2013-4].

Steinnes E, Berg T, Uggerud H T. 2011. Three decades of atmospheric metal deposition in Norway as evident from analysis of moss samples. Science of The Total Environment, 412: 351-358.

Szücs E. 1980. Similitude and Modeling: Fundamental Studies in Engineering（Balkay B, Trans.）. Amsterdam, the Netherlands: Elsevier Scientific Publishing Company.

Von Bertalanffy L. 1968. General System Theory: Foundations, Development, Applications. New York: George Braziller.

Xia Z G, Li Q. 2007. Structural phase transformation and electrical properties of （0.90-x）Pb （$Mg_{1/3}Nb_{2/3}$）O_3-xPbTiO$_3$-0.10Pb （$Fe_{1/2}Nb_{1/2}$）O_3 ferroelectric ceramics near the morphotropic phase boundary. Acta Materialia, 55（18）: 6176-6181.

Zackrisson M, Avellan L, Orlenius J. 2010. Life cycle assessment of lithium-ion batteries for plug-in hybrid electric vehicles-Critical issues. Journal of Cleaner Production, 18: 1519-1529.

Zhang B, Zhong J H, Li W J, et al. 2010. Transformation of inert $PbSO_4$ deposit on the negative electrode of a lead-acid battery into its active state. Journal of Power Sources, 195（13）: 4338-4343.

第 10 章 重金属人为迁移转化：概念、内涵与内容
Anthropogenic Transfer & Transformation of Heavy Metals in Anthrosphere: Concepts, Connotations and Contents*

10.1 Introduction

The rapid economic development of human society has been accompanied by increasingly intense disturbance of the environment. For example, research has convincingly shown that the emissions of heavy metals caused by human activities (anthropogenic emission) have become the main source of these pollutants in the environment (Mil-Homens et al., 2006; Wang et al., 2010). Because they are toxic substances, these metals greatly influence the quality of the ecological environment and human health after they enter the environment (Li et al., 2010). Especially in recent years, heavy metal pollution events have become more frequent in China due to the country's rapid rate of socioeconomic development (Zheng et al., 2010), and this has increased the urgency of finding ways to effectively manage heavy metals to control the pollution they cause.

Given the magnitude of human impacts on the environment, human activities have become an important component of Earth's ecological systems. In particular, the hybrid system comprising human society and economy has become an important subsystem that shapes the transfers and transformations of heavy metals in the anthrosphere (Gómez-Álvarez et al., 2011; Thevenon et al., 2011). The impacts on the environment exerted by humans involve transfers of heavy metals between locations, transfers of these substances between phases and transformation of these substances among different forms. However, previous research has traditionally focused on what happens after the heavy metals enter the environment (Ngoc et al., 2009; Yan et al., 2008), and there has been little research on the various changes they undergo within the hybrid socioeconomic system under the influence of human activities. Thus, it is difficult to solve problems related to changes in the properties, patterns, and locations of heavy

---

* Mao J S, Ma L, Niu J P. 2012. Anthropogenic Transfer & Transformation of Heavy Metals in Anthrosphere: Concepts, connotations and Contents. International journal of Earth Sciences and Engineering, 5 (5): 1129-1137.

metals under the influence of the human activities, the internal relationships among different patterns of transfer and transformation, how human activities drive these processes, and the environmental consequences.

Due to the large differences between hybrid socioeconomic systems and traditional Earth sciences studies, research on the transfer and transformation of heavy metals under the influence of the human activities, which can be described as "anthropogenic transfer and transformation (ATT)", the concepts and the research framework differ greatly from the traditional concepts and framework of Earth sciences and environmental chemistry. To better understand the ATTs of heavy metals in hybrid socioeconomic system, it is necessary to rigorously define the concepts, contents, and connotation of ATTs of heavy metals. The goal of the present study was to provide a new theoretical basis for understanding these transfers and transformations, thereby permitting more effective protection or remediation of environmental quality.

10.2 Concepts

10.2.1 Anthropogenic Transfers

In environmental chemistry, the transfers (or "transport", which is usually used for atmospheric pollutants) of pollutants relate to their spatial and temporal displacement in the environment and the processes that lead to enrichment, dispersion, and disappearance of these substances. There are three main categories of displacement mechanisms: mechanical, physicochemical, and biogenic (Wang and Zhao, 2009). Significant research has been done under this framework in recent years. For example, there has been considerable research about the transfers of heavy metals such as cadmium and lead in farming areas from the soil to the plant or from the soil into bodies of water (Ngoc et al., 2009; Yan et al., 2008; Li et al., 2010). The driving force for these transfers comes from natural forces and the occurrence place is the environment that is strongly disturbed by human activities.

The spatial component of anthropogenic transfers includes both the physical space that is familiar to us from traditional studies and a virtual space, which reflects how the attributes of the materials intersect with human needs, and which can be called the anthropogenic phase. Fig. 10.1 shows a conceptual framework for the transfers among the components of the anthropogenic phases. In addition to natural transfers, we can define four primary types of anthropogenic transfers: ①P-type transfers relate to metal production activities. ②U-type transfers relate to meeting human utilization needs. ③R-type transfers relate to resource conservation. ④E-type transfers relate to meeting environmental needs.

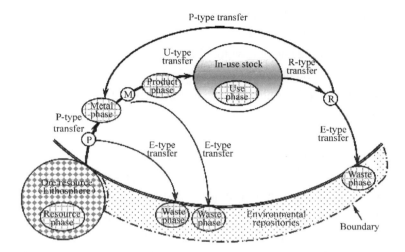

Fig. 10.1 The PURE conceptual framework for anthropogenic transfers of heavy metals in anthro-sphere

P-type transfers relate to metal production activities; U-type transfers relate to meeting human utilization needs; R-type transfers relate to resource conservation; and E-type transfers relate to meeting environmental needs.

These four transfers can be easily remembered in the form of an acronym: PURE. Any system for describing the transfers of heavy metals in a hybrid human socioeconomic system must provide a framework that accounts for these additional transfers. To successfully model how these substances move through the system, it will be necessary to parameterize each transfer and obtain reliable data on its quantities and its spatial and temporal patterns.

The spatial component also includes the three-dimensional physical space in which these materials exist. For example, during mineral exploration and subsequent smelting to obtain metals from the minerals, heavy metals transfer from the lithosphere into the industrial and economic system. During this transfer from the resource (e.g., the ore that is the original source of the metal), transformation also occurs (i.e., the physical attributes change from raw ore to metals suitable for meeting human needs). If the ores are not smelted at the same site where they are extracted, there will also be a spatial transfer from the mining location to the smelting location. If the demand for the ores and metals changes over time, there will also be temporal transfers. The driving force for both of these anthropogenic transfers is the desire to meet human needs.

From a chemical and physical perspective, a "phase" refers to a specific state of matter whose properties (including the content of each element in the substance and the nature and strength of the bonds between atoms) differ from those in any other phase. Other important properties of phases are whether the component elements are reactive (e.g., easily exchangeable) or stable, the physical phase of the substance (i.e., gas, liquid, or solid), and the position within the ecosystem (e.g., soil vs. water) (Gao and Li, 2007). The same phases exist in hybrid human-natural systems, but the four above mentioned transfer types also become important because they represent additional anthropogenic phases that relate to how a substance is moved and transformed to meet human needs. Four additional categories can be defined: ①Metal phase: the raw material that will be transformed into products. ②Product phase: the transformed material inside a product that will be used by humans. ③Application phase: use of the transformed material. ④Waste phase: loss or recycling of the substance during and after use.

Table 10.1 summarizes the characteristics of these additional phases. These four phases must be integrated with the four transfer types (P, U, R, and E) to fully describe the transfers and transformations of heavy metals as they pass through the hybrid human socioeconomic system.

Table 10.1 Explanations and examples of the proposed anthropogenic phases for heavy metals

Phase type	Explanation	Common examples of metal phases
Resource phase	The heavy metal substance in the mineral resources, which is the state that has not yet been exploited by humans.	Galena, towanite, lead zinc ore, greenockite, chromite, cinnabar
Metal phase	The metallic raw material after smelting to remove the non-metal components of the ore.	Lead, copper, zinc
Product phase	The heavy metal substances after they have become part of an industrial product, including semi-finished products that will undergo further transformation or inclusion in other products, but before the product enters service.	Welding rods, copper pipes, alloys, electroplating materials, plastic stabilizers, fluorescent powder, insecticides, chrome alum, plastics, paints, dyestuffs
Application phase	The heavy metal substances after they enter service to meet human needs.	Lead-acid batteries, radiators, cables, bearings, circuit boards, TV screens, fluorescent lamps, control rods in nuclear reactors, lead blocks
Waste phase	The heavy metal substances after they are no longer usable by humans. This phase includes the substances after they have been discharged into the environment before they have undergone physical, chemical, or biological reactions with the environment.	Gangue, waste residues from smelting, waster and gases from smelting, liquid wastes from electroplating, welding slag, incinerator ashes

The heavy metals present in mineral resources if not yet been disturbed by human activities, they seem to not belong to an anthropogenic phase. This state represents the initial state in the relationship between humans and heavy metals. Because this state greatly influences the subsequent anthropogenic transfers and transformations of the heavy metal (e.g., by determining the kind of mining and smelting processes that can be used), we have chosen to include the resource phase in the list of anthropogenic phrases in Table 10.1.

10.2.2　Anthropogenic Transformations

Definitions of "transformation" differ among the different scientific disciplines. In environmental chemistry, the transformation of pollutants refers to a process in which the existing form of a pollutant is changed or the pollutant itself is changed into another substance as a result of some combination of physical, chemical, or biological reactions. For example, mercury can be oxidized into Hg^{2+} after Hg-contaminated waste is discharged into water, after which it may be methylated into methylmercury and dimethylmercury by microbial processes (Wang and Zhao, 2009). In biogeochemistry, transformation of a heavy metal generally refers to a change from the existing form of the metal, which is usually classified into a soluble form in the water phase or a sedimentary form in rock. Furthermore, a metal in a specific form can be further classified into various species according to the nature of the bonds between the metal element and its associated elements. For instance, the solubility of a metal in water may include forms such as the metal's free ion (e.g., Zn^{2+}) and inorganic ligands (CrO_4^{2-}, $CuCO_3^0$, $CdCl^{2-}$), and so on (Wang, 2011). In addition, the heavy metal and its compounds may exist in exchangeable or stable forms, among others (Liao, 2005; Shi et al., 2010).

In the study of the ATTs of heavy metals, researchers must also focus on the properties of the metallic substances that let them meet specific human needs. These properties depend strongly on the form and the chemical structure of the heavy metal. Therefore, transformation of the heavy metal is an engineering process, in which the original form, in which the metal is found, with its own intrinsic physicochemical properties, is transformed through a series of production processes into a new substance with different properties that are better able to meet specific human needs. For example, lead's chemical structure can be transformed into an alloy with improved electrical performance to improve the performance of ceramic materials (Xia and Li, 2007), and $PbSO_4$ can be activated in the cathode of a lead-acid battery by means of X-ray diffraction to improve the energy efficiency of the battery (Zhang et al., 2010). Therefore, changes in the form of the heavy metal are anthropogenic transformations. Table 10.2 defines and provides examples of the anthropogenic forms and related species of heavy metals.

Table 10.2 Definition and examples of the anthropogenic forms and species of heavy metals

General species	Explanation	Subdivisions	Explanation	Example
Metal species	The chemical structure of the metallic material after smelting	None	The result of smelting a raw material (e.g., mineral ore) to produce the elemental metal.	Pb, Cu, Hg
Product species	The chemical structure of the heavy metal substance in the industrial final products	Dissipation products species	The chemical structure of the heavy metal in dissipation products (i.e., products whose components are released into the environment during use of the product) *	Cr_2O_3, $Na_2Cr_2O_7$, $Hg(NO_3)_2$, $HgSO_4$, $CuSO_4$
		Stable product species	The chemical structure of the heavy metal in stable products (i.e., products with a long service lifespan and for which the metal content remains stable during the product's service life) *	PbO, Pb_2O_3, $PbSb_2$, $PbSb_6$
Waste species	The chemical structure of the heavy metal substance when it is discharged into the environment after use	Waste gas species	The chemical structure of the heavy metal discharged into the environment in waste gas	PbO, Pb_2O_3, HgO, C_8H_2OPb, $(CH_3CH_2)_4Pb$
		Liquid waste species	The chemical structure of the heavy metal discharged into the environment in liquid wastes	$PbCrO_4$, $Cr_2(SO_4)_3$, Cr_2O_3, $2PbO \cdot PbSO_4$
		Waste residue species	The chemical structure of the heavy metal discharged into the environment in solid wastes	PbS, $CuFeS_2$, $PbCO_3$, $FeCr_2O_4$

* source: Mao and Graedel, 2009.

To incorporate these transformations within a model, it is necessary to define a separate parameter for each major form of the heavy metal. These parameters should include at least the following categories: ①The quantities of each form (e.g., raw form, smelted form, product or utilization form, and waste form). ②The quantities of each phase (e.g., solid, liquid, gas, mobile, stable).

Using these parameters, equations can then be developed that describe the transformations of a heavy metal among these various forms. This detailed description provides deeper insights than simply accounting for total masses of an element without accounting for its forms and its transformations among different forms.

10.2.3 Essential Characteristics

The ATTs of a heavy metal start with the natural mineral resource and end with discharge of wastes containing the metal into the environment before it can react with the environmental medium for the subsequent transformations as it undergoes physical, chemical, and biological reactions in the environment, and includes all the changes that occur between these two points. Compared with the concepts of transfer and transformation in traditional environmental sciences, anthropogenic transfers have the following unique characteristics.

(1) The changes within the human socioeconomic system are emphasized. This means that in addition to the physical, chemical, and biological transformations that have been previously and traditionally studied, the production (P), use (U), resource conservation (R), and environmental (E), and transfers shown in Fig. 11.1 and the metal, product, application, and waste phases shown in Table 11.1 must be considered.

(2) Human needs are taken as the original driving forces that cause the transfers and transformations. In environmental geology, the driving forces are natural forces, such as concentration differences, chemical bonding forces between elements, and physic-chemical processes such as dissolution by water. In human socioeconomics systems, each need or category of needs may define a process or set of related processes that must be considered.

(3) The status of the heavy metal is defined according to its relevance to humans. Accordingly, the spatial distribution of the metal also relates to human needs, and is therefore related to administrative (virtual) regions rather than physical regions. This has important implications such as trans-boundary and jurisdictional issues. This can be made explicit in environmental science, but is usually left implicit. Similarly, temporal patterns must be carefully considered using techniques such as life-cycle analysis.

10.3 Scientific Connotations

The anthropogenic transfers and transformations of heavy metals not only change their phase and form, but also their spatial and temporal distribution. Therefore, research on ATTs should aim to reveal the rules that govern these processes, clarify the impacts of these processes on the environment and on human socioeconomic development, and identify the evolution of these processes. Thus, this field of research has the following important scientific connotations:

The dependencies between humans and the environment from a systems perspective, anthropogenic transfers and transformations of heavy metals are inextricably linked with the environment and with natural transfers and transformations of these metals, but these processes are also driven by human needs that govern the transfers in the hybrid human socioeconomic system. Because the "anthropogenic environmental reservoir" (the part of the environment into which emissions are released but have not yet reacted with the environmental medium) is the container for environmental releases and accumulation of heavy metals, it is both the final destination of the metals within the human phases and the starting point where the heavy metals re-enter the natural environment.

The impacts of human activities on the environment, the processes that cause anthropogenic transfers and transformations of heavy metals have direct and indirect environmental impacts. During production activities, pollutants in various specific forms (e.g., see Table 10.2) will be emitted into the environmental reservoir, producing direct impacts on the environment. Indirect impacts include processes related to the effects of efforts to deal with these impacts (e.g., the creation of greenhouse-effect gases during production of the energy required to operate waste treatment or recovery facilities).

Human consumption preferences and required quantities as we have noted, all transfers and transformations within the human phases relate to the satisfaction of human needs. They are therefore strongly affected by human preferences. In addition, continuing expansion of the human population and the increased consumption that results from improvement of the standard of living will significantly increase the demand for heavy metals, along with all the associated processes from mineral extraction to the release of wastes. Finally, the complexity of modern technological civilization has created a huge diversity of human needs, which increase the complexity of the anthropogenic phases and the complexity of the transfers and transformations required to support this complexity and satisfy human consumption preferences.

The dynamic coupling between socioeconomic development and the environment, the ATT of heavy metals involve dynamic interactions between humans and our natural environment. The resulting spatial and temporal changes in the distribution and forms of heavy metals therefore represent a dynamic balance between the natural storage location, the resource quality (e.g., the chemical form of the heavy metal), and the locations where human needs arise and are satisfied. The needs that drive human activities and the activities themselves define the required product performance, product type, and production technologies. All of them change in response to changes in the relationships among the social and economic factors and natural factors that result from socioeconomic development. These changes lead to the redistribution of the phases and forms of heavy metals in different spheres of activity, regions, and periods. The overall transfers and transformation result from dynamic coupling among all of the factors previously discussed in this chapter: the heavy metal phase (Table 10.1), form, temporal and spatial distribution, required quantity, nature of the transfer (the P-, U-, R-, and E-type transfers in Fig. 10.1), as well as the dynamic interactions between humans and our environment. The anthropogenic transfers and transformations of heavy metals therefore reflect a dynamic coupling between human socioeconomic development and the natural environment.

Variations in the spatial and temporal distribution of human needs and natural elements, the spatial and temporal patterns for the ATT of heavy metals in various forms and phases among the various regions and spheres result from differences in the spatial and temporal distribution of mineral resources and human needs. These factors determine the direction and intensity of the transfers of heavy metals. For example, a refinery in a region that lacks sufficient mineral resources must purchase the metal ore from regions that are rich in the resource. Similarly, a customer who lives in a cultural or residential district must purchase paint containing heavy metals in its additives from an industrial production area in the same city or must import the paint from another city to decorate their house. The intensity of transport of daily consumption goods to a heavily populated region will obviously be higher than that to a sparsely populated region. The transportation of heavy metals between countries is usually termed "trade", and trade is strongly affected by human factors such as domestic trade policy, international regulations on the transport of goods and materials, and so on.

The consequences of human activities for the environment as one of the most vigorous components of the Earth's ecological system, human activities have had a surprisingly large impact on the global environment, and this impact is reflected in the anthropogenic transfers and transformations of heavy metals. From the perspective of the environmental impacts, these processes lead to the global redistribution of heavy metals. This redistribution is visible in the gradual depletion of the mineral reserves of a given heavy metal and significant increases in the accumulation of the heavy metal in the in-use stock (Elshkaki et al., 2004a; 2004b; 2004c), accompanied by increases in the amount of the heavy metal released into the environment as pollutants. However, these changes produce both ecological risks and risks to human health. For example, excessive lead concentrations in food in some areas threaten human health (Mo et al., 2010). Therefore, the transfer and transformation processes have more than purely theoretical consequences; they can affect both humans and our environment.

10.4　Core Issues

According to the proposal titled the anthropogenic transfer and transformation of lead and the quantitative analysis on environmental accumulated lead emission, which was granted by NSFC of China in Sep. 2011 (grant 41171361), the core issues and research contents of ATTs of heavy metals are listed as following sections.

10.4.1　Core Issues

Anthropogenic transfers and transformations of heavy metals are complicated processes, with human needs as their main driving force. These needs result in transfers between phases, spatial transportation of the metals in these phases, and transformation of the form of the metal through production activities and the use of various products that contain the metal. All of these activities will be influenced by factors such as the socioeconomic development status, technology level, management methods, and government policies, in addition to natural factors such as the resource and environmental conditions. Therefore, future research on anthropogenic transfers and transformation of heavy metals must solve the following key problems.

(1) How do human activities affect the temporal and spatial variations in a heavy metal?

Variations in the temporal and spatial patterns of a heavy metal are strongly influenced by both the natural distribution and quality of the resources and by human activities that result from the desire to satisfy human needs. An integrated consideration of the interactions between humans and our environment, the relationships among human and natural factors, and the temporal and spatial patterns in these relationships is a prerequisite for understanding the driving mechanisms that underlie the transfers and transformations of heavy metals to satisfy human needs.

(2) What spatial and temporal dynamics affect the coupling between the anthropogenic phases and the forms of the metal?

The temporal and spatial patterns of heavy metals result from the redistribution of these substances among various anthropogenic phases, metal forms, spheres of activity, regions, and time periods. These processes contain dynamic couplings among many parameters, including the heavy metal's phase, form, temporal and spatial distribution, and quantity, as well as the dynamic interactions between humans and our environment. Considering the impacts of these couplings on the phase, form, and quantity of the metal, its relationships with human needs and the modes of production, and the interactions among anthropogenic and natural factors will reveal the key points that define the structure of models capable of creating and testing hypotheses about processes and future trends.

(3) How can we capture the temporal and spatial variation in models capable of predicting future trends?

The accumulation, transfers, and transformations of heavy metals result from the phase transfers and form transformations defined by the related production and usage activities, driven by long-term demand for these substances to meet human needs. Some heavy metals are concentrated in high-quality form within products used by the human socioeconomic system, whereas others are discharged into the environment in the form of low-quality wastes and pollutants. This will result in gradual depletion of the associated mineral resources and the release of heavy metal pollutants into the environment. These changes are closely related to changes in the mode of action, technology level, and consumption preferences, each of which contains great uncertainties in the information required to describe and to predict these changes.

Answering the previous questions in section 4 will allow the development of models capable of describing the ATTs that result from and that cause temporal and spatial variations in the presence and characteristics of heavy metals. Combining this with an understanding of the uncertainty in each of these factors and a good estimate of the current conditions (i.e., the starting point for any simulation) will let researchers predict future trends and look for ways to protect both supplies of these important resources and the environment.

10.4.2 Main Research Contents

To answer the questions in section 4.1, research should focus on the following contents.

(1) Classification of the anthropogenic phases and forms of the heavy metal.

Life-cycle analysis of heavy metals and their products before, during, and after their use to support human activities will let researchers analyze changes in a metal's attributes, forms, and temporal and spatial distributions, as well as the relationships between changes in the factors and our ability to meet specific human needs. Defining the basic concepts by building on the framework described in this chapter will let researchers obtain data on the anthropogenic phases and forms of a heavy metal and relate these to the many factors that affect these phases and forms.

(2) Analysis of the processes that underlie the transfers and transformations and their temporal and spatial variation.

Analyzing the basic processes that determine the temporal and spatial variation of anthropogenic transfers and transformations of heavy metals will let researchers develop a conceptual framework for this variation. The present chapter provides such a framework, though much additional research will be required to refine the preliminary framework and begin defining the processes that operate within that framework.

(3) Developing a dynamic model of the temporal and spatial variations of anthropogenic transfers and transformations of heavy metals.

Identifying the factors that affect the temporal and spatial variations in the anthropogenic transfers and transformations of heavy metals will let researchers quantify and parameterize these factors and analyze the effects of these factors. In turn, this will allow the development of a dynamic model of these processes that is capable of simulating temporal and spatial changes in the transfers and transformations as well as the dynamic coupling between human and natural processes. Such models will provide insights into the characteristics of the hybrid human-natural system that determines the fate of a heavy metal.

(4) Quantifying the transfers and transformations of a heavy metal in a specific region.

Using the understanding and models developed in the previous steps, it should become possible to take a specific region as an example, then combine historical data with models of the anthropogenic transfers and transformations and the temporal and spatial variation to predict patterns and trends for the heavy metal. Such models can be used to estimate the rate of change in the anthropogenic transfers and transformations and their temporal and spatial variation, and predict changes in the intensity of usage, accumulation, and anthropogenic release of the heavy metal. In addition, analysis of the uncertainties in these predictions and sensitivity analyses to determine the model's dependence on the quality of the data will allow improved hypothesis testing and model development.

10.5　Prospects and Conclusions

It is increasingly important to understand human impacts on the natural environment created through the industrial system that produces products and the socioeconomic systems that determine the creation, use, and disposal of these products. Research on the transfers and transformations of heavy metals through the hybrid human-natural system is necessary to let researchers answer the following questions.

（1）How do human activities affect the state of a heavy metal and changes in that state?

（2）Under the disturbance created by human activities, what changes will the substance undergo?

（3）What processes affect these changes?

（4）What consequences arise from these changes in a substance?

Learning the basic rules that determine the anthropogenic transfers and transformations of heavy metals will provide an important basis for managing the relationships between humans and our environment and reducing the impacts of human activities on the ecological environment. Understanding these rules will let us solve the problems related to temporal and spatial variation in the risk to the ecological environment created by these processes and decide what measures should be taken to reduce the risk.

The impacts of human activities on the environment are increasingly understood, but the relationship between human socioeconomic systems and the natural environment are less well understood. Therefore, it remains difficult to understand the primary causes and the mechanisms that determine the occurrence of heavy metal pollution. We hope that the framework for the anthropogenic transfers and transformations of heavy metals proposed in this chapter will stimulate research on the dependency between socioeconomic systems and the natural environment, and will support efforts to avoid irrational use of resources that leads to exhaustion of the resource, accompanied by declining environmental quality. The concepts and connotations described in this chapter provide a framework that will help researchers to evaluate the use, discharge, accumulation, and changes in the transfers and forms of heavy metals. Combined with existing basic geographical and environmental knowledge and the available data, this research will provide objective evidence that can be used to guide management decisions to improve the relationship between human and ecological systems.

Acknowledgements

This project was financially supported by the National Natural Science Foundation of China (General Program) under grant 41171361.

References

Elshkaki A, Van der Voet E, Van Holderbeke M, et al. 2004a. Development of a dynamic model for substance flow analysis: Part 1-General stock model report No 2004/IMS/R/292. Belgium: Flemish Institute for Technological Research (VITO).

Elshkaki A, Van der Voet E, Van Holderbeke M, et al. 2004b. Development of a dynamic model for substance flow analysis: Part 2-Integration of stock and flow model report No 2004/IMS/R/293. Belgium: Flemish Institute for Technological Research (VITO).

Elshkaki A, van der Voet E, Van Holderbeke M, et al. 2004c. The environmental and economic consequences of the developments of lead stocks in the Dutch economic system. Resources, Conservation and Recycling. 42: 133-154.

Gao P Y, Li J B. 2007. Physical chemistry. Beijing: Science Press (In Chinese).

Gómez-Álvarez A, Valenzuela-García J L, Meza-Figueroa D, et al. 2011. Impact of mining activities on sediments in a semi-arid environment: San Pedro River, Sonora, Mexico. Applied Geochemistry. 26 (12): 2101-2112.

Li F L, Ni L J, Yuan J, et al. 2010. Cultivation practices affect heavy metal migration between soil and *Vicia faba* (broad bean). Chemosphere. 80 (11): 1393-1398.

Li Y L, Liu J L, Cao Z G, et al. 2010. Spatial distribution and health risk of heavy metals and polycyclic aromatic hydrocarbons (PAHs) in the water of the Luanhe River Basin, China. Environmental Monitoring and Assessment. 163: 1-13.

Liao G L. 2005. Research on the pollution assessment and the transport characteristics of heavy metals in a typical nonferrous metal mine. Changsha: Central South University (In Chinese with English abstract).

Mao J S, Graedel T E. 2009. Lead in-use stock: a dynamic analysis. Journal of Industrial Ecology. 13 (1): 112-126.

Mo W L, Chen Y Q, Sun X H. 2010. Study on determination of lead in various processing stages of rice by GF-AAS. Journal of Cereals & oils. 8: 3.

Mil-Homens M, Stevens R L, Boer W, et al. 2006. Pollution history of heavy metals on the Portuguese shelf using 210Pb-geochronology. Science of the Total Environment. 367 (1): 466-480.

Ngoc M N, Dultz S, Kasbohm J. 2009. Simulation of retention and transport of copper, lead and zinc in a paddy soil of the Red River Delta, Vietnam. Agriculture, Ecosystems & Environment. 129 (1-3): 8-16.

Shi Y Z, Han W Y, Ma L F, et al. 2010. Influence of exogenous lead on change of lead forms and bioavailability in different tea garden soils. Journal of Agro-environment Science. 29 (6): 1117-1124.

Thevenon F, Graham N D, Chiaradia M, et al. 2011. Local to regional scale industrial heavy metal pollution recorded in sediments of large freshwater lakes in central Europe (lakes Geneva and Lucerne) over the last centuries. Science of the Total Environment. 412-413: 239-247.

Wang H Y, Zhao L J. 2009. Environmental chemistry. Beijing: Journal of Chemical Industry and Engineering. 3: 68. (In Chinese)

Wang W X. 2011. Ecotoxicology and biogeochemistry for trace metals. Beijing: Science Press. (In Chinese)

Wang X P, Yang H D, Gong P, et al. 2010. One century sedimentary records of polycyclic aromatic hydrocarbons, mercury and trace elements in the Qinghai Lake, Tibetan Plateau. Environmental Pollution. 158 (10): 3065-3070.

Xia Z G, Li Q. 2007. Structural phase transformation and electrical properties of (0.90-x) Pb ($Mg_{1/3}Nb_{2/3}$) O_3-xPbTiO$_3$-0.10Pb ($Fe_{1/2}Nb_{1/2}$) O_3 ferroelectric ceramics near the morphotropic phase boundary. Acta Materialia. 55 (18): 6176-6181.

Yan S, Ling Q C, Yan S, et al. 2008. Behaviors of heavy metals in soil-rice system around in industry area around city. Environmental Chemistry, 27 (2): 226-230.

Zhang B, Zhong J H, Li W J, et al. 2010. Transformation of inert $PbSO_4$ deposit on the negative electrode of a lead-acid battery into its active state. Journal of Power Sources. 195 (13): 4338-4343.

Zheng N, Liu J S, Wang Q C, et al. 2010. Heavy metals exposure of children from stairway and sidewalk dust in the smelting district, northeast of China. Atmospheric Environment. 44 (27): 3239-3245.

第11章 铅元素人为服务归趋中的变化：功能、形态与位置

Changes in the Functions, Species and Locations of Lead during Its Anthropogenic Flows to Provide Services*

11.1 Introduction

Along with the implementation of sustainable development strategies, the development of a services-oriented society is becoming an important measure to conserve natural resources and reduce environmental pollution (Stahel, 1997). A variety of studies of this topic have been carried out in recent years, and have focused mainly on the following aspects: ①In terms of industrial economics, the goal is to increase the proportion of the economic structure accounted for by service industries, so that these sectors play a more important role in the economy (Henriques and Kander, 2010). ②For products and services, it is important to improve the product's performance or the service's quality by means of ecological innovation, particularly if this can increase services and decrease the importance of the least-efficient or most polluting products (Prettenthaler and Steininger, 1999). ③To guarantee that these processes occur, policies and regulations have been adopted to promote the implementation of a services-oriented society (Schmidt, 2001). These researches have emphasized the services that materials can provide to humanity. In other words, the fate of these materials in the anthrosphere (i.e., the human socioeconomic system and the parts of the environment that are affected by humans) is as final products with specific functions (ISO, 2006) that are capable of meeting human needs for specific services.

In hybrid human-natural systems, most materials that flow through the anthrosphere originate in nature (Graedel and Allenby, 1995; Manahan, 2005). When a material flows from its natural state into the anthrosphere to satisfy human needs, the substance undergoes three main processes: ①extraction from nature; ②transformation into products with specific functions that meet human needs by means of a series of processes

---

* Mao J S, Liang J, Ma L. 2014. Changes in the functions, species and locations of lead during its anthropogenic flows to provide services. Transactions of Nonferrous Metals Society of China, 24 (1): 233-242.

that extend from design to fabrication and manufacturing (ISO, 2006); and ③ performance of these functions to provide human services in the form of final products. During these processes, materials inevitably undergo a series of important changes (Mao et al., 2012), which involve redistribution within the anthrosphere at different temporal and spatial scales, forming different service patterns (van Beers and Graedel, 2007; Mao et al., 2008a; 200b; Eckelman et al., 2012). The chemical and physical properties of the substance change, as do its states, the services it provides, its functions, and the locations where it performs those functions, all are closely related to the substance's form (Mao et al., 2012). The quantitative changes have been extensively studied for various anthropogenic cycles (Mao et al., 2008a; 2008b; 2009; Mao and Graedel, 2009), but changes related to the form of a substance have also been studied by experts in engineering, technology, and basic sciences such as physics and chemistry (Zhang et al., 2010). However, due to a lack of communication among these disciplines, it is difficult to create a thorough picture of the main changes in a substance that result from human activities, thereby making it difficult to systematically analyze all the changes to a substance that occur during its flow through the anthrosphere to providing human services.

Lead is a good example of a material that flows through the anthrosphere and links it with the natural environment. Lead originates as a nonrenewable resource, in the form of lead ore (Mao et al., 2008a), and its emission into the environment creates a high level of ecological risk, including threats to human health (Zheng et al., 2010). From the perspective of the human services that it provides, lead is mainly used to manufacture lead-acid batteries (LABs), which store and deliver electrical energy, and because good records are maintained for these products, they can be used to quantitatively describe the services provided by lead (Mao et al., 2006). Lead also enters the economic components of the anthrosphere through trading of materials and marketing of products. Lead offers an additional advantage for such studies: its flows are easier to trace than those of compounds such as polymers and complex molecules. Furthermore, many previous studies on the anthropogenic lead cycle provide a sound foundation for a study of lead's life-cycle processes, uses, and service patterns, which make it easier to trace lead through the stages of its life cycle and, thus, to identify the functions, forms, and locations of lead. Such research represents a useful approach to understanding the interactions between humanity and nature that result from flows of materials to providing human services.

11.2 Analogy between Anthropogenic Flows and Environmental Flows

11.2.1 Factors That Influence Environmental Lead Flows

In natural systems, a substance such as lead has a certain temporal and spatial distribution patterns. The places where the substance is concentrated are termed "reservoirs", and the flows of the substance between reservoirs are called "fluxes" (Manahan, 2005). A substance generally flows in a certain direction, which is determined by the direction of the natural forces responsible for these flows (e.g., geological processes that concentrate minerals in certain locations, downhill flows of water that carries dissolved substances). These flows produce trends in the accumulation of the substance in specific locations, such as in deposits of lead-containing ore or in products.

Traditional studies on the fate of a material as a result of these processes originate in the fields of geochemistry and environmental chemistry, and include research that focuses on environmental pollutants and the sources, pathways, fates, and possible ecological risks of these substances in the environment (Manahan, 2005). Although there are some natural sources of pollution, most concentrated pollutants are anthropogenic, and enter the environment as gaseous emissions, wastewater discharges, solid wastes, and residues produced by human activities (Li et al., 2010; Niisoe et al., 2010). The pollutants then follow pathways between environmental components such as soils, plants, and water (Li et al., 2010; Niisoe et al., 2010). The environmental fates of the pollutants reflect the results of their flows through the environment.

Many studies have shown that the main factors influencing the flows of pollutants through the environment are their physical and chemical properties, forms and locations (Manahan, 2005). The physical and chemical properties of a material reflect its internal atomic or molecular structure, which in turn determines the material's environmental behavior (Gao and Li, 2007). The existing forms of a substance affect the chemical bonds that form between the substance and various other substances in the environment (Liu et al., 2010). The location of a substance determines its surrounding environment and how it contacts the environment, which in turn determines the kinds of natural forces it will experience. The forms and locations of a substance change in responses to changes in the state of the environment. Thus, all flows of a material through the environment result from interactions between the natural attributes of the substance and those of the environment, which also determine the eventual fate of the material.

11.2.2 The Relationships between the Anthropogenic and Biogeochemical Lead Cycles

In nature, lead flows between the lithosphere, atmosphere, hydrosphere, pedosphere, and biosphere; these flows are referred to as the "biogeochemical" or "natural" cycle of lead (Liu et al., 2010). Humans are the most active component of the biosphere, and both accelerate flows of lead into the anthrosphere and promote flows through the hybrid socioeconomic-environmental system; these flows are called the "anthropogenic" lead cycle (Mao, 2008; Eckelman et al., 2012). The environment acts as both the source of lead to the anthrosphere and a sink for lead released from the anthrosphere, leading to important interactions between the anthropogenic and biogeochemical lead cycles (Fig. 11.1). In this chapter, we will focus on the anthropogenic part and how lead changes its role in response to providing human services.

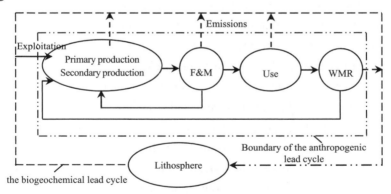

Fig. 11.1 Illustration of the conceptual relationship between the anthropogenic and biogeochemical lead cycles

F&M=fabrication & manufacture, WMR=waste management & recycling

11.2.3 Comparing the Anthropogenic and Environmental Flows of Lead

From the perspective of system theory (von Bertalanffy, 1968), both the anthropogenic lead cycle and the flows of lead after it enters are important components of the overall biogeochemical cycle of lead; that is, the human system is a component of the natural system that contains it, as shown in Fig. 12.1. Thus, the two cycles should exhibit similar phenomena, behaviors, processes, essential characteristics, and factors that influence the flows; this can be deduced from the homologous similarity principle in similarity theory (SzÜcs, 1980). The anthropogenic lead cycle extends from extraction of lead from ore to the final use of lead products that provide specific services, including the recycling of wastes to provide inputs for other services; the biogeochemical lead cycle extends from the release of lead from a source into the environment (analogous to human extraction from a source) to the subsequent flows of lead through the various components of that environment and the associated chemical transformations (analogous to human transformations of lead to provide services).

The anthropogenic lead flows result from human activities that are designed to meet specific human needs. This observation suggests three main differences from the environmental components of the lead cycle.

(1) Different dynamic mechanisms: Human needs are the primary driving forces of anthropogenic flows which purpose to provide specific human service. To reach such a target, lead flows result from the extraction of lead ore, which occurs more rapidly and on a larger scale than biogeochemical processes. This is followed by processing of the ore into intermediate materials and fabrication of products using these materials; in contrast with natural processes, these processes reverse entropy as a result of external energy inputs. In addition, the flows include artificial processes that do not occur in nature, such as trade and transportation of both primary and secondary materials, and the use and disposal of lead products (Mao et al., 2012). In contrast, environmental flows once lead enters the environment are driven primarily by natural forces.

(2) Different sources, pathways, and fates: In the environmental, lead pollutants may be focused and they may pass through air, water and so on, and finally be dispersed in the pedosphere, bodies of water, and other components of the ecosystem (Manahan, 2005). However, in the anthrosphere, lead ore is the original source, supplemented by lead scraps generated by obsolete lead products or recovery of lead-contained wastes as the secondary source. The in-use stock of lead represents the immediate fate of this material (Elshkaki et al., 2004) before the end of a product's life cycle and acts as the place where its anthropogenic functions and final services are provided to humans. Thus, the two main anthrosphere pathways flowing from extraction of lead ore and from the recovery of lead scraps will lead to the production and subsequent use of lead products to provide services to humans (Mao et al., 2012).

(3) Different factors influencing the flows and fates of lead: In the environment, the main factors that influence the fate of lead are its property, forms and locations in the environment. In the anthrosphere, the lead is used to meet human needs, so the primary factors influencing its fate are how the lead is used to meet these needs and what happens to the lead at the end of a product's useful life (i.e., after it becomes obsolete). Because the function of a material depends on its physical and chemical structure (Gao and Li, 2007), the anthrosphere forms of lead differ from their natural forms. The new functions and forms are created through design, processing, manufacturing, and engineering activities (Zhang et al., 2010), all of which are directly shaped by the required human services the lead will provide in its new form. In this chapter, we consider the functions, chemical structures or compositions, and locations where lead is used as the main factors that influence the flows and fate of lead in the anthrosphere.

11.3 Factors That Influence Anthropogenic Lead Flows

11.3.1 Tracing Flows of Lead through Its Life Cycle

Before we can identify the specific function, chemical structure, and location of lead within the anthrosphere, it is necessary to know what service a product provides and the stage of lead within its life cycle for that product. This information drives us to trace the flows of lead through the anthrosphere.

In our analysis, we paid attention to the following life cycle stages (Fig. 11.2): lead production from ore, manufacturing and use of lead products, treatment of wastes containing lead, and recycling of scrap lead (Mao et al., 2008a; 2008b). During primary and secondary lead production, lead metal will be produced through smelting, concentration, and refining (Song, 2011; Wang, 2012). During manufacturing and fabrication, lead metal is transformed into lead products with new forms that can provide specific services. Lead will sometimes be combined with other materials to meet a specific need, resulting in a specific chemical structure that supports that function (Zhang et al., 2010). The lead metal will be allocated to various products during the manufacturing and fabrication stage, creating different usage patterns (Mao et al., 2008a).

During use stage, lead products with different functions and life-spans flow into the hands of customers, where they provide services until they become obsolete at the end of a product's lifespan. The balance between the amount of lead entering and leaving use stage determine the accumulation of lead in-use stocks (Elshkaki et al., 2004). The amount of lead in-use stock may increase as products are in growing periods and decrease in dropping periods (Mao et al., 2009).

During the waste treatment and recycling stages, obsolete lead products are collected, disassembled, and sorted, and then the useful parts are returned to upstream processes for reuse or sent to secondary production processes as raw materials. Components that cannot be recovered are discharged into the environment after various treatments, including burning and disposal in landfill sites.

Throughout these stages, lead flows between regions (e.g., from mines to cities) and between components of the anthrosphere (e.g., from a battery manufacturing plant to a car assembly plant). Differences in the trends for these flows result from differences in the distribution of sources of a material that becomes the input for a process, and in the location of the industry that will use that input or the population that will use a given product. These differences change in response to changes in socioeconomic conditions, which determine the trade in materials and products and the resulting transportation needs. Tracing lead through these stages of its life cycle provides a framework that can be used to support subsequent analysis of the quantitative changes in flows of lead through the anthrosphere.

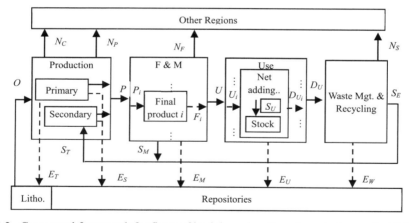

Fig. 11.2 Conceptual framework for flows of lead through anthrosphere during product life cycle

Symbols and definition

Symbol	Definition	Symbol	Definition
O	Lead from ore	E	Lead in emissions to the environment
P	Refined lead (P_i, for that in product i)	E_T	Lead in tailings
U	Lead entering its Use phase (U_i, for use of product i)	E_S	Lead in slag

Continued

Symbol	Definition	Symbol	Definition
D_U	End-of-life lead flows to waste management & recycling (WMR) (D_{Ui}, for product i)	E_M	Lead in fabrication and manufacture (F&M) emissions
S_U	Lead entering the in-use stock	E_U	Lead in emissions from use stage
N	Lead in net exports (exports minus imports)	E_W	Lead in WMR emissions
N_C	Lead in net exports of concentrates	S_T	Lead in total scrap flowing to production
N_P	Lead in net exports of refined lead	S_M	Lead in new scrap from F&M
N_F	Lead in net exports of products	S_E	Lead in end-of-life scrap
N_S	Lead in net exports of lead scraps	F	Final products (F_i, for product i)

11.3.2 The Factors That Influence the Flows of Lead to Providing Human Services

In this section, we will describe how the functions, forms (chemical structure or composition), and locations of lead (the factors influencing the flows) can be identified in the flows of lead through the anthrosphere. We will discuss this from the perspective of the services that the lead provides as a result of these flows.

1. Functions

In the course of lead flowing for providing anthropogenic services, the basic property of lead that attracts humans is its ability to satisfy some specific human demands, which includes further production needs and immediate use needs. To meet the first type of need, the function of lead is performed as that it can be further processed by transforming its chemical structure or components based on its physical and chemical nature, through technological processes such as designing and engineering (Liu et al., 2010). For example, the boiling point of lead is around 1740°C, and it can thus be extracted from lead ore or scrap lead by controlling the range of smelting temperature (Feng and He, 2010). Lead is ductile and malleable, thus it can be used to make products such as lead pipes and sheets that provide services such as transport of fluids and protection against radiation (Liu et al., 2010). Various references describe the physical and chemical properties of lead compounds, and these can be compared with the requirements for specific engineering or chemical applications that will provide such services. To meet the second type of needs, a final product which requires no further processing is put in to direct use. The functions of lead to meet human use need specifically the performance characteristics of a product (ISO, 2006). For example, the lead acid battery taking lead as its main components has the function to store and transform electricity (Liu, 2011). All the functions of lead that meets the human use needs are identified in professional manuals related to different lead products. The source of the main function related information is listed in Table 11.1.

2. Forms

The main properties of lead that make it useful for providing human services are related to its chemical structure, electrovalence structure and its ability to form other compounds when combined with other substances. This lays a foundation for its specific functions. All the forms in the anthrosphere are mainly obtained through a series of human activities such as designing and engineering. For example, Pb exists in galena but can be transformed in to elemental Pb after smelting and refining it (Feng and He, 2010). Industrial processing technology manuals provide forms' related information for designers or engineers in metal smelting and other processing factories. A summary of the sources for different forms of Pb are given in Table 11.1.

Table 11.1 Sources of information for the functions, forms, and locations of lead

Type	Name and example	Information source
Function	galena, cerussite, wulfenite, obsolete lead products[a]	Feng and HE, 2010
	Lead ingots, bars, powder, pipes, and sheets	Liu et al., 2010
	Lead-acid batteries	Shenyang Research Institute of Storage Battery, 2010
	Cables	Cang, 2010
	Lead alloys	National Technical Committee for the Standardization of Nonferrous Metals, 2008
	Others[b]	Shen, 2008
Forms	galena, cerussite, wulfenite	Wang, 2012
	obsolete lead products[a]	Song, 2011
	Ingots, bars, powder, pipes, and sheets	Liu et al., 2010
	Lead-acid batteries	Liu, 2011
	Cables	Liu et al., 2010
	Lead alloys	National Technical Committee for the Standardization of Nonferrous Metals, 2008
	Others[b]	SzÜcs, 1980

Continued

Type	Name and example	Information source
Location	galena, cerussite, wulfenite	Editorial Staff of the Yearbook of the Nonferrous Metals Industry, 2012
	Lead in service, end-of-life scrap lead[c]	National Bureau of Statistics of China, 2012
	Lead metal[d]	Editorial Staff of the Yearbook of the Nonferrous Metals Industry, 2012
	Lead products[e]	China battery industry association, 2012

a mainly obsolete lead-acid batteries, but also including obsolete pipes and sheets, cables, and lead residues;
b refers to lead in paints, polyvinyl chloride (PVC), and gasoline as components or additives;
c represented by lead-acid batteries used in vehicles;
d represented by the location of refineries, including primary and secondary production;
e represented by the location of production of lead-acid batteries.

In most products, the form of lead does not change during use. However, some forms of lead change during use, for instance, the lead-tin alloys that are used in soldering of metal parts, the initial form is $Sn_{60}Pb_{40}$ or $Sn_{63}Pb_{37}$ before use, but during soldering, Pb_2O_3, Pb_3O_4, or PbO is generated and lead is emitted into the environment during the joining. In this article, we will use the forms of lead in the final products to represent the forms of in-use lead.

3. Locations

Differences in the regional distribution of natural resources such as lead ore and in human factors determine the sources of supply for these resources. However, the needs for services are determined by human factors such as the locations of population centers and industries, and differences between regions in the economic development level and the industrial structure. As a result of these differences, lead-containing materials must be transported from source areas to destination areas that will use these materials to provide a service. These flows take two types: transportation between regions (e.g., exports of lead-acid batteries from China to Europe) and flows between components of the biogeochemical and anthropogenic systems (e.g., the flow of lead from the lithosphere into the anthrosphere as a result of mining). Trends in the flows between sources and destinations result from differences in the availability of and demand for services that require a given form of lead. Flows between components of the anthrosphere arise from pathways between primary and secondary industries and from the creation and emission of products and wastes. Although it is theoretically possible to describe the changes in the location of lead by tracking lead from mining and processing through product manufacturing to use and disposal, this is difficult in practice because of the diversity in lead use patterns, the distribution of enterprises, and the locations where lead products are used. In addition, precise statistics are not available for all locations of lead flows.

To solve these problems, we chose lead-acid batteries as our case for examining the regional changes in the anthropogenic flows, since batteries represents the main use of lead in China, accounting for 80% of the total lead consumption (Mao et al., 2006). We used China's provinces, autonomous regions, and municipalities as the regional units, and used government statistics to define the five most important locations for mining of lead ore, smelting of the ore, and manufacture and use of the batteries as representative locations. Because batteries are mainly used in vehicles, we used regional vehicle ownership statistics to represent the amount of in-use lead to provide a service in the form of electricity storage and delivery via batteries, and identified the three locations with the highest vehicle ownership to represent the main locations where final use of the batteries to provide this service occurs. On this basis, we determined trends in the directions of the lead flows between the main locations where lead is produced and used.

11.4 Results

11.4.1 Changes in the Functions of Lead

The properties and functions of lead change as it flows through the stages from the extraction of ore to manufacturing of the final products (Table 11.2). For example, raw ore has few direct uses, but can provide services as an input for the smelting and refining processes; because the resulting lead metal is both ductile and malleable, it can be transformed into clothing that provides the service of protection against radiation. As another example, paint that contains lead may appear red or yellow as a result of adding Pb_3O_4 or PbO, respectively (Liu et al., 2010). Other functions include storage and transfer of electrical energy in batteries and the provision of resistance to corrosion and radiation.

Table 11.2 The main functions of lead in providing human services in the anthrosphere

Type	General function	Examples	Specific function description
Lead ore and scrap lead	Sources of lead metal in the anthrosphere to meet production needs and be further processed into refined lead metal	Ore: galena, cerussite, wulfenite	Primary sources of lead metal in the anthrosphere. For example, galena can be made into lead metal through mining, concentration, smelting, and refining.
		Scraps: obsolete lead products, such as lead-acid batteries, cables, and lead residues	Secondary sources of lead metal in the anthrosphere. For example, lead-calcium and lead-antimony electrode plates and lead sulfate in obsolete lead-acid batteries can be made into lead metal through collection, disassembly, melting, and refining.
Lead metal	Raw material for industrial production. It can meet production needs and can be further processed into industrial lead products.	Lead ingots	Lead ingots can be used to manufacture lead-acid batteries, lead pipes, tubes, sheets, and radiation-proof shielding.
		Lead bars	Lead bars can be used to manufacture cable sheathings, printed circuit boards, and cathode-ray tubes.
		Lead powder	Lead powder can be used to manufacture lead alloys, pigments, and chemicals, and can be further used to manufacture parts of lead-acid batteries, paints, and lead-based components such as stabilizers in polyvinyl chloride (PVC) and tetraethyl lead as an antiknock additive in gasoline.
Lead products	To provide specific services and meet specific usage needs.	Lead-acid batteries	Lead-acid batteries can store and transfer electric energy.
		Lead pipes	Lead pipes can transport corrosive liquids.
		Lead sheets	Lead sheets can defend against nuclear radiation and corrosive chemicals.
		Cables	Lead sulfate in cable sheathings can protect cables from radiation and corrosion.
		Lead alloys	Lead alloys such as tin-lead solder can be used to repair and join metal pieces.
		Paints, pigments	Lead compounds in paints and pigment can be used to create colors in ceramic glazes and paints; for example, Pb_3O_4 for red and PbO for yellow.
		Televisions	Lead oxide in cathode-ray tubes TV can protect against radiation and stabilize optical effects.
		PVC	Lead is used as a stabilizer in PVC.
		Gasoline	Tetraethyl lead is used as an anti-knock additive for aviation fuel in piston-driven aircraft.
		Photovoltaic cells	Lead can store and transfer energy in lead-based semiconductors, such as lead telluride, lead selenide, and lead antimonide, which are important components of photovoltaic cells.

11.4.2 Changes in the Forms of Lead

Lead also changes form in response to the need for different functions, since different forms offer different properties. For example, the form of lead in galena is PbS, and this is transformed into lead metal (Pb) by smelting and refining. The lead can also be transformed into PbO_2, Pb, and $PbSO_4$ when it is incorporated into batteries. Table 11.3 summarizes the main forms of lead and the corresponding chemical structures. The various forms that lead takes in a battery, which represent the major use of lead, include Pb metal, PbO_2 in fully charged batteries, and $PbSO_4$ in fully discharged batteries. The lead metal (Pb) in lead plates, pipes, and sheets is the next most common form. The next most common forms are lead oxides such as PbO_2 and PbO.

Table 11.3　The main forms of lead in the anthrosphere

Form of lead	Materials	Main chemical structure or components
Lead ore	Galena, cerussite, wulfenite	PbS, $PbCO$, $Pb[MoO_4]$
Lead scraps	Obsolete lead products, such as lead-acid batteries, cables, and lead residues	$PbSO_4$, Pb, PbO_2, Pb_xSb_y, Pb_xCa_y
Lead metal	Lead ingots, bars, powder	Pb
Lead products	Lead-acid batteries	$PbSO_4$, Pb, PbO_2, $Pb_xSb_y^*$, $Pb_xCa_y^*$
	Lead pipes	Pb
	Lead sheets	Pb
	Cables	$3PbO \cdot PbSO_4 \cdot H_2O$, $2PbO \cdot PbHPO_3 \cdot H_2O$
	Lead alloys	$Pb_{96}Sn_4$, $Pb_{93}Sn_7$, $Sn_{60}Pb_{40}$, $Sn_{63}Pb_{37}$, $Sn_{30}Pb_{50}Zn_{20}$
	Paint, pigment	Pb_3O_4, PbO
	Televisions	PbO
	PVC	$3PbO \cdot PbSO_4 \cdot H_2O$, $2PbO \cdot PbHPO_3 \cdot H_2O$
	Gasoline	$(CH_3CH_2)_4Pb$
	Photovoltaic cells	PbTe, PbSe, PbSb

* lead alloys

11.4.3　Changes in the Locations of Lead

The changes in the location of lead start with the mining of lead ore. China

Nonferrous Metals Industry Yearbook reported that the total production of mine concentrates containing lead in China was 1.85 Mt of lead in 2010, mainly from the Inner Mongolia autonomous region and Hunan Province. Fig. 11.3 shows the locations of the five largest producers of lead ore (Inner Mongolia, and Hunan, Sichuan, Guangxi, and Guangdong provinces), which together contributed about 64% of China's total domestic production of lead concentrates. An additional 1.6 Mt of lead is imported from Australia and North America to meet the domestic demand.

A total of 4.20 Mt of refined lead was produced in 2010 in China, of which 2.84 Mt represented primary lead and 1.36 Mt represented secondary lead (Editorial Staff of the Yearbook of the Nonferrous Metals Industry, 2012). The primary lead was mainly produced in Henan, Hunan, and Yunnan provinces, which contributed around 75% of China's total primary lead. The secondary lead was mainly produced in Anhui and Jiangsu provinces, which contributed around 70% of the total secondary lead. These five provinces were the largest sources of refined lead, and accounted for 79% of China's total refined lead.

A total of 144 Gigavolt ampere hours of batteries are produced in 2010 of China (China battery industry association, 2012). The main battery production centers are Zhejiang, Guangdong, and Hebei provinces. The top 5 provinces (Fig. 11.3) accounted for about 66% of the total lead-acid battery production.

Lead-acid batteries are mainly used in automobiles (Guberman, 2012). It was reported that the number of automobiles totaled about 110 million in China in 2010 (National Bureau of Statistics of China, 2012). Fig. 11.3 shows that the main provinces based on automobile numbers were Guangdong, Shandong, Jiangsu, Zhejiang, and Hebei provinces, which accounted for around 28% of the total automobiles in China and represent an important conceptual final destination of lead in the anthrosphere.

In general, totally about 4Mt of lead was transferred from lithosphere to anthrosphere to meet China's production need in 2010, in which China itself contributes a rough half in lead concentrates. Meanwhile, 1/3 of this refined lead was transferred to other countries through trade of various lead products.

Fig.11.3 conceptualizes the trends in the overall flows of lead through the anthrosphere throughout its life cycle. In general, the lead moves from central to eastern China as it moves through anthrosphere to providing final service, companying a rough 2Mt of lead import of lead concentrates and 0.7Mt of lead export of various lead products.

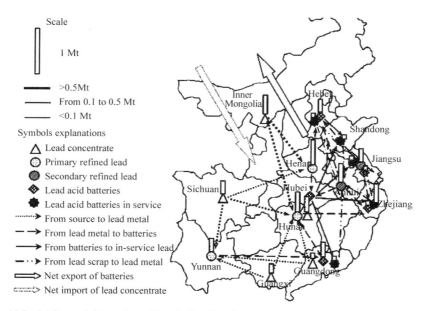

Fig. 11.3　Main spatial transfers of lead related to flows through anthrosphere in part of China

11.4.4　Characteristics of the Anthropogenic Flows to Providing Service

In contrast with the environmental flows of lead pollutants, the tendency of the anthropogenic service shows the following remarkable characteristics: ①All of the transformations of lead are driven by human activities that are performed to satisfy need specific human services within the anthrosphere; ②the flow of lead is from various components of the natural system into various components of the anthrosphere; ③to provide the desired services from lead, the natural forms are transformed into forms with more useful physical and chemical properties; and ④these transformations have the goal of providing services (i.e., serving different functions), and changes in the forms of lead required to provide these functions are stimulated by engineering and technology demands such as design, production, and transportation, which in turn drive the transformations of lead among its various forms.

11.5　Conclusions

(1) To meet human needs for specific services, lead flows through the anthrosphere,

undergoing transfers and transformations in response to design, production, transportation, trade, and utilization needs.

(2) The flows of lead through the anthrosphere resemble environmental flows of lead pollutants: both sets of flows include sources, pathways, and service fates. Ore and scrap are the main sources of lead, and the flows from lead ore extraction and collection of scrap leading to that the final use of lead products is the main pathways for providing human services. The in-service lead that provides these services represents the final fate of lead in the anthrosphere.

(3) The main changes of lead in the anthrosphere occur in its functions, forms, and locations. The changes in function relate to transformations from raw ore into useful products such as batteries, products that provide protection against corrosion, and products that provide protection against radiation. The form of the lead also changes to reflect the properties required by these functions. The changes in location occur in the form of flows of lithospheric ore reserves in central China and to regions with high densities of automobiles in eastern China, companying a rough 2 Mt of lead import of lead concentrates and 0.7 Mt of lead export of various lead products.

References

Cang B. 2010. Electrical cables. Beijing: China Electric Power Press. (In Chinese)

China battery industry association. 2012. China battery industry yearbook (2011). Beijing: China Battery Industry Association. (In Chinese)

Eckelman M J, reck B K, Graedel T E. 2012. Exploring the global journey of nickel with Markov chain models. Journal of Industrial Ecology, 16 (3): 334-342.

Editorial Staff of the Yearbook of the Nonferrous Metals Industry. 2012. China Nonferrous Metals Industry Yearbook 2011. Beijing: China Nonferrous Metals Industry Yearbook Press. (In Chinese)

Elshkaki A, van der Voet E, VAN HOLDERBEKE M. 2004. The environmental and economic consequences of the developments of lead stocks in the Dutch economic system. Resources, Conservation and Recycling, 42: 133-154.

Feng G L, He G P. 2010. The exploitation and fabrication technology of nonferrous mineral resource. Extraction and smelting part. Kunming: Yunnan Science and Technology Press. (In Chinese)

Gao P Y, Li J B. 2007. Physical chemistry. Beijing: Science Press. (In Chinese)

Graedel T E, Allenby B R. 1995. Industrial ecology. New Jersey: Prentice Hall.

Guberman D E. 2012. Minerals Yearbook. Lead. http://minerals.usgs.gov/minerals/pubs/commodity/lead/#pubs. Jan. 2012. (Accessed on Dec. 20, 2012)

Henriques S T, Kander A. 2010. The modest environmental relief resulting from the transition to a service economy. Ecological Economics, 70 (2): 271-282.

International Organization for Standardization Technical Committee TC-207 on Environmental Management Systems, 2006. ISO 14044-2006, Environmental management-Life cycle assessment-Requirements and guidelines. International Organization for Standardization, Geneva.

Li F L, Ni L J, Yuan J. 2010. Cultivation practices affect heavy metal migration between soil and *Vicia faba* (broad bean). Chemosphere, 80 (11): 1393-1398.

Liu C M, Lin G Y, Deng Y L. 2010. Fabrication of nonferrous metal materials. Changsha: Central South University Press. (In Chinese)

Liu G L. 2011. Introduction of the technology on lead acid battery. Beijing: China Machine Press. (In Chinese)

Liu Q, Wang H K, Li W D. 2010. The ecological effect and global chemical recycle of the heavy metal lead. Journal of Anhui Institute of Education, 23 (6): 97-100. (In Chinese with English abstract).

Manahan S E. 2005. Environmental chemistry, 8th ed., 2005, CRC Press, Boca Raton.

Mao J S, Cao J, Graedel T E. 2009. Losses to the environment from the multilevel cycle of anthropogenic lead. Environmental Pollution, 157: 2670-2677.

Mao J S, Dong J, Graedel T E. 2008a. The multilevel cycle of anthropogenic lead: I. Methodology. Resource Conservation & Recycling, 52: 1058-1064.

Mao J S, Dong J, Graedel T E. 2008b. The multilevel cycle of anthropogenic lead: II. Results and discussion. Resource Conservation & Recycling, 52: 1050-1057.

Mao J S, Graedel T E. 2009. Lead in-use stock: a dynamic analysis. Journal of Industrial Ecology, 13 (1): 112-126.

Mao J S, Lu Z W, Yang Z F. 2006. The eco-efficiency of lead in China's lead-acid battery system. Journal of Industrial Ecology, 10 (1/2): 185-197.

Mao J S, Ma L, Niu J P. 2012. Anthropogenic transfer & transformation of heavy metals in anthrosphere: concepts, connotations and contents. International Journal of Earth Sciences and Engineering, 5 (5): 1129-1137.

National Bureau of Statistics of China. 2012. China statistical yearbook (2011). Beijing. China Statistics Press.

National Technical Committee for the Standardization of Nonferrous Metals. 2008. Comprehensive collection of standards about lead and lead alloys. 5th ed. Beijing: Standards Press of China. (In Chinese)

Niisoe T, Nakamura E, Harada K. 2010. A global transport model of lead in the atmosphere. Atmospheric Environment, 44 (14): 1806-1814.

Prettenthaler F E, Steininger K W. 1999. From ownership to service use lifestyle: the potential of car sharing. Ecological Economics, 28 (3): 443-453.

Schmidt W P. 2001. Strategies for environmentally sustainable products and services. Corporate Environmental Strategy, 8 (2): 118-125.

Shen C L. 2008. Manual for chemicals and building raw materials. Beijing: Standards Press of China. (In Chinese)

Shenyang Research Institute of Storage Battery. 2010. Comprehensive collection of battery standards: volume of lead acid battery. Beijing: Standards Press of China. (In Chinese)

Song X C. 2011. Smelting of nonferrous heavy metals. Beijing: Metallurgical Industry Press. (In Chinese)

Stahel W R. 1997. The functional economy: Cultural and organizational change. In: The industrial green game, edited by D. J. Richards. Washington, DC: National Academy Press.

Szücs E, 1980. Similitude and Modeling: Fundamental Studies in Engineering (Vol. 2). Translated by Balkay B., Elsevier Scientific Publishing Company, Amsterdam, the Netherlands.

Van Beers D, Graedel T E. 2007. Spatial characterisation of multi-level in-use copper and zinc stocks in Australia. Journal of Cleaner Production, 15 (8-9): 849-861.

Von Bertalanffy L. 1968. General system theory: foundations, development, applications. New York: George Braziller.

Wang L K. 2012. Technical manual for lead and zinc melting. Beijing: Metallurgical Industry Press. (In Chinese)

Zhang B, Zhong J H, Li W J. 2010. Transformation of inert PbSO4 deposit on the negative electrode of a lead-acid battery into its active state. Journal of Power Sources, 195 (13): 4338-4343.

Zheng N, Liu J S, Wang Q C. 2010. Heavy metals exposure of children from stairway and sidewalk dust in the smelting district, northeast of China. Atmospheric Environment, 44 (27): 3239-3245.

第 12 章　中国铅元素的人为迁移与转变
Lead Anthropogenic Transfer and Transformation in China*

12.1　Introduction

Humans have increasingly altered the physical and chemical processes on the earth's surface over the last several centuries. However, it is only within the past several decades that the impact of humans on the earth has begun to widely attract the attention of scientists (Sen and Bernhard, 2012). Human activities have changed the mobilization and transformation processes of many elements in the nature, including the heavy metals such as lead (Rauch and Pacyna, 2009). Specifically, the scale of anthropogenic lead cycle is much greater than the natural cycle (Sen and Bernhard, 2012). As anthropogenic activities have increasingly disturbed the earth's natural system, lead anthropogenic transfer and transformation become integral parts of its biogeochemical cycle. As one of the most poisonous metals in human civilization, lead poses great threats to eco-security as well as human health (Smith and Flegal, 1995). Therefore, research on lead transfer and transformation is desperately pressing. Currently, many studies have outlined the characteristics of anthropogenic lead cycle by quantifying the flows of lead (Elshkaki et al., 2009; Mao et al., 2008a, 2008b, 2009). Those reports have established a foundation for a further study of lead cycle whereas some shortcomings still exist. Studies on transfer and transformation seem to be limited at present. Some studies on lead transfer have mainly focused on the transfer processes occurring in nature, such as the metal transfer in soil-plant system and the mass transfer between different media (Bi et al., 2010; Guillen et al., 2009; Kondo et al., 2013), and others focus on toxicology studies (Hallen et al., 1995; Lin et al., 2010). Lead transformation studies have been generally limited to the changes caused by technical innovations (Volpe et al., 2009; Zhu et al., 2013) or natural process such as that in the soil (Birkefeld et al., 2007; Cao et al., 2003). Therefore, lead redistribution and speciation after anthropogenic

---

* Liang J, Mao J S. 2015. Lead anthropogenic transfer and transformation in China. Transaction of Nonferrous Metals Society of China, 25 (4): 1262-1270.

input are not fully understood, and this knowledge is essential to clarify human interference as well as understand the complex processes of pollution formation.

We pay special attention to the anthropogenic lead flow in China in 2010. China has become one of the largest lead producer and consumer all over the world (Zhang et al., 2012). Additionally, with the booming economy and rising social demands, the scale of lead transfer and transformation in China will continue to increase in future (Editorial Staff of the China Nonferrous Metals Industry Yearbook, 2000-2012). Therefore, lead cycle in China is a representative case study to examine lead redistribution and speciation change after anthropogenic input.

12.2 Methodology

12.2.1 Basic Concepts

Practically, the changes such as product features, chemical species or locations (Mao et al., 2014) occur consistently in the anthropogenic lead flow. We will explain some key concepts and give a brief retrospect in the following part because those concepts are not widely understood now.

1. Anthropogenic Transfer

Anthropogenic transfer normally refers to the mobilization and corresponding virtual redistribution of substances in the anthrosphere. They are driven by the social demands and are accomplished under human activities (Mao et al., 2012). The natural transfer of pollutants refers to the moving processes of pollutant in the environment, which normally leads to the enrichment, dispersion or disappearance of the pollutants (Dai, 2006). The differences between the anthropogenic transfer and the natural transfer are as follows: ① the driving forces behind anthropogenic transfer are social demands other than the natural forces, which motivate the natural transfer; ② the anthropogenic transfer occurs in the hybrid socioeconomic system, although this system is often related to the natural environment.

Anthropogenic transfer can be classified into two categories: transfers in physical space and in virtual space. Physical transfer involves the transfers between different regions or the earth's spheres, i.e., the lithosphere, hydrosphere, atmosphere and anthrosphere. For example, lead is transferred from the lithosphere to the anthrosphere after lead ores are mined from the earth. The virtual space transfer refers to the transfer between different industrial sectors, which are conceptional but essential for knowing lead movement in the whole society. In this work, we will focus on the virtual anthropogenic transfer, regardless of the plentitude of studies on physical transfer (Niisoe et al., 2010; Xu et al., 2014).

In this work, we will focus on the transfer in the virtual space. Based on lead life cycle, lead transfer can be further classified into the transfer at production, fabrication and manufacture (F&M), use and waste management and recycling (WMR) stages. Each transfer at a certain life stage can be further subdivided into different industrial sectors. Also, according to different purposes, lead transfer can be divided into the flow to satisfy production needs, the flow to manufacture product and the flow to meet the demand of humans as a finished product. For example, lead flow from the lithosphere to refining industry belongs to the flow to satisfy production needs.

In the framework (Fig. 12.1), the detailed acronyms for the industrial sectors are cited based on the research of Mao et al. (2011). Lead goes through life stages while transferring in various industrial sectors. Moreover, lead losses are emitted into the earth's spheres. For example, tailings and slag are discharged from production. Lead dust and waste water are released into soil, atmosphere or water from F&M. Hibernating lead refers to lead that is not finally disposed of, but no longer serving any useful function (Mao et al., 2008a).

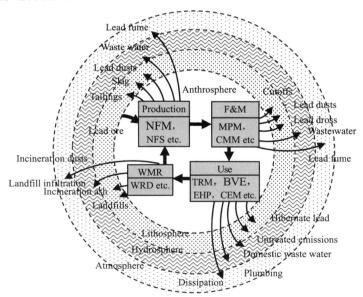

Fig. 12.1 Schematic diagram of the anthropogenic transfers of lead

Codes for industrial sectors: TRM, transportation manufacturing; BVE, building and civil engineering; EHP, electricity and heating power; CEM, computer and electronics manufacturing; and WRD, waste resources and discard recycling.

2. Anthropogenic Transformation

Anthropogenic transformation refers to the process where chemical species of a material are changed into another or change into several other species, and the new species are meant to satisfy specific human demands. Practically, these processes can be achieved by various engineering techniques (Mao et al., 2012). The natural transformation of pollutants refers to the physical, biological or chemical reactions where a substance changes into another (Dai, 2006). The differences can be figured out as follows: ① the natural one is affected by many factors, such as the temperature, pH, soil or water properties and other anions (Birkefeld et al., 2007), while anthropogenic flow is mainly affected by social factors; ② the natural one tends to form stable substances while anthropogenic transformation aims at serving human beings.

Generally, chemical species refer to the chemical species, but they can also refer to the mineralogical, molecular, or ionic type (Zheng et al., 2010). In this work, we analyze the chemical species although they are combined with other materials. For example, PbO can exist in acidic battery paste but is also widely used as a cosolvent in alloys.

12.2.2 Analysis of Lead Anthropogenic Transfer and Transformation

Overall, lead life cycle is traced with material flow analysis to understand the transfer and transformation. Then, lead transfer between different industrial sectors is studied. Finally, possible transformation is identified with a physicochemical analysis to find the possible changes in lead species during the life cycle.

1. Tracing lead life cycle

The boundary in this work is the system made of all the four lead life cycle stages. The geographical boundary is the mainland China and the reference year is 2010. Lead life cycle starts with the mining of lead resources, continues with the lead fabrication and manufacturing into products or semi-products, which enter the use stage serving humans, and ends up with lead deposited in landfill after recycling. In short, it contains all the stages that metal experiences from the state of resource to waste. Previous research on lead paid close attention to the quantities of the multilevel lead cycle such as lead in-use stock and lead losses to the environment (Mao and Graedel, 2009; Gerst and Graedel, 2008; Elshkaki et al., 2004), which can be useful in tracing lead. By material flow analysis, which is widely applied to characterizing the flow in a well-defined system (Glöser et al., 2013), we tried to find out the mass of each flow in lead cycle based on data from the database or estimation. The flows needed to be determined can be seen from the schematic diagram, which were established in STAF project in 2006 as Fig. 14.1 in the research of Mao et al. (2008a). Additionally, mass conservation law must be observed. Tracing lead life cycle helps to determine the exact stage where lead transfer occurs. Although the import and export cannot be excluded from the flow (they are important for the balance of the flows), we mainly focus on the domestic ones in China.

2. Analysis of the Anthropogenic Transfer

Lead transfers in different industrial sectors while flowing through the anthropogenic cycle. Principal lead products are lead acid batteries, pigments, alloys, ammunition, cable sheathing and additives. We identified where lead products fulfill their function based on the necessary information and the general knowledge. We followed lead transfer into different sectors and traced lead redistribution in the human society. In this work, we chose the industrial sectors according to the China's National Standard of Industrial Classification, GB/T 4754—2011.

3. Analysis of the Anthropogenic Transformation

Lead products cannot fulfill their functions without certain chemical species. It is already known that anthropogenic input acts as the dominant driving force behind the transformation. To analyze species transformation, we firstly identified the chemical species before they underwent the industrial processes, and then referred to handbooks or technical references to find out the probable chemical reactions that happened in the industrial processes, which helped us to predict the possible outcomes of reactions. Furthermore, during those chemical reactions, we also tried to analyze the emissions from anthrosphere. To conclude, we focused on the initial species of lead compounds as well as the endpoints of the industrial processes while analyzing the possible industrial emissions during those reactions.

12.2.3 Data Sources

Some basic data are essential to determine lead flows, which are already available. Their values and sources are shown in Table 12.1. Those data are the basis for the further calculation and estimations.

Table 12.1 Basic data for calculation of anthropogenic lead cycle and their data sources

parameter	value	data sources
Mass flow of refined lead production/kt	4200	USGS, 2013
Mass flow of primary refined lead/kt	2840	USGS, 2013
Mass flow of secondary refined lead/kt	1360	USGS, 2013
Mass flow of domestic lead ore/kt	1850	USGS, 2013
Mass flow of lead consumption/kt	3950	China nonferrous metals industry yearbook, 2000–2012
Mining efficiency	0.84	China nonferrous metals industry yearbook, 2000–2012
Mass flow of primary refining efficiency	0.96	China nonferrous metals industry yearbook, 2000–2012
Mass flow of lead in net import of lead ore/kt	1287	China nonferrous metals industry yearbook, 2000–2012
Mass flow of lead in net export of products/kt	50	China nonferrous metals industry yearbook, 2000–2012
Mass flow of lead in net import of scrap/kt	0	China nonferrous metals industry yearbook, 2000–2012

12.3 Determination of Lead Flows and Species

12.3.1 Lead Flow Quantities in Anthropogenic Cycle

Although some lead flows are accessible, some others cannot be acquired precisely for the moment such as the lead flow in China in 2010. In this work, the quantities of those flows were determined.

The recycling flows are important in anthropogenic lead cycle. In secondary production, the scraps entering secondary production are estimated as the secondary refined lead (1360 kt) divided by the recovery rate (98%) (Mao et al., 2008a). F&M scrap occupies approximately 4% of the total scrap (Liang and Mao, 2014). Apart from the recycled flows, lead products are manufactured with the fabrication efficiency of 91% (Mao et al., 2008a), and some lead products become in-use stock, for which the accumulation rate is 0.18 (Mao et al., 2008b). The net import of lead ore and net export of lead products are 1287 kt and 50 kt, respectively, while the net export or import of lead scraps is zero because the import and export of lead scraps are strongly restricted according to the Law of the People's Republic of China on the Prevention and Control of Environmental Pollution by Solid Waste and Basel Convention on the Control of Transboundary Movements of Hazardous Waste and Their Disposal. In addition, lead sludge and lead slag are listed among the Catalogue of Commodities Forbidden to Import (the third batch and the fourth batch). And also, the scarp import and export data are not included in the China Nonferrous Metals Industry Yearbook (2000–2010).

Although the life spans of lead products may affect the quantities at the use stage (Mao and Graedel, 2009), in this work, we mainly focused on a static analysis of lead transfer and transformation in a specific year, and a dynamic analysis of lead in-use stock is not included.

Lead losses from the anthrosphere differ greatly in every stage of the life cycle. Tailings are estimated from the lead ores multiplied by 1 minus the mining rate, and refining slag equals the primary or secondary refined lead multiplied by 1 minus the primary or secondary refining rate. Lead losses from F&M are less than 5% of the total losses (Mao et al., 2008b). The high proportion of dissipative lead is responsible for the heavy loss in use, which is estimated to be about 30% of the total losses (Liang and Mao, 2014a). For every kilogram of lead consumed, 0.5 kg is assumed to be lost to the environment (Smith and Flegal, 1995). Following previous study (Smith and Flegal, 1995; Liang and Mao, 2014a), the total losses to environment are estimated to be 2.1 Mt. Finally, with mass conservation law, we can work out all the data, such as the lead entering WMR and the losses to the environment from WMR. The lead flow in 2010 in China is shown in Fig. 12.2.

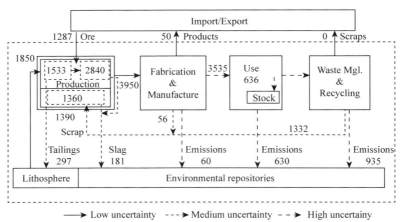

Fig. 12.2 Anthropogenic flow of lead in 2010 in China (unit: kt)

12.3.2　Lead Transformation at Production Stage

At the production stage, for lead ore, galena (PbS), cerusite ($PbCO_3$) and anglesite ($PbSO_4$) are in the raw materials (Wang, 2012), and galena accounts for 75% (mass fraction), cerusite constitutes 15% (mass fraction), and the left are anglesite or other species (Yu and Chen, 2005).

Wastes are discharged during the production process, including primary refining, secondary refining and mining activities. In production, mining contributes a lot to the emissions into the environment (Zhang et al., 2011). The mass proportion of those wastes can be referred to from the cleaner production audit report (Zhao, 2007), on-site monitoring or the lead mass balanced sheet (Elshkaki et al., 2004). The primary species in wastes can be identified, with $PbSO_4$ being the main species in tailing, PbO and PbS in primary lead refining process and PbS and PbO_2 in secondary lead refining. We can get the exact amount of the species by multiplying the total quantities of tailings, primary lead slag and secondary lead slag by the ratio of the species in the wastes (Liang and Mao, 2014b), and then sum up the same species.

12.3.3　Lead Transformation at F&M Stage

At the F&M stage, the principal lead products and the species in them are focused on to get the species distribution. Lead acid batteries, lead oxidants, lead alloys, and rolled and extruded products altogether account for 94% of the total, of which batteries constitute 73%, and lead alloys occupy 8% (Wang, 2011). As for the chemical species, in lead acid batteries, the active material in the positive pole is mainly PbO_2 while the negative pole consists of spongy lead; in lead oxidants, the species are mainly PbO, Pb_3O_4 and Pb_2O_3; lead alloys, and rolled and extruded products are mainly metallic lead. With the proportion of species, an approximate speciation can be obtained at F&M stage in the anthrosphere.

Lead losses are discharged during the manufacturing. For lead acid batteries, the emissions are largely PbO, Pb and $PbSO_4$ (Liu, 2011). For lead oxidants production, lead oxidants are usually produced with the metal oxidation method and PbO, Pb, Pb_2O_3 and Pb_3O_4 are lost. For lead alloy smelting, PbO and Pb are emitted in dross or fumes (Editorial Board of Non-ferrous Metals Industry Analysis, 1992).

12.3.4　Lead Transformation at Use and WMR Stage

A similar calculation was carried out for speciation at use and WMR stages. The species at those two stages may change and differ from those at production and F&M stage. In lead acid batteries, $PbSO_4$ is formed when the battery is discharged. A large part of lead oxidants consumed is PbO (massicottite) and Pb_3O_4 (red lead). Lead pigments and additives are mainly lead salt compounds, such as $PbCO_3$ and $PbCl_2$. The lead species at WMR stage show the species of discarded products whereas the losses from WMR show the species in landfill.

Use losses to the environment are dissipative in nature. Batteries discarded without treatment are considered to be the losses during use. The proportions of Pb, PbO_2 and $PbSO_4$ are 30%, 30% and 40%, respectively. For lead oxidants, losses in PbO, Pb_2O_3 and Pb_3O_4 are 60%, 30% and 10%, respectively. For lead additives, $PbCO_3$ and $PbCl_2$ constitute about 30% of all emissions, with the left being other species (Lang and Mao, 2014b). In waste batteries, $PbSO_4$, PbO_2, Pb and PbO account for approximately 50%, 20%, 17% and 13% of the total lead in batteries, respectively (Zhou, 2002). Lead oxidants will release complex compounds when they are treated: Pb_3O_4 is released as the ingredient of red lead paint; $PbCO_3$ in lead subcarbonate is also emitted; chemicals such as Pb_2O_3 are released from incineration. In China, with the high content of chlorine and sulfur in wastes, during incineration lead residues and fly ash are usually discharged with $PbCl_2$, $PbSO_4$ and lead oxidants, and the mass ratio of chloride to sulfide is 2∶1 (Shi et al., 2008).

12.4　Results and Discussion

12.4.1　Anthropogenic Transfer in Anthropogenic Lead Cycle

Lead anthropogenic transfers play an important role in the social-economic interaction. Lead transfers in a multitude of industrial sectors. In production, lead is transported from the industrial sector of NFM to NFS. At F&M stage, refined lead is manufactured into products after being transmitted to the industrial sector of MPM or CMM (the acronyms can be referred to in Fig. 12.1). During use, lead products, including lead batteries, lead alloys and building materials, realize their functions after entering into the industrial sectors, such as TRM, BE and EHP. Finally, discarded products are recycled or disposed of in WRD.

In conclusion, lead firstly moves into primary industrial sectors of raw material processing, such as lead mining and refining sectors, and then enters secondary manufacturing sectors. Then, lead products provide services mainly in such industrial sectors as transportation, electrical power and buildings and construction. Finally, lead ends its transfer process when discarded lead is recycled or put into landfilling.

Lead transfer is reflected directly by the change of flux. The refined lead production was 4.20 Mt in 2010 (China nonferrous metals industry yearbook, 2000-2012), with 1.85 Mt lead ore being mined domestically. Then, 3.95 Mt was consumed, with 3.53

Mt entering end services and approximately 2.10 Mt lost into environmental repositories (Fig. 12.3), namely the atmosphere, hydrosphere and pedosphere. WMR emissions are higher in mass than other emission flows. Based on all the data of lead flow, it can be concluded that in the anthropogenic cycle for every 1 kg of lead consumed, 0.89 kg lead comes into the domestic market, 0.47 kg lead needs to be mined from the lithosphere and 0.53 kg lead is lost to the environment.

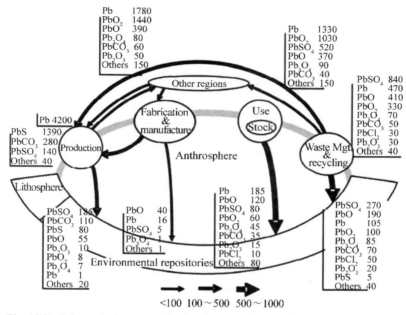

Fig. 12.3 Schematic diagram of lead anthropogenic transfer and transformation of China in 2010 (unit: kt/year)

12.4.2 Implications on Resources and Environment

Overall lead in the anthrosphere transfers in the following direction: resources–end servicees–environmental repositories. Lead is mined from natural resources, fabricated into lead products, which enters end services for human welfare, and finally discarded into the environment after being used. It can be seen from Table 12.2 that there is a huge difference among the lead species in resources, end services and environmental losses. PbS, PbCO$_3$ and PbSO$_4$ are the primary chemical species in resources, in which PbS accounts for as high as 75%, with PbCO$_3$ and PbSO$_4$ occupying 15% and 8%, respectively. After being transformed into various products, lead serves human society largely in the chemical species of Pb (38%), PbO$_2$ (26%) and PbSO$_4$ (15%). Lead losses are released from the anthropogenic cycle as $PbSO_4$ (26%), PbO (19%), Pb (15%) and $PbCO_3$ (10%), among which $PbSO_4$ probably comes from tailings and waste lead acid batteries, and PbO is from lead chemicals and batteries. The toxicity of those species varies a great deal and one of the most suitable hazard score ranking was given by the Indiana Relative Chemical Hazard (IRCH) Ranking System (Eckelman and Graedel, 2007).

Table 12.2 Distribution of lead chemical species in lead resources, end services and losses

Item	Chemical species	Mass flow/kt	Percent of total/%
Resource	PbS	1390	75
	$PbCO_3$	280	15
	$PbSO_4$	150	8
	Others	30	2
	Total	1850	100
End service	Pb	1330	38
	PbO_2	1030	29
	$PbSO_4$	520	15
	PbO	370	10
	Pb_3O_4	90	3
	$PbCO_3$	45	1
	Others	150	4
	Total	3535	100
Environmental loss	$PbSO_4$	540	26
	PbO	405	19
	Pb	307	15
	$PbCO_3$	215	10
	PbO_2	168	8
	Pb_3O_4	138	7
	PbS	85	4
	Pb_2O_3	45	2
	$PbCl_2$	60	3
	Others	141	7
	Total	2104	100

Rules of lead transfer and transformation could identify the adverse influences on the resources or environment and allow a better control of lead inflows and outflows. Overdoses may compromise the future availability for human utilization and improper disposal could result in a huge health risk in potential, such as the negative effect on developmental intelligence quotient (IQ) and blood lead levels (Carrington and Bolger, 2014). Therefore, knowledge of lead life cycle could help to improve or even provide guidance for human activity so as to alleviate the negative effects on lead flow. For example, efficiencies, such as the mining efficiency or fabrication efficiency, should be improved to ensure that less is lost to the environment and more enters the products. Human should also be sparing when exploiting the resources, because more mining may lead to more environmental losses (Liang and Mao, 2014a). At last, the mass of lead flow can be less if the demand for lead is reduced. Therefore, human should actively seek for lead substitution to reduce the amount of lead consumed and thus reduce the pollution or help resource conservation.

12.4.3 Data Uncertainty

Due to uncertainty, the lead cycle is not entirely correct and complete. There is an inconsistency in data source with varying reliability-data for the refined lead production, primary or secondary lead and mined ore are from the USGS, while the others are from the CNMIY (China Nonferrous Metals Industry Yearbook). Some flow mass may be not accurate enough, especially the emission flows, as they are based on previous research work. And some empirical information is used, such as fabrication efficiency, the mass ratio of F&M scrap to total scrap.

Although the best estimations are made to study lead anthropogenic transformation, the species analysis is far from accurate. The absent information was estimated, such as the species percentage of lead oxidants, lead pigments and additives. Therefore, high uncertainty exists except for lead acid batteries, which have available information like the species in F&M losses, and in new or waste battery.

12.5 Conclusions

(1) Lead anthropogenic transfer and transformation greatly differ from the traditional one in geochemistry as it focuses on the relationship between anthropogenic input and material redistribution or species change.

(2) Lead transfers in different industrial sectors. Lead products provide services mainly in such industrial sectors as transportation, electrical power and construction. Also, lead transfer can be shown by the change of flux. For every 1 kg of lead consumed, 0.89 kg comes into the domestic market, 0.47 kg lead needs to be mined from the

lithosphere and 0.53 kg lead is lost to the environment.

(3) Lead anthropogenic transformation refers to the chemical species change in the anthropogenic cycle. For China, lead ores are primarily in the specie of PbS (75%). Lead products in end services are largely in chemical species of Pb, PbO_2 and $PbSO_4$ (over 80% of the total). Also, lead losses emitted into the environment are primarily in such species as $PbSO_4$ (26%), PbO (19%) and Pb (15%).

References

Bi X Y, Ren L M, Gong M, et al. 2010. Transfer of cadmium and lead from soil to mangoes in an uncontaminated area, Hainan Island, China. Geoderma, 155 (1-2): 115-120.

Birkefeld A, Schulin R, Nowack B. 2007. In situ transformations of fine lead oxide particles in different soils. Environmental Pollution, 145 (2): 554-561.

Cao X D, Ma L Q, Chen M, et al. 2003. Lead transformation and distribution in the soils of shooting ranges in Florida, USA. Science of the Total Environment, 307 (1-3): 179-189.

Carrington C D, Bolger P M. 2014. Toxic metals: Lead. Encyclopedia of Food Safety, 2: 349-35.

Dai S G. 2006. Environmental chemistry. Beijing: Higher Education Press, 14. (in Chinese)

Eckelman M J, Graedel T E. 2007. Silver emissions and their environmental impacts: A multilevel assessment. Environmental Science and Technology, 41 (17): 6283-6289.

Editorial Board of Non-ferrous Metals Industry Analysis. 1992. Series of non-ferrous metals industrial analysis //Chemical phase analysis for ore and industrial products. Beijing: Metallurgical Industry Press, 255-258. (in Chinese)

Editorial Staff of the China Nonferrous Metals Industry Yearbook. 2000-2012. China nonferrous metals industry yearbook, 2000-2012. Beijing: China Nonferrous Metals Industry Yearbook Press. (in Chinese)

Elshkaki A, van der Voet E, Holderbeke M V, et al. 2009. Long-term consequences of non-intentional flows of substances: Modeling non-intentional flows of lead in the Dutch economic system and evaluating their environmental consequences. Waste Management, 29 (6): 1916-1928.

Elshkaki A, van der Voet E, van Holderbeke M, et al. 2004. The environmental and economic consequences of the developments of lead stocks in the dutch economic system. Resources Conservation and Recycling, 42 (2): 133-154.

Gerst M D, Graedel T E. 2008. In-use stocks of metals: Status and implications. Environmental Science and Technology, 42 (19): 7038-7045.

Glöser S, Soulier M, Tercero Espinoza L A. 2013. Dynamic analysis of global copper flows: Global stocks, postconsumer material flows, recycling indicators, and uncertainty evaluation. Environmental Science and Technology, 47 (12): 6564-6572.

Guillén J, Baeza A, Ontalba M A, et al. 2009. 210Pb and stable lead content in fungi: Its transfer from soil. Science of the Total Environment, 407 (14): 4320-4326.

Hallén I P, Jorhem L, Oskarsson A. 1995. Placental and lactational transfer of lead in rats: A study on the lactational process and effects on offspring. Archives of Toxicology, 69 (9): 596-602.

Kondo A, Yamamoto M, Inoue Y, et al. 2013. Evaluation of lead concentration by one-box type multimedia model in Lake Biwa-Yodo River Basin of Japan. Chemosphere, 92 (5): 497-503.

Liang J, Mao J S. 2014a. A dynamic analysis of environmental losses from anthropogenic lead flow and their accumulation in China. Transactions of Nonferrous Metals Society of China, 24 (4): 1125-1133.

Liang J, Mao J S. 2014b. Speciation analysis of lead losses from anthropogenic flow in China. Environmental Sciences, 35 (3): 1191-1197. (in Chinese)

Lin C M, Doyle P, Wang D L, et al. 2010. The role of essential metals in the placental transfer of lead from mother to child. Reproductive Toxicology, 29 (4): 443-446.

Liu G L. 2011. Introduction of the technology on lead acid battery. Beijing: China Machine Press. (in Chinese)

Mao J S, Cao J, Graedel T E. 2009. Losses to the environment from the multilevel cycle of anthropogenic lead. Environmental Pollution, 157 (10): 2670-2677.

Mao J S, Dong J, Graedel T E. 2008a. The multilevel cycle of anthropogenic lead: I: Methodology. Resources Conservation and Recycling, 52 (8-9): 1058-1064.

Mao J S, Dong J, Graedel T E. 2008b. The multilevel cycle of anthropogenic lead: II: Results and discussion. Resources Conservation and Recycling, 52 (8-9): 1050-1057.

Mao J S, Du Y C, Xu L Y, et al. 2011. Quantification of energy related industrial eco-efficiency of China. Frontiers of Environmental Science and Engineering in China, 5 (4): 585-596.

Mao J S, Graedel T E. 2009. Lead in-use stock. Journal of Industrial Ecology, 13 (1): 112-126.

Mao J S, Ma L, Liang J. 2014. Changes in the functions, species and locations of lead during its anthropogenic flows to provide services. Transactions of Nonferrous Metals Society of China, 24 (1): 233-242.

Mao J S, Ma L, Niu J P. 2012. Anthropogenic transfer & transformation of heavy metals in anthrosphere: Concepts, connotations and contents. International Journal of Earth Sciences and Engineering, 5 (5): 1129-1137.

Niisoe T, Nakamura E, Harada K, et al. 2010. A global transport model of lead in the atmosphere. Atmospheric Environment, 44 (14): 1806-1814.

Rauch J N, Pacyna J M. 2009. Earth's global Ag, Al, Cr, Cu, Fe, Ni, Pb and Zn cycles. Global Biogeochemical Cycles, 23 (2): 1-16.

Sen I S, Bernhard P E. 2012. Anthropogenic disturbance of element cycles at the earth's surface. Environmental Science and Technology, 46 (16): 8601-8609.

Shi D Z, Wu W X, Lu S Y, et al. 2008. Effect of MSW source-classified collection on the emission of PCDDs/Fs and heavy metals from incineration in China. Journal of Hazardous Material, 153 (1-2): 685-694.

Smith D R, Flegal A R. 1995. Lead in the biosphere: Recent trends. Ambio, 24 (1): 21-23.

Statistic Design Management Department of National Bureau of Statistics of PRC. GB/T 4754—2011 for national economic activities industrial classification. (in Chinese)

USGS (U.S. Geological Survey). 2013. Minerals yearbook–Lead [EB/OD]. 2013-10-09. http://minerals.usgs.gov/minerals/

pubs/commodity/lead/.

Volpe M, Oliveri D, Ferrara G, et al. 2009. Metallic lead recovery from lead-acid battery paste by urea acetate dissolution and cementation on iron. Hydrometallurgy, 96 (1-2): 123-131.

Wang J K. 2012. Handbook of lead and zinc smelting production technology. Beijing: Metallurgical Industry Press, 30. (in Chinese)

Wang Y. 2011. Development and perspectives of Chinese lead industry. Nonferrous Metals Engineering, (1): 27-29. (in Chinese)

Xu H, Sun Y Y, Ding Z H, et al. 2014. Lead contamination and transfer in urban environmental compartments analyzed by lead levels and isotopic compositions. Environmental Pollution, 187: 42-48.

Yu X, Chen H. 2005. The experimental study of flotation on improving recovery of a lead ores. Non-Ferrous Mining and Metallurgy, 21 (6): 18-20. (in Chinese)

Zhang X W, Yang L S, Li Y G, et al. 2011. Estimation of lead and zinc emissions from mineral exploitation based on characteristics of lead/zinc deposits in China. Transactions of Nonferrous Metals Society of China, 21 (11): 2513-2519.

Zhang X W, Yang L S, Li Y H, et al. 2012. Impacts of lead/zinc mining and smelting on the environment and human health in China. Environmental Monitoring and Assessment, 184 (4): 2261-2273.

Zhao X S. 2007. Cleaner production evaluation of innovations on ISP lead and zinc smelting. Engineering Design and Research, 123: 11-19. (in Chinese)

Zheng G D, Fu B H, Takahashi Y, et al. Chemical speciation of redox sensitive elements during hydrocarbon leaching in the Junggar basin, northwest China. Journal of Asian Earth Sciences, 39 (6): 713-723.

Zhou Z H. 2002. Pollution-free pyrometallurgical regeneration of lead and its alloys from scrapped batteries. Shanghai Nonferrous Metals, 23 (4): 157-163. (in Chinese)

Zhu X F, Yang J K, Gao L X, et al. 2013. Preparation of lead carbonate from spent lead paste via chemical conversion. Hydrometallurgy, 134-135: 47-53.

第 13 章 铅元素人为循环环境释放物形态分析
Speciation Analysis of Lead Losses from Anthropogenic Flow in China[*]

从岩石圈中开采出的矿物铅，经过富集、冶炼成精铅，然后加工、制造成产品为人类社会服务，使用后又返回到自然环境中，整个过程称为铅的人为循环过程（Mao et al., 2008）。研究表明，生物圈中 95%以上的铅是由人为因素造成的（Smith and Flegal, 1995），铅释放物进入大气、土壤、水体等环境介质中会对生态系统和人类健康造成危害。值得注意的是，铅的最终归宿以及健康风险除了与数量有关外，更大程度上取决于其形态和暴露水平（Shah et al., 2009；Xie et al., 2013）。形态的差异直接导致污染物在环境中的迁移转化、生物有效性和潜在毒性的不同（章骅等，2011）。正确识别人为循环释放到环境中的铅的起始形态，可以为其健康风险研究奠定基础，更重要的是，弄清铅释放物的形成特点，有利于对污染物实行有效的源头控制，从而大大提高管理的科学性和有效性。

人为循环的铅释放物不同于铅污染物，是铅与环境介质发生复杂作用生成污染物之前的状态（Mao et al., 2012）。本研究中形态分析的对象是人为循环产生的、尚未和环境介质发生作用的铅释放物。目前，形态分析在环境领域的应用已成为分析化学的趋势和热点（Szpunar and Obiński, 1999）。现有铅形态的研究多集中在环境化学和工程技术领域。其中，环境化学关注的是释放物进入环境后的状态，通过典型化学试剂提取来分析铅的不同形态（Liu et al., 2007；Cui et al., 2011），这种分析难以对所有环境释放的产生源头进行测量，同时，采样结果易受到采样点位置、测试方法等人为因素的影响。工程技术领域则关注技术革新中铅形态的变化（Volpe et al., 2009；Zhu et al., 2013；Sonmez and Kumar, 2009），这些研究虽然为研究某环节铅形态的变化提供了方法，但无法了解铅整个生命周期中发生的转化以及释放到环境中的形态种类。本章通过应用物理化学分析方法，追踪铅元素人为循环流动过程，辨识生命周期各阶段环境释放物的形态，并以 2010 年为例，定量分析中国人为循环中铅释放物的形态分布特征。

---

[*] 梁静，毛建素. 2014. 铅元素人为循环环境释放物形态分析. 环境科学, 35（3）: 1191-1197.

13.1 研究方法

13.1.1 环境释放物研究框架

铅的人为循环包括生产、加工制造、产品使用和废物处置与回收四个阶段。由于各阶段主要的生产过程和生产工艺存在差异，产生的环境释放物在化学结构、赋存状态上也有很大不同。本研究在追踪铅人为循环的过程中，着重分析人类社会-环境界面上环境释放物的形态，按照铅的生命周期的划分，建立铅环境释放物形态的研究框架，如图13.1所示。

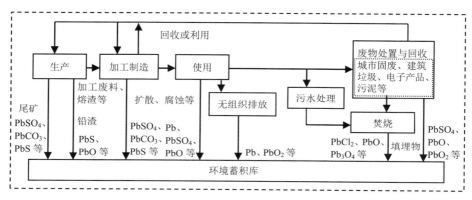

图 13.1 人为循环中铅释放形态分析研究框架

Fig. 13.1 Framework of lead speciation analysis for anthropogenic lead losses

铅元素的化学形态包含形态种类和数量分配特征两方面。其中，形态种类包括元素赋存状态和化学结构（Mao et al.，2012），本研究中主要讨论铅元素环境释放物的化学结构，不考虑铅与其他物质复杂的结合状况。如 PbO 既包含在冶炼的铅渣中，也和其他物质结合起来作为铅蓄电池中铅膏的成分，无论怎样的结合状况，铅渣和铅膏中所含铅是相同的化学结构；在数量分配方面，文中所涉及的量均指铅释放物中的含铅量。在生命周期的某阶段，形态 X 的铅释放物产生量为

$$X = \sum_{j=1}^{n} K_j Y_j \tag{13.1}$$

式中，j 表示铅金属生产阶段中某过程（包括采选、原生铅冶炼和再生铅冶炼），或其他生命周期三个阶段中某特种铅产品（如铅酸电池、电缆包皮等）；n 表示生产阶段中生产过程总数，或其他阶段中铅产品类别总数；Y 为该阶段中第 j 个过程或第 j 种铅产品所对应的铅环境释放总量；K 为铅释放物中 X 形态的比例。

整个铅人为循环过程中，铅释放物表现为 X 形态的总量 E_X 为

$$E_X = \sum_{i=1}^{4} X_i \qquad (13.2)$$

式中，i 为生产、加工制造、产品使用和废物处置与回收 4 个阶段中的第 i 个阶段；X_i 为第 i 阶段释放物表现为 X 形态的量。

将式（13.1）和式（13.2）结合可得

$$E_X = \sum_{i=1}^{4} \sum_{j=1}^{n} K_{ij} Y_{ij} \qquad (13.3)$$

13.1.2 生命周期各阶段铅环境释放物形态分析

1. 金属生产阶段

铅矿石一般含铅 3%～9%，富集后可获得达到冶炼要求的铅精矿，没有进入铅精矿的铅留存于尾矿中。方铅矿（PbS）、白铅矿（$PbCO_3$）、和铅矾（$PbSO_4$）是目前炼铅的主要原料，选矿过程会产生尾矿。我国尾矿中铅矿物以氧化矿为主，$PbSO_4$ 是矿物的主要形态，氧化率高达 85%（郭翠香和赵由才，2008），综合我国铅矿石实际情况，$PbSO_4$、$PbCO_3$、PbS 与其他形态的比例分别取 50%、35%、10% 和 5%。

国内原生铅冶炼多以烧结鼓风炉传统工艺为主（蒋继穆，2004），然后以柏兹电解法（Betts electrolytic process）精炼得到含铅 99%以上的精铅。炉渣中包含来不及发生还原反应而产生的一系列铅氧化物（如 PbO、Pb_2O_3、Pb_3O_4）、硫化物的共熔体铅锍（PbS）以及少量 $PbSO_4$ 等。电解精炼过程中 Pb^{2+} 在阴极上放电，形成含有 PbO、PbFeCl 等的阳极泥，电解前加硫除铜的工艺产生含有 PbS 的浮渣，之后碱性精炼产生含有 PbO 的氧化渣。原料和熔炼条件不同会使铅渣和铅锍的组分存在差异。综合企业清洁生产审计（赵晓声，2007）和现场调研结果，PbO、PbS、$PbSO_4$、Pb_2O_3、Pb_3O_4 及其他形态的比例可分别取 40%、30%、12%、10%、6%和 2%。

中国再生铅的冶炼多采用传统的反射炉（王吉坤，2012），冶炼的原料中废铅蓄电池占 85%（郭学益等，2009）。含铅的废电池壳和隔板上粘有 $PbSO_4$、PbO_2、Pb、PbO 等含铅物质（Zhu et al.，2013）。熔炼过程一般加入铁屑、苏打和碎焦进行还原、固硫和造渣反应，生成的熔炼铅渣中一般含有 PbS、PbO、$PbSO_4$ 等。同时，冶炼的铅尘为 PbS、PbO 等形态的释放物，其比例分别为 51% 和 49%（Lewis and Beautment，2002）。结合再生铅企业铅平衡表（杨继东等，2012），可得 PbS、PbO、$PbSO_4$、PbO_2 和 Pb 所占的比例分别为 34.6%、22.0%、18.4%、23.3%和 1.7%。

2. 产品加工制造阶段

精铅主要用于铅蓄电池、铅氧化物产品、铅合金及铅材等，其中铅蓄电池的使用量占70%以上（王晔，2011）。铅蓄电池生产过程分为铅粉与板栅制造、和膏、固化组装等多种工序，加工制造中的含铅废物主要为板栅冶炼的铅渣以及废弃极板、板栅等，根据成分可划分为铅合金损失和铅膏损失。铅合金主要为铅锑合金（Pb_xSb_y）或铅钙合金（Pb_xCa_y），铅膏的成分为 PbO、$3PbO·PbSO_4·H_2O$ 和 $4PbO·PbSO_4$（李志明等，2009）。制造产生的铅合金和铅膏废物的比例约为1∶2（Dalodwalla and Heart，2000），由此，该阶段释放的 Pb、PbO 和 $PbSO_4$ 的比例约分别为33.4%、53.3%和13.3%。

铅氧化物的制备通常采用金属氧化法，通过金属熔融、磨粉后焙烧氧化。温度在330～450℃范围内熔融过程中会形成 Pb_2O_3，450～470℃的温度范围内则形成 Pb_3O_4，温度再高时变为 PbO。铅氧化物产品在制造阶段产生 PbO、Pb、Pb_2O_3 和 Pb_3O_4 的比例分别取55%、30%、10%和5%。

铅合金熔炼时熔融铅的表面覆盖有氧化物浮渣，其成分为 PbO，同时有少量铅蒸气和铅尘的产生。铅渣的具体成分因合金种类不同而不同，如在铅钙合金配制生产过程中，浮渣中含有 $CaPb_3$ 和少量铅金属（郭占金等，2013）。铅合金熔炼中 PbO、Pb 的比例分别取85%、10%，其余形态比例取5%。

3. 产品使用阶段

铅使用的类型通常可以划分为铅蓄电池、铅材等可回收性铅以及弹壳、涂料、添加剂等耗散性铅。产品使用阶段的铅损失主要包含以下三种：一是耗散型产品，如涂料、弹药、塑料添加剂、焊料等，这些产品中的铅在其使用中将全部释放进入环境（Mao et al.，2009）；二是部分电缆难以就地回收或回收成本过高，常留存于远洋海底，这部分铅也被称作"休眠铅"（hibernating lead）（Mao et al.，2009）；三是铅产品使用中经风化和磨损产生环境扩散、腐蚀等。Lohm（Lohm et al.，1994）曾对第三种释放过程进行深入分析。本研究沿用其中环境释放系数的概念，采用环境释放系数表示特定时间段内向环境中发生自由迁移的量占使用中铅产品的总量的比值。中国和瑞典的铅释放情况存在差异，首先中国的铅产品消费结构中耗散型产品的比例大，其次，中国的技术水平与瑞典有所差异。考虑到这些因素，对2010年中国使用阶段的环境释放系数进行估值，具体如表13.1所示。

不同的铅产品产生不同形态的环境释放物，铅蓄电池、铅氧化物产品和铅盐等使用过程中释放到环境中的铅的形态往往存在差异。铅蓄电池中正极活性物质为 PbO_2，负极为海绵状铅，铅蓄电池中 Pb、PbO_2、$PbSO_4$ 比例分别取30%、

30%和40%；铅氧化物产品中PbO、Pb$_3$O$_4$（红丹）和Pb$_2$O$_3$的环境释放量根据其使用情况分别记为65%、25%和10%；含铅涂料和添加剂多为铅盐，如2PbCO$_3$·Pb（OH）$_2$（铅白）、PbCl$_2$、PbCrO$_4$等，文中PbCO$_3$和PbCl$_2$占铅盐释放物的比例分别取25%和5%，其余为PbCrO$_4$、硬脂酸盐等其他形态；而焊料、电缆护套、子弹壳释放到环境中的形态主要表现为Pb。

表13.1 铅使用阶段的环境释放系数

Table 13.1 Loss coefficients of lead products in use stage

铅产品	铅释放系数	
	瑞典	中国
铅蓄电池	0.01	0.09
铅氧化物产品	0.10	0.45
其他涂料和添加剂	0.50	0.75
电缆包皮	0.01	0.25
金属制品	0.005	0.008
焊料	—	0.90
其他	0.01	0.05

4. 铅废物处置与回收阶段分析

我国回收的废铅主要包括铅酸电池、电缆包皮、耐酸器皿衬里、印刷合金及轴承合金等（郭学益等，2009），含铅垃圾80%以上进行填埋，其余进行焚烧处理。含铅垃圾的种类包括城市固体废弃物、建筑垃圾、危险废弃物（如阴极射线管）和污水处理的污泥等。

部分化学添加剂和铅氧化物产品经过焚烧被释放到环境中。焚烧产生的含铅物质主要为底渣和飞灰，其成分与氯、硫组分的含量、燃烧环境和金属特性等有关。我国垃圾中含有大量的厨余垃圾、PVC塑料等含氯物质和纺织、橡胶等含硫物质，焚烧中会产生PbCl$_2$、PbSO$_4$和铅的氧化物。根据垃圾中氯、硫含量，化合物PbCl$_2$和PbSO$_4$的比例约为2∶1（Shi et al.，2008）。废铅蓄电池的成分区别于新电池，PbSO$_4$、PbO$_2$、Pb和PbO所占比例分别为50%、20%、17%、13%（周正华，2002）；铅氧化物产品在废物处置与回收阶段释放的形态种类比较复杂，Pb$_3$O$_4$（铅丹）为涂料的主要成分，占环境释放的25%；PbCO$_3$作为铅白的成分所占比例为20%；假设通过焚烧产生的Pb$_2$O$_3$所占比例为5%，其余为PbCl$_2$、PbSO$_4$和PbO形态的铅。废物处置与回收阶段环境释放系数的取值汇总后可见表13.2。

表 13.2 废物处置与回收阶段的铅环境释放系数

Table 13.2 Loss coefficients of lead products in waste management & recycling

铅产品	环境释放系数
铅蓄电池	0.15
铅氧化物产品	0.55
其他涂料和添加剂	0.15
电缆包皮	0.08
金属制品	0.10
其他	0.70

13.2 结果与讨论

13.2.1 各阶段铅释放物形态比例

人为循环中环境释放物形态及其所占比例汇总整理后见表 13.3。可以看出，生命周期的某些过程产生的释放物形态种类较多，如焚烧过程可产生 $PbSO_4$、Pb_3O_4、$PbCl_2$、$PbCO_3$ 等 7 种形态的释放物，而选矿过程释放物的种类较少，主要为 $PbSO_4$、$PbCO_3$ 和 PbS。

表 13.3 生命周期各阶段铅环境释放物的形态及比例

Table 13.3 Speciation and contribution of the lead losses in product life-cycle stages

生命周期阶段	产生过程	环境释放物名称	释放物形态	所占比例[1]/%
生产	选矿	尾矿	$PbSO_4$∶$PbCO_3$∶PbS∶其他	50%∶35%∶10%∶5%
	原生铅冶炼	炉渣、铅锍、阳极泥、浮渣、氧化渣	PbO∶PbS∶$PbSO_4$∶Pb_2O_3∶Pb_3O_4∶其他	40%∶30%∶12%∶10%∶6%∶2%
	再生铅冶炼	铅渣、废电池隔板、除尘灰	PbS∶PbO_2∶PbO∶$PbSO_4$∶Pb	34.6%∶23.3%∶22.0%∶18.4%∶1.7%
加工制造	电池制造	铅膏、铅渣、废弃极板、板栅	PbO∶Pb∶$PbSO_4$	53.3%∶33.4%∶13.3%
	铅氧化物产品	铅渣	Pb∶PbO∶Pb_3O_4∶Pb_2O_3	55%∶30%∶10%∶5%
	合金冶炼	浮渣	PbO∶Pb∶其他	85%∶10%∶5%
产品使用	铅氧化物产品	铅渣	Pb∶PbO∶Pb_3O_4∶Pb_2O_3	55%∶30%∶10%∶5%
	合金冶炼	浮渣	PbO∶Pb∶其他	85%∶10%∶5%
	铅蓄电池	腐蚀和无组织排放	Pb∶PbO_2∶$PbSO_4$	30%∶30%∶40%
	铅氧化物产品	腐蚀和扩散	PbO∶Pb_2O_3∶Pb_3O_4	60%∶30%∶10%
	其他涂料和添加剂	腐蚀和扩散	$PbCO_3$∶$PbCl_2$∶其他	25%∶5%∶70%

续表

生命周期阶段	产生过程	环境释放物名称	释放物形态	所占比例[1]/%
废物处置与回收	铅蓄电池	固体废弃物	$PbSO_4 : PbO_2 : Pb : PbO$	50% : 20% : 17% : 13%
	铅氧化物产品	灰渣、飞灰	$PbO : Pb_3O_4 : PbCO_3 : PbCl_2 : Pb_2O_3 : PbSO_4 :$ 其他	31.3% : 25% : 20% : 13.3% : 5% : 3.3% : 2%
	其他涂料和添加剂	灰渣、飞灰	$PbO : PbCl_2 : Pb_2O_3 : Pb_3O_4 : PbSO_4 :$ 其他	63.3% : 13.3% : 10% : 10% : 3.3% : 10%

1) 各数据均以含铅量计算。

13.2.2 生命周期铅释放物形态构成

生产阶段的铅释放来自选矿过程产生的尾矿、原生和再生冶炼的熔炼铅渣。尾矿中的含铅量可用 2010 年铅精矿的使用量（USGS，2006）乘以 1 减去采矿回收率算得，熔渣中的含铅量可以用精铅量乘以 1 减去冶炼回收率算得，其中精铅包含原生铅和再生铅，采矿效率、原生和再生铅冶炼效率分别为 83.97%、95.74%（中国有色金属工业年鉴编辑委员会，2012）和 98%（Mao et al.，2008）。简单计算可得铅生产阶段选矿、原生和再生铅冶炼过程中环境释放物总量，又已知铅各形态释放物所占的比例（表 13.3），由式（13.1）可得铅释放物在生产阶段的形态构成，结果见表 13.4。为简化研究结果和使之便于比较，文中研究精铅的生产量为 1 kt 所对应的环境释放情况。

中国 2006 年起限制精铅出口，2010 年精铅为净进口，铅矿进口将是长期趋势，研究中假设我国生产的精铅全部用于国内消费。铅的消费结构中铅蓄电池约占 73%，铅氧化物产品约占 13%，铅合金及铅材约占 8%，上述铅使用量在铅产品的总制造中占 94% 左右（王晔，2011），文中制造阶段主要考虑以上 3 种产品。制造过程中没有进入铅产品的铅，即释放到环境中铅的数量采用铅产品量乘以 1 减去制造效率可得。由于铅酸电池和化工制品的制造效率为 85%~98%，铅材的铅利用率较低（毛建素和陆钟武，2003），铅蓄电池制造效率取值为 0.98（t/t），铅氧化物产品为 0.97（t/t），铅合金及铅材为 0.9（t/t）。同理，由表 13.3 和式（13.1），可得加工制造阶段各形态铅环境释放物的量。

以 2010 年中国铅的消费情况为例，除铅蓄电池和铅氧化物产品外，铅材占铅消费量的 3%（王晔，2011），焊料 3%，电缆护套 1%，铅弹 1%。每消费 1 kt 精铅，根据铅的消费结构以及产品使用、废物处置与回收阶段的铅产品释放物各形态的比例（表 13.3），由式（13.1）可求得使用阶段和废物处置与回收阶段各形态铅的量，结果如表 13.4 所示。

表 13.4 生命周期各阶段铅环境释放物的形态构成

Table 13.4 Constitution of lead speciation for the lead losses in product life-cycle stages

生命周期阶段	环境释放物形态的数量/t									
	PbSO$_4$	PbO	Pb	PbCO$_3$	PbO$_2$	Pb$_3$O$_4$	PbS	PbCl$_2$	Pb$_2$O$_3$	其他
生产	42.6	13.1	0.1	29.4	1.7	1.7	23.8	—	2.9	4.8
加工制造	1.9	16.7	6.8	—	—	0.4	—	—	0.2	0.4
产品使用	26.3	38.0	59.5	11.3	19.7	14.6	—	1.9	5.9	24.9
废物处置与回收	56.1	40.1	22.4	14.3	21.9	18.3	0.7	13.2	4.0	8.5
总计	126.9	107.9	88.8	55.0	43.3	35.0	24.5	15.0	12.9	38.5

注：—表示可忽略不计。

由式（13.2）和式（13.3）可得整个生命周期释放的各形态铅的量。从表 13.4 可以看出，2010 年中国铅人为循环的环境释放物主要形态为 PbSO$_4$，当精铅消费量为 1 kt 时其产生量为 128.2 t，约占总释放量的 23.4%。PbSO$_4$ 主要来自尾矿中以硫酸盐存在的铅（郭翠香和赵由才，2008）以及废旧铅蓄电池的组成成分。PbSO$_4$ 还可用作颜料、草酸生产催化剂、纤维增重剂、涂料、油漆等。其次释放量较大的是 PbO、Pb 和 PbCO$_3$，共占释放量的 46.2%。PbO 作为铅化合物中最重要的一种，广泛用于蓄电池、涂料、陶瓷、玻璃、铅盐制备、聚氯乙烯和橡胶等工业领域，由于多为耗散性使用而难以回收，PbO 常在使用或者焚烧处置时释放到环境中。PbO 具有氧化性，通常不能稳定存在，在环境介质中逐渐与 CO$_2$ 等发生反应，这是进一步研究铅释放物在环境中迁移转化所要考虑的重要内容。铅和铅合金形式的弹壳、焊料等在产品使用阶段几乎全部以金属或合金形式释放出来，故 Pb 形态的释放物也较多。Pb 在潮湿和含有 CO$_2$ 的空气中会失去光泽而变成暗灰色，转变成碱式碳酸盐。PbCO$_3$ 主要来源于生产阶段的选矿后的尾矿，PbCO$_3$ 在环境中一般比较稳定。

13.2.3 生命周期铅释放物来源构成

2010 年，国内每消费 1kt 精铅产生的环境释放物的总量为 547.9t，如表 13.5 所示。考虑中国的经济发展情况及研究时间的不同，这一结果和相关研究中世界范围内每消费 1 kg 铅就有 0.5 kg 铅进入环境中的结果一致（Smith and Flegal，1995；Mao et al.，2009）。可以看出，铅释放物形态基于生命周期各阶段的分布存在很大差异，PbSO$_4$ 主要分布在生产阶段和废物处置与回收阶段；PbO 虽然在生命周期各个阶段都产生较多，但源于废物处置与回收阶段的量最多；Pb 形态的释放物主要来自产品使用阶段的无组织的排放和腐蚀扩散，而 PbCO$_3$ 主要来自生产阶段如图 13.2 所示。

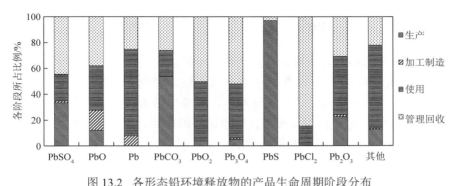

图 13.2 各形态铅环境释放物的产品生命周期阶段分布

Fig. 13.2 Distribution of lead species fractions among product life-cycle

表 13.5 铅释放物在各生命周期阶段的数量分布

Table 13.5 Distribution of total lead losses among life-cycle stages

生命周期阶段	释放量/t	所占总量比例/%
生产	120.1	21.9
加工制造	26.5	4.8
使用	201.9	36.9
废物处置与回收	199.3	36.4
总计	547.9	100.0

就整体而言，铅释放物来自使用、废物处置与回收阶段的数量远大于生产、加工制造阶段，使用阶段和废物处置与回收阶段的产生量占总量的 73.3%。即使生产和加工制造阶段的数量相对较少，但是由于职业暴露和职业风险，采矿业生产和电池制造回收、喷漆、焊接等加工车间的工人血铅含量较高，相关研究中也较多涉及铅生产和加工制造业中工人的血铅含量（United Nations Environment Programme，2013）。我国铅产品使用阶段环境释放偏高，这可能与以下原因有关：一是我国耗散性铅在消费结构中所占的比例高（毛建素和陆钟武，2003）；二是我国公民的环保意识比发达国家落后，随意丢弃现象严重；三是缺乏全国性的回收网络（王红梅等，2012），废铅酸蓄电池回收处于无序状态，很多铅产品无法进入管理回收的下一阶段。另外，废物处置回收阶段的环境释放量也较大。这可能是由于国内再生铅技术设备较落后、生产规模较小、回收率低、环境污染严重（陈永桥和王冬，2013），同时，铅管理回收的相关法律法规还有待完善。我国废物处置与回收阶段的回收效率与发达国家相比低 10 个百分点，仅为 85%左右（王红梅等，2012）。

13.2.4 讨 论

中国铅人为循环的环境释放受到多种因素的影响，如资源利用情况、再生铅的比例、铅的消费结构等。中国虽然是世界上铅资源最丰富的国家之一，但随着近年来大规模的开采，高品位矿逐年减少，后备储量难以满足未来发展需求（王淑玲，2004）。铅冶炼原料供应不足，将使铅精矿长期处于净进口状态。未来国内铅矿石开采和尾矿的产生量将减少，相应地环境释放物中 $PbCO_3$、$PbSO_4$ 和 PbS 的数量也将减少。

未来再生铅的比例将对铅释放物的产生造成影响。目前，工业发达国家再生铅的产量占精铅总产量的比例可达 65%，而我国再生铅所占比例不超过 30%（刘毅，2012）。随着可回收性铅使用量的增加，我国再生铅原料将稳定增长，再生铅的比例将不断提高。废物处置与回收阶段铅释放物总量将减少，该阶段产生的释放物如 PbO、Pb、$PbCl_2$ 等也将减少。但在未来十几年内，如果中国铅市场出现供过于求的现象，则有可能导致铅回收动力不足，回收量减少，铅的环境释放较以前反而增加（Elshkaki et al，2004）。

铅的市场消费结构对于环境释放物的形态分布的影响很容易理解，如禁止向汽油中添加铅极大地减少了环境中的四乙基铅等有机铅。现阶段，无铅化已成为一种趋势，在未来十几年内，耗散性铅如阴极射线管、铅弹、塑料稳定剂和铅焊料的应用将进一步减少。同时，电子产业和汽车市场的发展必将带来铅蓄电池需求量的快速增长。在不改变现有技术的情况下，电池制造过程中 PbO、Pb 的释放量和使用阶段及废物处置与回收阶段 $PbSO_4$、PbO_2 等的释放量将增加。

13.3 结 论

（1）2010 年中国国内每消费 1kt 精铅，铅人为循环产生的环境释放物总量约为 547.9t，主要源于产品使用阶段和废物处置与回收阶段（73.3%）。

（2）生命周期中所产生的环境释放物主要形态表现为 $PbSO_4$，约占总量的 23.4%。然后是 PbO、Pb 和 $PbCO_3$，共占总释放量的 46.2%。

（3）$PbSO_4$ 主要分布在生产阶段和废物处置与回收阶段，PbO 源于废物处置与回收阶段的量最多，Pb 形态的释放物主要来自产品使用阶段，而 $PbCO_3$ 主要来自生产阶段。

参 考 文 献

陈永桥，王冬. 2012. 废铅酸蓄电池的再生利用及其污染控制. 环境科学与技术，35（S1）：439-441.

郭翠香, 赵由才. 2008. 我国含铅废物现状及铅回收技术研究进展. 有色冶金设计与研究, 28（2-3）: 46-49, 54.

郭学益, 钟菊芽, 宋瑜, 等. 2009. 我国铅物质流分析研究. 北京工业大学学报, 35（11）: 1554-1561.

郭占金, 吴国庆, 赵振波, 等. 2013. 铅钙合金生产过程中浮渣控制的生产实践. 世界有色金属,（2）: 72-73.

蒋继穆. 2004. 我国铅锌冶炼现状与持续发展. 中国有色金属学报, 14（1）: 54-62.

李志明, 谭晓波, 吴贤章, 等. 2009. 和膏工艺对铅膏性能的影响. 蓄电池, 46（1）: 18-20.

刘毅. 2012. 废铅蓄电池回收提取氧化铅粉循环利用新技术. 科技与企业,（14）: 304-305.

毛建素, 陆钟武. 2003. 关于我国废铅实得率低下的原因的研究. 世界有色金属,（11）: 24-28, 32.

王红梅, 刘茜, 王菲菲, 等. 2012. 中国铅酸蓄电池回收处理现状及管理布局研究. 环境科学与管理, 37（6）: 51-54.

王吉坤. 2012. 铅锌冶炼生产技术手册. 北京: 冶金工业出版社: 30.

王淑玲. 2004. 世界铅资源形势分析. 国土资源情报,（6）: 28-36.

王晔. 2011. 中国铅行业发展及展望. 有色金属工程, 1（1）: 27-29.

杨继东, 刘佳泓, 徐建京, 等. 2012. 铅蓄电池生产企业的清洁生产审核. 化工环保, 32（3）: 264-268.

章骅, 何品晶, 吕凡, 等. 2011. 重金属在环境中的化学形态分析研究进展. 环境化学, 30（1）: 130-137.

赵晓声. 2007. ISP炼铅锌技改清洁生产评价. 工程设计与研究,（123）: 11-19.

中国有色金属工业年鉴编辑委员会. 2012. 2011中国有色金属工业年鉴. 北京: 中国有色金属年鉴编辑部: 602.

周正华. 2002. 从废旧蓄电池中无污染火法冶炼再生铅及合金. 上海有色金属, 23（4）: 157-163.

Cui Y S, Fu J, Chen X C. 2011. Speciation and bioaccessibility of lead and cadmium in soil treated with metal-enriched Indian mustard leaves. Journal of Environmental Sciences, 23（4）: 624-632.

Dahodwalla H, Heart S. 2000. Cleaner production options for lead-acid battery manufacturing industry. Journal of Cleaner Production, 8（2）: 133-142.

Elshkaki A, Van der Voet E, Van Holderbeke M, et al. 2004. The environmental and economic consequences of the developments of lead stocks in the Dutch economic system. Resource, Conservation and Recycling, 42（2）: 133-154.

Lewis A E, Beautement C. 2002. Prioritising objectives for waste reprocessing: a case study in secondary lead refining. Waste Management, 22（6）: 677-685.

Liu Y S, Ma L L, Li Y Q, et al. 2007. Evolution of heavy metal speciation during the aerobic composting process of sewage sludge. Chemosphere, 67（5）: 1025-1032.

Lohm U, Anderberg S, Bergbäck B. 1994. Industrial Metabolism at the National Level: A Case Study on Chromium and Lead Pollution in Sweden, 1880-1980. Tokyo: United Nations University Press, 101-118.

Mao J S, Cao J, Graedel T E. 2009. Losses to the environment from the multilevel cycle of anthropogenic lead. Environmental Pollution, 157（10）: 2670-2677.

Mao J S, Dong J, Graedel T E. 2008. The multilevel cycle of anthropogenic lead: I. methodology. Resource Conservation and Recycling, 52（8-9）: 1058-1064.

Mao J S, Ma L, Niu J P. 2012. Anthropogenic transfer & transformation of heavy metals in anthrosphere: concepts, connotations and contents. International Journal of Earth Sciences and Engineering. 5（5）: 1129-1137.

Shah P, Strezov V, Nelson P F. 2009. X-ray absorption near edge structure spectrometry study of nickel and lead speciation in coals and coal combustion products . Energy Fuels, 23 (3): 1518-1525.

Shi D Z, Wu W X, Lu S Y, et al. 2008. Effect of MSW source-classified collection on the emission of PCDDs/Fs and heavy metals from incineration in China. Journal of Hazardous Materials, 153 (1-2): 685-694.

Smith D R, Flegal A R. 1995. Lead in the biosphere: recent trends. Ambio, 24 (1): 21-23.

Sonmez M S, Kumar R V. 2009. Leaching of waste battery paste components. Part 2: leaching and desulphurisation of $PbSO_4$ by citric acid and sodium citrate solution. Hydrometallurgy, 2009, 95 (1-2): 82-86.

Szpunar J, Obiński R. 1999. Speciation in the environmental field-trends in analytical chemistry. Fresenius Journal of Analytical Chemistry, 363 (5-6): 550-557.

United Nations Environment Programme. 2013. Draft final review of scientific information on lead. http: //www. chem. unep.ch/Pb_and_Cd/SR/Draft_final_reviews/Pb_Review/Final_UNEP_Lead_review_Nov_2008.pdf.

USGS (U. S. Geological Survey). 2006. Mineral commodity summaries. http: // minerals. usgs. gov/ minerals/ pubs/ commodity/lead/lead mcs05. pdf.

Volpe M, Oliveri D, Ferrara G, et al. 2009. Metallic lead recovery from lead-acid battery paste by urea acetate dissolution and cementation on iron. Hydrometallurgy, 96 (1-2): 123-131.

Xie X, Ding G D, Cui C, et al. 2013. The effects of low-level prenatal lead exposure on birth outcomes. Environmental Pollution, 175 (4): 30-34.

Zhu X F, Yang J K, Gao L X, et al. 2013. Preparation of lead carbonate from spent lead paste via chemical conversion. Journal of Hazardous Materials, 134-135: 47-53.

第14章 全球铅元素人为释放物源头数量与形态分析
Source Analysis of Global Anthropogenic Lead Emissions: Their Quantities and Species[*]

14.1 Introduction

In recent years, lead demand and consumption have seen a continuous increase due to economic development and an increasing population (Ordoñezand and Rahe, 2013; DiFrancesco et al., 2012). Lead-containing wastes are a huge threat to ecological safety and human health (Tukker et al., 2006; Chen et al., 2012). It is well known that many pollutants are from anthropogenic sources and the scale of the lead anthropogenic cycle is much greater than the natural lead cycle (Sen and Bernhard 2012). Therefore, it is vital to know how much lead was emitted in the past and to obtain more information on emission characteristics, such as their species type, because these characteristics greatly affect the hazard potential of lead (Shah et al., 2009). Source abatement decreases the overall pollution level; therefore, better source control makes the creation of an emissions inventory a priority. In this study, lead emissions refer to the release of lead into various environments, such as inputs to the atmosphere, the terrestrial environment and the aquatic environment. The definition of 'emissions' is the same as the definition used by Mao et al (2008a).

Many studies on lead emissions, both on local and global scales, have been conducted (Zhang et al., 2011; Rauch and Pacyna, 2009), but all of these studies lack information on the species of lead that are being emitted or their changes over time. Pacyna and colleagues have estimated lead emissions to various environmental reservoirs based on emissions factors (Rauch and Pacyna, 2009; Nriaguand Pacyna, 1988; Pacyna et al., 1995). Studies on lead species mostly focus on detecting the species present in soil or dust (Manceau et al., 1996; Sakata, 2014). These data can then be

---

[*] Liang J, Mao J S. 2015. Source analysis of global anthropogenic lead emissions: their quantities and species. Environmental Science and Pollution Research, 22 (9): 7129-7138.

used to identify possible environmental pollution sources. Although the previous studies have provided detailed species analyses at the local scale, information at the global level is still missing. In this study, we have estimated the 2010 lead emissions and the species types that have been released by the anthropogenic lead cycle. In addition, we have simulated how the lead cycle evolves as a function of technological progress and quantified and analyzed the total emissions between 1930 and 2010.

In previous research, we analyzed lead emissions using China as a case study (Liang and Mao, 2014a). This study investigates the distribution of lead among the different life cycle stages and includes a dynamic analysis of temporal evolution. Although a worldwide study on lead emissions is already available for 2000 (Mao et al., 2009), it appears to be a static study and is a snapshot of lead emissions in 2000. This drawback means that the investigation of further temporal features and the creation of a dynamic model are required. A species analysis (Liang and Mao, 2014b) has already been undertaken, which showed the feasibility of this kind of analysis. All of these studies have laid a sound foundation for a more comprehensive analysis of the anthropogenic lead cycle, which is the aim of this study.

14.2　Methodology

14.2.1　Estimation of Lead Emissions

1. Estimation of lead emissions in 2010

The lead life cycle consists of four principal stages: production, fabrication and manufacture (F&M), use and waste management and recycling (WMR). The scope and system boundary of the anthropogenic lead cycle is defined by the STAF project by Yale University in 2006. We have only considered intentional lead use driven by human demands of lead and have not included some non-intentional applications of lead that arise from their natural occurrence in fossil fuels, other metal ores, and phosphate ores (Elshkaki et al., 2009)

This framework (Fig. 14.1) is also consistent with our previous studies (e.g., Mao et al., 2008a; Mao et al., 2009a, 2009b). In this study, the framework is applied to identify how the different chemicals are formed while previous studies focused on the mass of the flows. This framework is revised based on a summary research (Chen and Graedel, 2012). The difference between the global anthropogenic cycle and the country-level cycle is that there are no imports and exports at the global level.

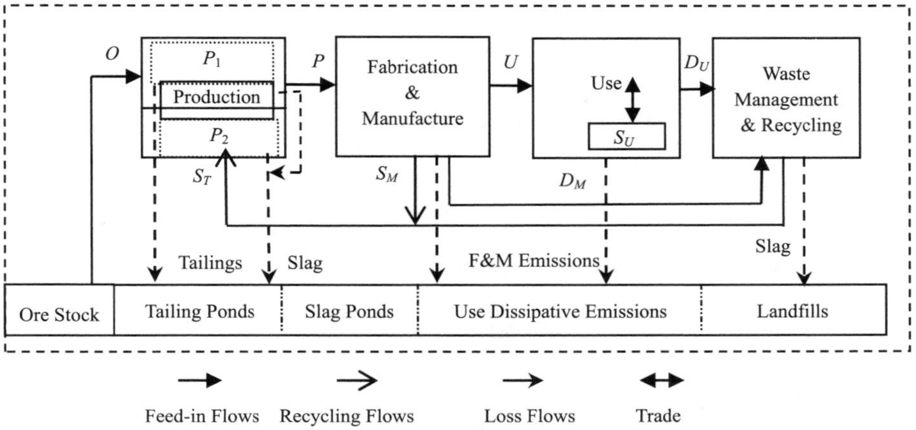

Fig. 14.1 Lead emissions framework for the anthropogenic lead cycle

Symbols and definition: O, lead from ore; P_t, refined lead in year t; U, lead entering the 'use' category; S_U, lead entering the in-use stock; D_U, end-of-life lead to waste management and recycling (WMR); D_M, lead in the F&M discards to WMR; S_T, lead in the total scrap to production processes; S_M, lead in new scrap from manufacturing.

The lead cycle was quantified by substance flow analysis, which is the most widely applied metal flow analysis method (Glöser et al., 2013). We managed to combine the emissions with lead production because refined lead production has the most direct influence on all lead fluxes. Namely, we considered the emissions as simply being production-driven.

Lead emissions are affected by many factors, such as technological advancement and social progress-related parameters. These parameters are empirical and have been verified by previous studies (Liang and Mao, 2014b; Mao et al., 2008b). Production or manufacturing efficiencies, such as mining efficiency, η_1, primary refining efficiency, η_{21}, secondary refining efficiency, η_{22} and fabrication efficiency, f_m, can affect emission levels. Lead in the in-use stock eventually goes into the environment so the in-use stock accumulation ratio, α, also influences the environmental emissions. Rises in the secondary lead ratio, λ and the secondary supply ratio, θ, lead to an increase in the amount of recycled and secondary refined lead and a decrease in the amount of lead waste released into the environment. Other parameters used in the model are presented in Table 14.1.

Table 14.1 Parameters used in the estimation of anthropogenic lead emissions

Parameter	name	Definition	Formula definition
α	Accumulation ratio	Proportion of in-use stock to lead entering use	$\alpha = \dfrac{S_U}{U}$
β	New scrap ratio	Proportion of new scrap from manufacturing to the total scrap	$\beta = \dfrac{S_M}{S_T}$
μ	Discard-management ratio	F&M discards entering WMR as a fraction of the sum of the discards and emissions from F&M	$\mu = \dfrac{D_M}{D_M + E_M}$
λ	Secondary lead ratio	Ratio of the secondary lead to the total production	$\lambda = \dfrac{P_2}{P}$
θ	Secondary supply ratio	Scrap supplied to production as a fraction of refined lead input to F&M	$\theta = \dfrac{S_T}{P}$
f_m	Fabrication efficiencies	Fraction of lead entering manufacturing that leaves as a part of the finished products	$f_m = \dfrac{U}{P}$

The lead life cycle follows the mass conservation law, which is reflected by the equivalence relationships among the lead flows. In addition, some of the lead flows can be formulated as a combination of the parameters provided in Table 14.1. Therefore, through the use of appropriate mathematical substitutions, the lead emissions from the lead life cycle can be formulated into equations that include lead production (P). The lead emission rate refers to the lead emission quantity during a specific period of time, normally a year. At the production stage, lead in tailings and slag can be described by equation (14.1) and equation (14.2) respectively. At the F&M stage, the E_M emission rate is described by equation (14.3).

$$T = \frac{\lambda(1-\eta_1)}{\eta_1} P \tag{14.1}$$

$$S_a = \left[\frac{(1-\lambda)(1-\eta_{21})}{\eta_{21}} + \frac{(1-\lambda)(1-\eta_{22})}{\eta_{22}} \right] P \tag{14.2}$$

$$E_M = (1-\mu)(1-f_m-\beta\theta) P \tag{14.3}$$

The meanings of the terms in the equations are outlined in Table 16.1. The E_U emissions rate during the use stage can be calculated from the total dissipation of the different products in use. E_U and E_M at the WMR stage are described by equation (14.4) and equation (14.5) respectively.

$$E_U = \sum_{i=1}^{N} \varepsilon_i k_i P \tag{14.4}$$

$$E_W = \left[f_m(1-\alpha) - \varepsilon_i k_i - (1-\beta)\theta - \mu(1-f_m-\beta\theta) \right] P \tag{14.5}$$

where i is the product group index, N is the number of product groups, ε_i is the emission factor for product i and k_i is the contribution product i makes the total consumption. The meanings of the other parameters can found in Table 14.1.

2. Temporal Evolution of Lead Emissions

1) Total quantity of lead emissions between 1930 and 2010

There seems to be a delay between the release of lead emissions and them becoming hazardous. The emissions are discharged and then they enter the environment and interact with the environmental compartments. After that, the exposure to the emissions will occur before they finally pose a risk. Therefore, the estimation of source emissions is the starting point of the hazard, and it is important to estimate the emissions quantity from the source.

One can determine the quantity of the lead emitted from the lead cycle by integrating the annual emissions for 1930–2010. Lead emissions during the period can be described by equation (14.6):

$$Q_T = \int_{t_0}^{t} Q_t dt = \sum_{t_0}^{t} Q_t$$

$$= \sum_{t_0}^{t} \left[1 - \alpha f_m - \theta + \frac{\lambda(1-\eta_1)}{\eta_1} + \frac{(1-\lambda)(1-\eta_{21})}{\eta_{21}} + \frac{(1-\lambda)(1-\eta_{22})}{\eta_{22}} \right] P \quad (14.6)$$

where t_0 is 1930 and t is 2010 and all the parameters vary with time.

2) Parameter values

During production, the mining efficiency, η_1, ranges from 0.75 to 0.91; the primary refining efficiency, η_{21}, varies from 0.89 to 0.99 and the secondary refining efficiency, η_{22}, varies within the range 0.86–0.985 (Mao et al., 2008a). The mining efficiency and the primary refining efficiency data can be found in the reports of Nonferrous Metals Industry Yearbook (China Nonferrous Metals Industry Yearbook, 2010) and the unreported efficiencies are estimated based on the factors that may impact the efficiency, such as the consistency of technology and the related technological advance (CNIA, 2009). All the efficiencies needed for this study are listed in Table 14.2. China is referred to here because it is one of the largest producers in the world, ranking first since 2003 (Zhang et al., 2011). The secondary lead ratio, λ, estimates were based on information from the USGS (US Geological Survey) and ILA (International Lead Association) (DiFrancesco et al., 2012; International Lead Association, 2014) and increased between 1930 and 2010. The secondary supply ratio, θ, can be obtained from its relationship with λ, which indicates that the secondary supply ration is the secondary lead ration divided by the secondary refining efficiency.

For F&M, the fabrication efficiencies f_m, are estimated based on the fabrication efficiencies in the 20th century (Mao and Graedel, 2009). The following information was applied for the estimation of β values (the global proportion of new scrap to total scrap): β was 0.0416 in 2000 (Mao et al., 2008), was approximately 0.103 for China in 2006 (Guo et al., 2009) and was 0.05 for the United States in 1998 (Smith et al., 2004). The values of β for the global level should be between the United States and China. The global discard-management ratio, μ, is estimated based on the overall trend of technological development inferred from the efficiency changes during production. A value is given as 0.659 worldwide in 2000 (Mao et al., 2008b). All the estimations of parameter values are summarized in Table 14.2.

Table 14.2 Values for the parameters related to lead environmental emissions

Year	1930–1939	1940–1949	1950–1959	1960–1964	1965–1969	1970–1974	1975–1979	1980–1984	1985–1989	1990–1994	1995–1999	2000–2004	2005–2010
f_m	0.77	0.79	0.80	0.81	0.83	0.84	0.85	0.86	0.89	0.91	0.92	0.93	0.94
α	0.010	0.010	0.010	0.010	0.014	0.021	0.038	0.052	0.069	0.094	0.110	0.120	0.173
β	0.075	0.070	0.065	0.060	0.055	0.052	0.049	0.048	0.046	0.045	0.042	0.042	0.041
μ	0.58	0.59	0.60	0.61	0.61	0.62	0.62	0.63	0.64	0.64	0.65	0.70	0.73
λ	0.34	0.35	0.36	0.37	0.38	0.41	0.42	0.43	0.45	0.48	0.47	0.5	0.52
θ	0.395	0.405	0.414	0.423	0.432	0.461	0.467	0.470	0.484	0.508	0.490	0.520	0.538
η_1	0.750	0.760	0.780	0.800	0.820	0.830	0.840	0.860	0.880	0.880	0.890	0.900	0.910
η_{21}	0.855	0.860	0.865	0.870	0.875	0.885	0.895	0.910	0.925	0.940	0.955	0.970	0.980
η_{22}	0.860	0.865	0.870	0.875	0.880	0.890	0.900	0.915	0.930	0.945	0.960	0.980	0.985

At the use stage, the in-use accumulation ratio, α, in 2000 was 0.112. What is clear is that the in-use stock is influenced greatly by social development (Graedel and Cao, 2010). The accumulation ratio, α, can be estimated based on the worldwide GDP per capita. In addition, the dissipation varies depending on the type of lead product. There is hardly any dissipation for lead acid batteries during use (Elshkaki et al., 2004) whereas 76% of lead additives in petrol are emitted into the environment (Li et al., 2012). It has also been reported that 50%–75% of cable sheathing is recycled (Tukker et al., 2006; Elshkaki et al., 2004). Furthermore, lead emissions from pigments or additives, ammunition and solder alloys are all considered to be lost during use. The emission factors for lead products are presented in Table 14.3.

Table 14.3　Environmental emission factors for various lead products at the use stage

Item	1930–1939	1940–1949	1950–1959	1960–1964	1965–1969	1970–1974	1975–1979	1980–1984	1985–1989	1990–1994	1995–1999	2000–2004	2005–2010
Cable Sheathing	0.35	0.33	0.34	0.32	0.31	0.3	0.29	0.28	0.27	0.26	0.25	0.23	0.22
Alloys	0.4	0.41	0.42	0.4	0.39	0.38	0.37	0.35	0.33	0.31	0.25	0.22	0.2
Gasoline	0.88	0.86	0.85	0.85	0.84	0.83	0.82	0.82	0.79	0.78	0.77	0.76	0.75
Others	0.47	0.46	0.45	0.44	0.43	0.42	0.41	0.4	0.46	0.45	0.44	0.42	0.4

14.2.2　Lead Emissions Species Analysis

In this study, the lead species refer to the chemical forms. The lead species can be identified by tracing their physiochemical transformations during the lead cycle. Specifically, the basic species before any physiochemical reactions took place were first determined. Then the possible reactions were identified during the performance of industrial processes by referring to handbooks or technical references (Mao et al., 2014). Finally, the possible emissions species can be deduced based on the identified transformations.

The transformations vary significantly from one stage of the lead cycle to another. Mining activities represent a major source of environment contamination (Alberto et al., 2012). At the production stage, galena, cerusite and sardinianite account for 95% of the total lead ore (Woodward, 2006), of which, PbS constitutes approximately 45% (Ostergren et al., 1999) because the chemical reactions that take place in a blast furnace and during pyrorefining suggest that PbS and PbO will be released (Woodward, 2006).

In secondary refining, waste lead acid batteries are the dominant raw material (Jolly and Rhin, 1994). PbO accounts for 65%–70% of the emissions and $PbSO_4$ accounts for 30%–35% (Wang and Li, 2003) during this process. At the F&M stage, each lead product has a different emission rate during their manufacturing processes. In battery production, lead alloys and lead paste are normally discharged at a ratio of 1∶2 (Dahodwalla and Heart, 2000). The lead alloys and the lead paste composition data (Li et al., 2009) suggests that the proportions of PbO, Pb and $PbSO_4$ are 55%, 35% and 10%, respectively. In addition, PbO and Pb are usually released during the fusion stage, which is part of lead sheet and cable sheathing production process. At the use stage, tetraethyl lead is emitted as it is a petrol additive and metallic Pb is discharged when alloys or cable sheathing are used. Furthermore, the dissipative emissions from pigments and additives, which refer to the waste materials not recycled or re-used (Lohm et al., 1994), contain various chemical species. The most common forms in pigments are lead chromate ($PbCrO_4$), white lead [$2PbCO_3 \cdot Pb(OH)_2$], and red lead (Pb_3O_4). The common industrial additives include lead stabilizer in plastics (e.g., tribasic lead sulphate and dibasic lead phosphite) and catalyst (e.g., lead acetate and naphthenic acid). Pigments and additives may undergo complex reactions during use. For example, $2PbCO_3 \cdot Pb(OH)_2$ may decompose into PbO or react with SiO_2 in ceramics. Therefore, the amounts of specific forms are hard to define. Due to their complexity, we classified them into the "others" category. At the WMR stage, a discarded battery consists of $PbSO_4$ (60%), PbO_2 (25%), PbO (13%) and Pb (2%) (Zhu et al., 2013). $PbCl_2$, $PbSO_4$ and PbO are considered to be discharged when the batteries are incinerated along with municipal solid waste (Shi et al., 2008).

The emissions species undergo changes because the factors that affect them vary with time. The secondary lead ratio results in different lead species proportions during production, while changes in the consumption structure lead to speciation changes in all of the other stages. Although pollution disposal technology may also play a role, we specifically discuss the impacts of secondary lead and consumption in this chapter.

14.2.3 Data Sources

A significant amount of data on lead production and on the lead consumption structure exists. The lead production data used in this study were based on data from ILZSG (International Lead and Zinc Study Group) and USGS or ILA (DiFrancesco et al., 2012; International Lead Association 2014; ILZSG, 1992; 2005a; 2005b; 2008), which were slightly modified for use in this study (See Uncertainty section). The lead products were classified into six categories: batteries, sheet/pipe, cable sheathing, alloys, gasoline additives and others. The consumption structure from 1960 (ILZSG, 1992, 2005a, 2005b, 2008) is shown in Fig. 14.2 as data prior to 1960 were not available.

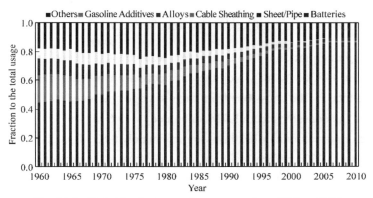

Fig. 14.2 Global lead consumption patterns from 1960 to 2010

14.3 Results and Discussion

14.3.1 Global Lead Emissions

1. Global lead emissions in 2010

The emissions produced by each stage were determined using equation (14.1) – equation (14.5). The total emissions from the lead anthropogenic cycle in 2010 was calculated as being 3.56 Mt (Table 14.4) and the emissions from WMR amounted to 2.32 Mt, which was 65.2% of the total. The emissions from the production (including the tailings and the slag) and use stages were 0.68 Mt (19.0%) and 0.47 Mt (13.1%), respectively, and the lowest emissions were released by F&M (2.8%). The emissions in 2010 were used as a case study in the emissions rate over time analysis.

Table 14.4 Global environmental emissions from the anthropogenic lead flow in 2010

Emissions	Symbol	Lead flow /kt	percentage of total loss/%
Tailings	T	505	14.2
Slag	S_a	170	4.8
F&M emissions	E_M	99	2.8
Use emissions	E_U	466	13.1
WMR emissions	E_W	2319	65.2
Total		3559	100

2. Temporal evolution of lead emissions between 1930 and 2010

According to equation (14.6), the total lead emissions from 1930 to 2010 were calculated to be 173.8 Mt, of which the emissions from WMR amounted to 84.9 Mt (48% of the total), the emissions from the use stage amounted to 33.95 Mt (20% of the total) and the emissions from F&M amounted to 10.6 Mt (6% of the total).

The total emission rate in 1930 was approximately 0.83 Mt/year, which increased rapidly after the 1960s before leveling out at 3–4 Mt/year between 1970 and 2010 (Fig. 14.3). After the 1980s, the total emissions declined. The phase out of petrol additives and reduced non-battery uses, such as solder and pipes (Schwarz et al., 2012; Genaidy et al., 2008), may be responsible for the decrease. Since 1980, the use emissions, F&M emissions and slag decreased, while the tailings and WMR emissions increased.

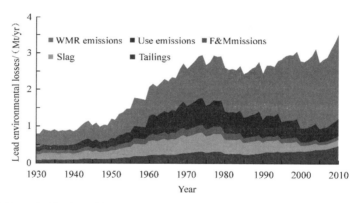

Fig. 14.3 Distribution of lead environmental emissions based on the life cycle stages between 1930 and 2010

In addition, the proportion of specific emissions flows to the total emissions changed during this period. The WMR emissions increased the most, about 6.2 times 1930 to 2010. The F&M emissions proportion dropped most from 9.6% in 1930 to 2.8% in 2010. The proportion of the production emissions (including the slag and tailings together) also dropped, from 32% in 1930 to only 19% in 2010.

3. Production-emissions relationship and pollution assessment

Fig. 14.4 shows the total lead production from 1930 to 2010 and the corresponding emissions to the environment. It can be seen that their trends are not exactly the same.

We can deduce that the emissions for one unit of production decreased continuously over the 70 years period studied in this investigation. In 1930, as much as 0.88 kg was emitted for every 1 kg of refined lead, whereas in 2010, 0.37 kg was emitted. In particular, after 1970, the emissions per unit production remained almost constant despite the overall increase in production. The fact that lead emissions did not rise in proportion to the increased amount of lead produced indicates that improvements in technology have occurred.

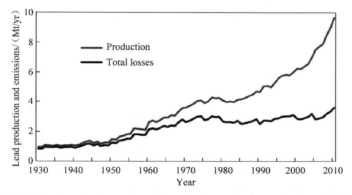

Fig.14.4 Total lead production and emissions to the environment between 1930 and 2010

14.3.2 Species Present in the Global Lead Emissions

1. Lead emission species in 2010

Lead speciation for 2010 is shown in Table 14.5. The main contributing lead species are different for each stage of the life cycle. The primary species emitted by the production stage are PbS, $PbCO_3$ and PbO, whereas the emissions are mainly PbO and Pb during the F&M stage. At the use stage, Pb and tetraethyl lead are typically emitted and $PbSO_4$, PbO and $PbCO_3$ are the major species produced during the WMR stage (Table 14.5). During use, 63% of the emissions are unspecified. These dissipative emissions may include many unspecified species, such as $PbCrO_4$, $2PbCO_3 \cdot Pb(OH)_2$, and Pb_3O_4 in pigments and organic lead compounds in additives. This proportion can be explained from the lead consumption patterns (occupying more than 10% of the total in 2010, see in Fig. 14.2) and the high emission factors (see in Table 14.3.)

By summing up the same species from all four stages, we can predict the lead speciation across the whole anthropogenic cycle. In 2010, $PbSO_4$ emissions were

1.51 Mt, which is 42.5% of the total. The other two main species emitted were PbO_2 (0.58 Mt, 16.2%) and PbS (0.30 Mt, 8.3%). These three species account for approximately 67.1% of the total.

Table 14.5 Lead speciation during different stages of the lead life cycle for 2010

Life cycle stage	Proportion of different lead species/%									
	$PbSO_4$	PbO	Pb	PbO_2	$PbCO_3$	PbS	$PbCl_2$	TEL[b]	Others	Total
Production	16.3	8.6	0.2	0.3	25.9	43.9	—[a]	—[a]	5.0	100
F&M	8.2	51.6	34.0	—[a]	—[a]	—[a]	—[a]	—[a]	6.2	100
Use	—[a]	—[a]	33.7	—[a]	—[a]	—[a]	—[a]	2.9	63.4	100
WMR	60.2	4.1	2.0	24.8	2.7	—[a]	1.4	—[a]	4.8	100

a Negligible; b TEL, tetraethyl lead.

2. Lead species in the 1960–2010 emissions

The species in the emissions between 1960 and 2010 were analyzed. The results indicated that the $PbSO_4$ emissions were 39.6 Mt, which was 28.5% of the total. The PbO emissions were 15.7 Mt (1.3%) and the Pb and PbS emissions were 15.3 Mt (11.0%) and 14.6 Mt (10.5%), respectively. These four species account for 61.2% of the total. The distribution of species in the total emissions between 1960 and 2010 tends to vary over time and the dataset is very heterogeneous.

The emissions rate for each species, i.e., the emission of a species over a single year, changes over time. The emissions for different species were plotted at 10 year intervals to show the changes in emission rates (Fig. 14.5). Tetraethyl lead and Pb declined, whereas PbO_2, $PbSO_4$ and $PbCO_3$ increased. No detailed estimations are given for $PbCl_2$ because its emissions were very low (they are included in the "Other" category in Table 14.5). The tetraethyl lead emissions in 1960 were approximately ten times higher than in 2010. Despite the fact that lead additives in petrol have been banned (the UK in 2000, the U.S. in 1996 and China in 2001), the ILZSG statistics suggest that lead additives were still used in a few countries, such as Mexico (ILZSG, 2008). PbO_2 emissions increased the most, with the emissions in 2010 75 times those of 1960. This increase was possibly caused by the increase in the use of lead batteries, which drove up the PbO_2 emissions during the secondary lead refining and landfilling processes. The $PbSO_4$ and $PbCO_3$ emissions remained relatively high because of the increased amounts of lead acid battery and tailings.

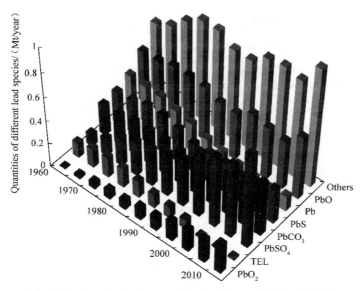

Fig. 14.5 Lead emission species rates between 1960 and 2010

TEL, tetraethyl lead.

14.3.3 Discussion and Uncertainty

1. Discussion

It is often considered surprising that more than 25% of the lead production in 2010 and more than 50% of the accumulated lead production between 1930 and 2010 were lost to emissions. Indeed, the recycling ratio seems insignificant. However, the implicit ratios indicated in this work are the true recycling ratios rather than the ratio of recycled lead only to total lead waste, which is actually the end of life recycling ratio (Mao et al., 2008b).

It is useful to compare the results of this analysis with previous reports. Smith found that nearly 50% of the lead produced over all of human history was discharged to the environment as a contaminant (Smith and Flegal, 1995). Our previous study produced a similar result. We found that for every 1 kg of lead used, 0.5 kg was lost to the environment (Mao et al., 2009). Many global anthropogenic cycles for metals have been established and a comparatively large amount is lost in emissions, such as two-thirds for selenium (Kavlak and Graedel, 2013a) and 80% for tellurium (Kavlak and Graedel, 2013b). A report on silver showed that more than 13 kt was emitted annually to the environment, which amounted to about half of the entire production of silver

(Eckelman and Graedel, 2007).

We have only considered the direct lead flows driven by human demands of lead without considering the emissions from the fossil fuels, other metal ores, and phosphate ores. The emissions from fossil fuels and ores are only part of the emissions from the non-intentional flows, which occupy less than a half of the total emissions (Elshkaki et al., 2009). Therefore, the emissions from fossil fuels and ores may not in comparison to those that are included in this study.

In this study, we mainly discuss lead losses as emissions to the environment, but we do recognize that studies about lead's impacts on the environmental compartments are equally important. In one of our following studies, some work on the interaction between lead emissions and the environment is attempted.

2. Uncertainty

Uncertainty in the estimations resulted from two main sources of error: ①missing data for some years; the statistics for 1960-2007 were from ILZGS, whereas the data for 1930-1960 were extrapolated from USGS data and the data for 2007-2010 were rectified by subtracting the average difference between the ILZGS and ILA data. No data were available for some years, such as lead production between 1940 and 1944. In this case, we extrapolated the data from other years. ②Parameter biases; The emission factors for petrol were based on the estimations for China (Li et al., 2012), which are higher than average global emissions. The emission factors for cable sheathing were estimated using data for developed countries, which were lower than for China. Furthermore, factors for alloys and others are unknown and were deduced by comparison with the pigments, ammunition or solder data set.

The precision of the lead speciation results for the different stages of the life cycle was low. Most estimations were empirical, such as estimations for the production stage, the $PbCO_3$ rate and $PbSO_4$ rate in the tailings, the PbS and PbO proportions in primary lead refining and the species rates in secondary lead refining. Moreover, the species rates in secondary lead refining were estimated from the amount of lead acid battery waste, which is the main source of secondary refining emissions (Wang and Li, 2003). In F&M, PbO and Pb were both assessed as being approximately 45% during the sheet/pipe and cable sheathing manufacturing processes and this was based on just a basic knowledge of the industrial processes involved. During use, various species, such as Pb_3O_4 and $PbCrO_4$, exist in pigments and additives, but only the primary species, such as Pb and tetraethyl lead, were considered. Recycling rate differences at the WMR stage were also not considered. The recycling rate for batteries was the highest, so the emissions from the species contained in batteries, such as $PbSO_4$ and PbO_2, should be higher, while the other rates may be lower than the true values.

14.4　Conclusions

Total lead emissions in 2010 were 3.56 Mt, of which 65% was from the waste management and recycling stage. The main species emitted from each life cycle stage were very different. The primary species emitted during production were PbS, $PbCO_3$ and PbO, while the emissions from the F&M stage were mainly PbO and Pb. Lead and tetraethyl lead were typically emitted during the use stage and $PbSO_4$, PbO and $PbCO_3$ were the major species emitted during the waste management and recycling stage.

The total lead emissions from 1930-2010 amounted to 173.8 Mt, which were primarily from the waste management and recycling (48%), production (26%) and use (20%) stages. The main species emitted were $PbSO_4$, PbO, Pb and PbS, which accounted for 61.2% of the total emissions. Between 1960 and 2010, tetraethyl lead and Pb emissions declined, whereas the species in the others category, such as PbO_2, $PbSO_4$ and $PbCO_3$ increased, in particular PbO_2, which increased the most.

Acknowledgment

This work was supported by the National Natural Science Foundation of China (General Program) under grant No. 41171361, entitled *The Anthropogenic Transfer and Transformation of Lead and the Quantitative Analysis on Environmental Accumulated Lead Emissions*.

Conflict of Interest

All of the authors declare that they have no conflicts of interest. The authors confirm that this is an original submission which has not been published previously or submitted to any other journal.

References

Alberto S, Marcello V, Luca R, et al. 2012. Combination of beehive matrices analysis and ant biodiversity to study heavy metal pollution impact in a post-mining area (Sardinia, Italy). Environmental Science & Pollution Research,

19（9）：3977-3988.

Chen J M，Tong Y P，Xu J Z，et al. 2012. Environmental lead pollution threatens the children living in the Pearl River Delta region，China. Environmental Science & Pollution Research，19（8）：3268-3275.

Chen W Q，Graedel T E. 2012. Anthropogenic cycles of the elements：A Critical review. Environmental Science & Technology，46（16）：8574-8586.

CNIA（China Nonferrous Metals Industry Association）. 2009. China Nonferrous Metals Industry 60 Years. Changsha：Central South University Press.（in Chinese）

Dahodwalla H，Heart S. 2000，Cleaner production options for lead-acid battery manufacturing industry. Journal of Cleaner Production，8（2）：133-142.

DiFrancesco C A，Smith G R，Gabby P N，et al. 2012. Lead statistics（1900-2011），http：//minerals.usgs.gov/ds/2005/140/ds140-lead.pdf.（Accessed 8 September 2013）

Eckelman M J，Graedel T E. 2007. Silver Emissions and their Environmental Impacts： A Multilevel Assessment. Environmental Science & Technology，41（17）：6283-6289.

Editorial Staff of the Yearbook of the Nonferrous Metals Industry. 1990-2010. China Nonferrous Metals Industry Yearbook（1990-2010）. Beijing：China Nonferrous Metals Industry Yearbook Press.（in Chinese）

Elshkaki A，Van der Voet E，Van Holderbeke M，et al. 2004. The environmental and economic consequences of the developments of lead stocks in the Dutch economic system. Resources Conservation &Recycling，42：133-154.

Genaidy A M，Sequeira R，Tolaymat T，et al. 2008. An exploratory study of lead recovery in lead-acid battery lifecycle in US market：An evidence-based approach. Science of the Total Environment，407（1）：7-22.

Glöser S，Soulier M，Tercero Espinoza L A. 2013. Dynamic analysis of global copper flows. Global stocks，postconsumer material flows，recycling indicators，and uncertainty evaluation. Environmental Science & Technology，47：6564-6572.

Graedel T E，Cao J. 2010. Metal spectra as indicators of development. Proceedings of the National Academy of Sciences of the United States of America，107（49）：20905-20910.

Guo X Y，Zhong J Y，Song Y. 2009. Substance flow analysis of lead in china. Journal of Beijing University of Technology，35（11）：1554-1561.（in Chinese）

ILA（International Lead Association）. 2014. Lead Statistics World Totals，http：//www.ila-lead.org/lead-facts/statistics.（Accessed 1 January 2014）

International Lead and Zinc Study Group（ILZSG）. 1992. Principal uses of lead and zinc 1960-1990，London.

International Lead and Zinc Study Group（ILZSG）. 2005. Lead and zinc： End use industry statistical supplement，1994-2003，London.

International Lead and Zinc Study Group（ILZSG）. 2005. Principal Uses of Lead and Zinc 2005，London.

International Lead and Zinc Study Group（ILZSG）. 2008. Principal Uses of Lead and Zinc 2005，London.

Jolly R，Rhin C. 1994. The recycling of lead-acid batteries：production of lead and polypropylene. Resources Conservation & Recycling，10（1-2）：137-143.

Kavlak G, Graedel T E. 2013a. Global anthropogenic selenium cycles for 1940-2010. Resources Conservation & Recycling, 73: 17-22.

Kavlak G, Graedel T E. 2013b. Global anthropogenic tellurium cycles for 1940-2010. Resources Conservation & Recycling, 76: 21-26.

Li Q, Cheng H G, Zhou T, et al. 2012. The estimated atmospheric lead emissions in China, 1990-2009. Atmospheric Environment, 60: 1-8.

Li Z M, Tan X B, Wu X Z, et al. 2009. The Influence of mixing process on leady paste properties. Battery, 46 (1): 18-20. (in Chinese)

Liang J, Mao J S. 2014. A dynamic analysis of environmental losses from anthropogenic lead flow and their accumulation in China. Transactions of Nonferrous Metals Society of China, 24 (4): 1125-1133.

Liang J, Mao J S. 2014. Speciation analysis of lead losses from anthropogenic flow in China. Environmental Science, 35 (3): 1191-1197. (in Chinese)

Manceau A, Boisset M, Sarret G, et al. 1996. Direct determination of lead speciation in contaminated soils by EXAFS spectroscopy. Environmental Science & Technology, 30 (5): 1540-1552.

Mao J S, Cao J, Graedel T E. 2009. Losses to the environment from the multilevel cycle of anthropogenic lead. Environmental Pollution, 157: 2670-2677.

Mao J S, Dong J, Graedel T E. 2008a. The multilevel cycle of anthropogenic lead: I. Methodology. Resources Conservation & Recycling, 52 (8-9): 1058-1064.

Mao J S, Dong J, Graedel T E. 2008b. The multilevel cycle of anthropogenic lead: II. Results and discussion. Resources Conservation & Recycling, 52 (8-9): 1050-1057.

Mao J S, Dong J, Graedel T E. 2009. Lead in-use stock. Journal of Industrial Ecology, 13 (1): 112-126.

Mao J S, Ma L, Liang J. 2014. Changes in the functions, forms and locations of lead during its anthropogenic flows to provide services. Transactions of Nonferrous Metals Society of China, 24 (4): 233-242.

Nriagu J Q, Pacyna J M. 1988. Quantitative assessment of worldwide contamination of air, water and soils by trace metals. Nature, 333: 134-139.

Ordoñez I, Rahe U. 2013. Collaboration between design and waste management: Can it help close the material loop? Resources Conservation & Recycling, 72: 108-117.

Ostergren J D, Brown Jr G E, Parks G A, et al. 1999. Quantitative speciation of lead in selected mine tailings from Leadville, CO. Environ Science & Technology, 33 (10): 1627-1636.

Pacyna J M, Scholtz M T, Li Y F. 2011. Global budget of trace metal sources. Environmental Reviews, 3 (2): 145-159.

Rauch J N, Pacyna J M. 2009. Earth's global Ag, Al, Cr, Cu, Fe, Ni, Pb, and Zn cycles. Global Biogeochemical Cycles, 23: 1-16.

Sakata K, Sakaguchi A, Tanimizu M, et al. 2014. Identification of sources of lead in the atmosphere by chemical speciation using X-ray absorption near-edge structure (XANES) spectroscopy. Journal of Environmental Sciences, 26 (2): 343-352.

Schwarz K, Pickett S, Lathrop R G, et al. 2012. The effects of the urban built environment on the spatial distribution of lead in residential soils. Environmental Pollution, 163: 32-39.

Sen I S, Bernhard P E. 2012. Anthropogenic disturbance of element cycles at the Earth's surface. Environmental Science & Technology, 46 (16): 8601-8609.

Shah P, Strezov V, Nelson F. 2009. X-Ray absorption near edge structure spectrometry study of nickel and lead speciation in coals and coal combustion products. Energy & Fuels, 23 (3): 1518-1525.

Shi D Z, Wu W X, Lu S Y, et al. 2008. Effect of MSW source-classified collection on the emission of PCDDs/Fs and heavy metals from incineration in China. Journal of Hazardous Materials, 153 (1-2): 685-694.

Smith D R, Flegal A R. 1995. Lead in the Biosphere: Recent Trends. Ambio, 24 (1): 21-23.

Smith G R. 2004. Lead recycling in the United States in 1998, http: //infohouse.p2ric.org/ref/45/44143.pdf. (Accessed 20 August 2013)

Tukker A, Buist H, van Oers L, et al. 2006. Risks to health and environment of the use of lead in products in the EU. Resources Conservation & Recycling, 49: 89-109.

Wang J M, Li R R. 2003. The refining technology at home and abroad. Nonferrous Resource Recycling, 3: 11-14. (in Chinese)

Woodward A. 2006. How products are made. http: //www.madehow.com/Volume-2/Lead.html#b#ixzz2pa2pAl6o. (Accessed 8 July 2014)

Zhang X W, Yang L S, Li Y H, et al. 2011. Estimation of lead and zinc emissions from mineral exploitation based on characteristics of lead/zinc deposits in China. Transactions of Nonferrous Metals Society of China, 21 (11): 2513-2519.

Zhu X F, Yang J K, Gao L X, et al. 2013. Preparation of lead carbonate from spent lead paste via chemical conversion. Hydrometallurgy, 134-135: 47-53.

Zimmermann T, Gößling-Reisemann S. 2013. Critical materials and dissipative losses: A screening study. Science of the Total Environment, 461-462: 774-780.

第 15 章 多尺度铅元素人为循环的环境排放
Losses to the Environment from the Multilevel Cycle of Anthropogenic Lead*

15.1 Introduction

Human activities dominate the global flows of lead (Klee and Graedel, 2004), a metal whose loss from the anthropogenic cycle is of particular concern because of its toxicity (Tukker et al., 2006). This property has stimulated considerable research on anthropogenic lead loss, which is known to have occurred for some three millennia (Shotyk et al., 1998). Anthropogenic lead has been distributed all over the planet (Rankin et al., 2005) and found even in Antarctic snow (Rosman et al., 1994). An archetypical study, for Sweden between 1880 and 1980, discusses lead emission from metal production, fabrication and manufacturing, dissipation from in-use stock, and loss following discard (Lohm et al., 1994). In addition to evaluating and capturing lead in these loss streams in order to minimize hazard, lead discards may in the future be viewed as critical resources. Indeed, lead today is recycled at rates greater than those for any other commonly-used metal (Henstock, 1996). (The United States, because of its emphasis on easily recovered vehicle batteries, produced 77% of its refined lead from recycling in 2000 (U.S. Geological Survey, 2000).

These issues of loss, hazard, and reuse cannot be addressed in a coherent manner unless the sources of loss are identified and their magnitudes estimated. To this end, we have constructed a model of the anthropogenic lead cycle (Mao et al., 2008a), and characterized that cycle for 52 countries for year 2000 (representing about 97%, 97%, and 90% of the total mine, lead production, and lead use in the world), eight world regions, and the planet as a whole (Mao et al., 2008a; 2008b). This model explicitly addresses seven loss streams: tailings, slag, fabrication and manufacturing, dissipation

---

* Mao J S, Cao J, Graedel T E. 2009. Losses to the environment from the multilevel cycle of anthropogenic lead. Environmental Pollution, (157): 2670-2677.

from use, hibernation, landfilling, and dispersion following product discard. We present herein our results for lead loss, compare countries and regions, and discuss the implications of our findings.

15.2　Losses of Lead and Potential Environmental Risk

The comprehensive model we employ for study of the anthropogenic lead cycle is shown in Fig. 15.1. It treats four principal life stages, Production, Fabrication and Manufacturing, Use, and Waste Management & Recycling (WMR), connected by the flows of lead among them. The lead stocks at each life stage are enhanced by imports and diminished by exports and by losses to the environment. The countries are grouped into larger regions based on their geographical locations (Mao et al., 2008a).

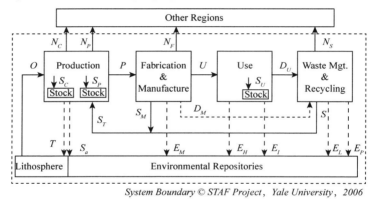

Fig. 15.1　A simplified schematic diagram of a comprehensive lead cycle, with successive life stages plotted from left to right

Symbols and definition

O	Lead from ore	S_a	Lead in slag
P	Refined lead	S_P	Lead entering stock of refined lead
U	Lead entering use	S_C	Lead entering concentrate stock
D_U	End-of-life lead to waste management & recycling	S_U	Lead entering in-use stock
D_M	Lead in F&M discards to WMR	S_T	Lead in total scrap to production
N	Lead in net export (export minus import)	S_M	Lead in new scrap from Manufacture
N_P	Lead in net export of refined lead	E	Lead in emissions to the environment
N_C	Lead in net export of concentrate or mine	E_M	Lead in F&M emissions
N_S	Lead in net export of scrap	E_I	Lead in dissipation from Use
N_F	Lead in net export of semi- or finished products	E_H	Lead to hibernation stock
T	Lead in tailings	E_P	Lead dispersed from WMR
S	Lead in scrap from WMR	E_L	Lead to landfill

In general, the lead flows at the regional level can be obtained by aggregating the lead flows in countries within the specific region. However, it is generally not possible to obtain complete information for every country within a region, especially for the less developed countries where few data are reported. In such cases, we use the reported data from other countries within the same region to estimate the missing data. For instance, data for lead use was reported for 30 of our 52 countries by the International Lead-Zinc Study Group (ILZSG, 2005); we generally use the average data from data-available countries to represent countries in the same region for which data were not available.

15.2.1 Lead Loss in Tailings

Tailings contain the lead extracted from ore but discarded with waste rock following milling and concentrating (Ostergren et al., 1999). The main lead-bearing constituents of this flow are fine particles of the lead minerals galena (lead sulfide), cerusite (lead carbonate), and anglesite (lead sulfate). Tailings have low potential for reuse and are usually discharged to holding ponds or directly to the environment. We estimated the amount of lead in tailings as the reported lead in concentrate (ILZSG, 2007) multiplied by one minus the recovery rate in milling and concentrating; the latter is estimated to vary within a range of 0.75 to 0.91 based on the technology employed (China Nonferrous Metals Industry Yearbook, 2001, 2002; Beijing General Research Institute of Mining and Metallurgy, 2000; Raw Materials Group, 2005).

15.2.2 Lead Loss in Slag

Lead emissions resulting from the smelting and refining of lead concentrate, or refining of lead scrap (secondary refining), are contained in slag, the waste byproduct of these purifying operations (Condor et al., 2001). The amount of lead in slag can be estimated as the reported amount of refined lead (ILZSG, 2007) divided by the recovery rate, where the refined lead includes both primary and secondary lead (11, 16). The recovery rate at primary and secondary refineries is estimated to range from 0.88 to 0.99 and 0.86 to 0.985 respectively (12, 17-20). Slag has little industrial use; it is usually retained on the refinery property or discharged into the environment.

15.2.3 Lead Losses from Fabrication & Manufacture

The fabrication & manufacture (F & M) life stage transforms refined lead into semi-finished products such as additives for paint, glasses, ceramics, PVC, and petrol (Thornton et al., 2001), and finished products such as lead-acid batteries, lead pipe and sheet, cable sheathing, and several lead alloys. Most lead that enters F & M but is not retained in products is either recycled as prompt scrap, discarded, or emitted. Emissions to the environment from F & M processes have historically been significant (Pacyna et al., 1995; Pechnikov et al., 1996; Karlsson, 1999).

A mass balance around the F & M life stage may be expressed as

$$P - U - N_F = S_M + D_M + E_M \quad (15.1)$$

where the symbols are defined in Fig. 15.3 of (Mao et al., 2008a). P, the rate of lead ingested into the F & M stages, was reported by ILZSG (2007) for countries. The sum of U (the lead entering Use) and N_F (the lead in net export of semi- or finished products) constitutes the lead entering semi- or finished products; it is estimated by multiplying P by the fabrication efficiency (i.e., the fraction of material entering fabrication that is embodied in the final products; Jiang, 2000; Mao et al., 2008a). The lead that is not retained as part of semi- or finished products becomes prompt scrap, discards to waste management, or emissions. The division among these three flows depends largely on the level and policies of environmental management; we estimated the branching ratios as 0.5, 0.3 and 0.2 of the total lead not entering finished products based on previous studies of lead flow analyses in China (Mao et al., 2006; 2007). For countries with F & M capacity but without secondary production capacity, lead entering F & M but not leaving F & M in products is treated entirely as emissions.

15.2.4　Lead Losses from the In-Use Stage

Most of the lead in the lead cycle at any one time is present in the Use phase, where it provides the services desired by individuals and corporations. Eventually most of the lead is discarded to WMR. However, some products such as cable sheathing are abandoned in place because it is often used in places where it is difficult or expensive to collect[e.g., the sea bed (Thornton et al., 2001)]. We refer to this loss as "comatose stock", and estimate its quantity as discussed in (Mao et al., 2008a). (Such material has sometimes been termed "hibernating stock", but that biological term refers to organisms expected to awake at a defined point in time. Because it is unlikely that lead cable sheathing will ever "awaken", "comatose" is the more appropriate term.)

A small but important portion of the lead in service is lost to the environment through corrosion [of lead pipes and roofing lead (Leygraf and Graedel, 2000)], dissipation[of lead in ammunition (Darling and Thomas, 2003; Giles, 2004)], of lead fishing weights (Scheuhammar and Norris, 1996), or of lead tire weights (U.S. Environmental Protection Agency, 2005), or by combustion of lead-containing motor fuels (Thomas, 1995; von Storch et al., 2002; Soto-Jimenez et al., 2006). Much of the dissipated lead is lost to soil, but water bodies (Ritson et al., 1994) and (eventually) sediments (Jonsson et al., 2002) also receive in-use emissions.

15.2.5 Lead Losses from the Waste Management & Recycling Life Stage

Lead discards from upstream life stages F&M and Use flow to waste management & recycling, where sorting and processing for reuse occurs. This "old scrap" may be sent to production facilities either in the country where the discard occurred, or exported for processing elsewhere. These reuse flows are generally well-characterized. Discarded lead that is not reused may be landfilled (incinerator ash, cathode ray tubes, consumer electronics, etc.) or "dispersed" [lost during transport, incinerated without landfilling (Walsh et al., 2001), incinerator ash used in pavement (Thornton et al., 2001), etc.]. Small (but widely publicized) amounts are lost in electronics discards or low-technology recycling. On the basis of discussions with the recycling industry, we estimate that the landfilled lead is 70% of the lead not reused at WMR; the remainder we treat as dispersed emissions.

15.2.6 Lead Losses Not Treated in This Study

We do not include in this work the anthropogenic lead emissions that arise from fossil fuel combustion in power plants, or in cement production. Pacyna et al. (1995) estimate lead emission from these processes to be 110–140 Gg Pb/year at the global level. As will be seen, this constitutes less than 5% of the total emissions that we estimate from all other lead losses.

15.2.7 Evaluation of Environmental Hazard of Lead Emissions

For each geographical region in our study, we rank the lead emissions in two different ways: by the relative magnitude (including that at the per capita level), and by the relative magnitude adjusted for potential toxicity. The former is calculated by adding the emissions throughout the life cycle, the latter by multiplying the lead emissions by the estimated total hazard score for the form or forms of lead in the emissions. The most suitable hazard score ranking was determined to be that given by the Indiana Relative Chemical Hazard(IRCH)ranking system(Green Media Toolshed, 2005), which is designed to indicate (on a scale of 0-100) how one chemical compares with others in terms of its capacity to impact human health, ecosystems, or environmental health. Additionally, the IRCH score takes into account the behavior of the compound in air, water, and land, and so is a mixed-media score. Compared with other toxic materials, lead compounds are of significant concern. It is not straightforward to characterize the physical and chemical forms of lead that are emitted in several of the life-cycle stages under the IRCH methodology, but the results of our efforts to do so are indicated in Table 15.1.

Table 15.1　Potential hazard rankings for lead emissions

Emission stream		IRCH*	Exposure	Mobility	Chemical	
Typical Potential					Hazard	Hazard
Tailings	PbS	50-75	Low	Low	Low	20
Slag	Pb, PbS	75-90	Low	Low	Medium	40
F&M emissions	Pb compounds	50-100	High	High	High	90
Comatose stock	Pb	75-90	Medium	Low	Medium	30
Dissipated from use	Pb, Pb$(C_2H_5)_4$	75-100	High	High	High	100
Landfilled	Lead wastes	75-90	Low	Low	Medium	30
Dispersed	Lead wastes	75-90	Medium	Medium	Medium	70

* Indiana Relative Chemical Hazard (Green Media Toolshed, 2005).

This approach to quantifying the risks from lead loss represents an attempt to take variable toxicities into account, and is clearly an approximation. Nonetheless, it recognizes that some forms of emission (e.g., gaseous lead compounds rather than metallic lead) are of more concern than others, and provides at least modest insight into the implications of that variability. The results should not be misconstrued as applicable to any particular facility, technology, or geographical location, or to have a high degree of rigor.

15.3　Results

15.3.1　Lead Loss at the Global Level

The combined losses of lead to the global environment in 2000 that are derived from our characterization of the cycle and its losses total about 3000 Gg Pb/year. When this result is compared with the global flow to use of about 6300 Gg Pb/year (Mao et al., 2008b), we see that for every kilogram of lead entering use during that year, about 0.48 kg (net) is lost to the environment. Not only do these flows carry with them the potential for adverse environmental impact, but the lead that is lost must be replaced by virgin material if the level of services that lead provides is to be maintained.

Table 15.2 shows the amounts of lead emission in the seven loss streams that we identify. The highest rate is landfilling (1115 Gg Pb/a, or 34% of the total), followed by dissipation from use (670 Gg Pb/a, or 21% of the total). The other streams are smaller, ranging from 16% of the total (tailings) to 2% (F & M emissions).

Table 15.2 Global lead losses and relative hazards in 2000 in the seven emission streams

Emission stream	Symbol	Global flow/ (Gg Pb/a)	Percentage of total/% Loss	Percentage of total/% Hazard
Tailings	T	530	16.4	6.5
Slag	S_a	230	7.1	5.5
F & M emissions	E_M	70	2.2	3.9
Dissipated from use	E_I	670	20.7	40.8
Comatose stock	E_H	145	4.5	2.7
Dispersed after discard	E_P	480	14.8	20.3
Landfilled	E_L	1115	34.4	20.3
Total		3240	100	100

It is of interest to consider the environmental media into which those flows occur, as best we can. A paucity of data prevents us from being highly accurate in this regard, but a rough estimate of the division of the flows among the media is shown in Fig. 15.2. There are five media repositories for the emitted lead: contained slurry on land, contained solids on land, water, air, and uncontained solids on land. In the case of five of the emission streams, the allocation to one of the media is definitive. The dissipative flow from Use (E_I) is divided into air emissions (largely from lead additives to motor vehicle fuel), emissions to water (corrosion of exposed metal products), and emissions to land (in ammunition and a variety of uses in lead compound additives). We estimate the fractions of the dissipative flow into these three substreams as 5%, 5%, and 90%, respectively (Nriagu and Pacyna, 1988; Tukker et al., 2006). The division of emissions from F&M (E_M) is also difficult to specify in the absence of detailed data, but on the basis of information in the literature (Nriagu and Pacyna, 1998), we divide the F&M emission stream as 86% to contained solids on land, 7% to water, and 7% to air.

Improvements in pollution control devices in the last decade or so have probably reduced the latter two fractions, although little supporting information is available. Nonetheless, the two flows as computed constitute only about 2% of total lead emissions, so are of only modest significance.

Fig. 15.2 demonstrates that the largest of the emissions to a specific environmental medium is to contained solids on land at about 1550 Gg Pb/a. followed by uncontained solids on land and contained slurry on land at about 1080 Gg Pb/year and 530 Gg Pb/year, respectively. Flows to air and water are much less (each is only about 1% of the overall total), though they could be important from the standpoint of environmental hazard. At the global level, it turns out that about 2/3 of the lead discards are recycled. Nonetheless, for every kilogram of lead entering use about 0.48 kg of lead on a net basis is lost to the various environmental repositories.

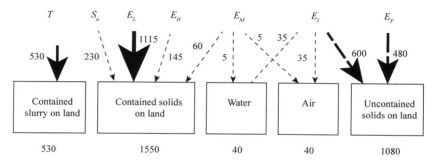

Fig. 15.2 Estimated global rates of loss of lead (Gg Pb/a) to different types of environmental media in 2000

The totals of all inputs to each medium is given below each box.

In the Supplemental Material, we calculate recycling and emission ratios at the different spatial levels, which reveal differences in cycle efficiencies.

15.3.2 Lead Loss at the Regional Level

The eight world regions are of very different importances for the lead cycle, as was shown in the regional cycles in Mao et al. (2008b). The importance of the emissions is quite different as well. In Fig. 15.3 we show the amount of lead recycled and emitted in each of the regions, and compare it to the amount entering the Use phase. Asia is seen to lead all regions in emissions, and North America in lead use and recycling. Europe as a region is quite important as well. In comparison, all of the flows for the other five regions are very small.

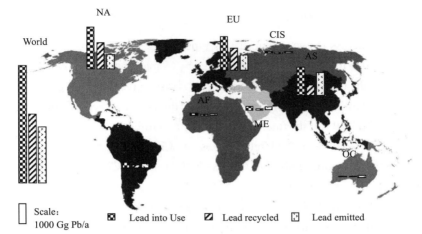

Fig. 15.3 Rates of flow of lead into use, recycled, and emitted within eight world regions

The regional codes are AF=Africa, AS=Asia, CIS=Commonwealth of Independent States, EU=Europe, LAC=Latin America and the Caribbean, ME=Middle East, NA=North America, OC=Oceania.

The relative magnitudes of the regional lead losses at the different life stages are shown in Fig. 15.4. We see readily that the most lead losses occur at the Use and WMR stages for the three largest emissions regions, i.e., Asia, Europe and North America. The largest losses from WMR occur in North America, and the largest losses from Processing and Use in Asia. In the Supplementary Material we perform a similar analysis for a selection of countries from different regions.

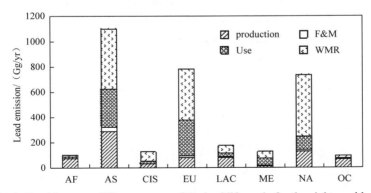

Fig. 15.4 Lead losses at different stages of the lead life cycle for the eight world regions

15.3.3 Comprehensive Lead Emission Patterns for Countries and Regions

Lead emission patterns for the entire set of countries and regions are shown in

Fig. 15.5(a). This diagram provides a semi-quantitative picture of the countries and regions most important for specific emission streams, as well as a picture of the relative emissions for each country across the spectrum of loss types. On a country level, the largest lead emission occurred in the United States, followed by China; these two countries contribute 17% and 15% of the total emissions of the world, respectively. On a regional level, the largest lead emissions occurred in Asia (34% of the total), followed by Europe and North America (24% and 23% of the total, respectively). Different countries are important to the different flows, although the United States and China dominate the picture. Some of the features of the individual emission streams are: ①Tailings–China dominates, Australia, Peru, the United States are important. ②Slag–China dominates, and the United States and Japan are important. ③Fabrication & Manufacturing Emissions–the United States is largest, then China. ④Comatose stock–Belgium-Luxembourg dominates, followed by China and United Arab Emirates. ⑤Dissipation from Use–China is largest. The United States is significant. ⑥Landfill–The United States is the largest. China, France, the United Kingdom and Mexico are significant. ⑦Dispersal from WMR–The United States is the largest, followed by China.

On a total basis, the lead emissions ranking for the United States is the highest, followed by China, Japan, France, the United Kingdom and Mexico.

The regional emissions, evaluated on the same basis, are shown at the right side of the diagram. Here it is seen that Asia is highest for most of the loss streams, while North America is the highest region in landfill and dispersal emissions. Europe is the highest region in hibernate as a consequence of its extensive use of lead sheeting and lead roofing. For total lead losses to the environmental repositories, Asia is highest, Europe next, and then North America.

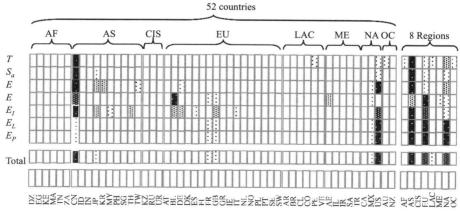

(a)

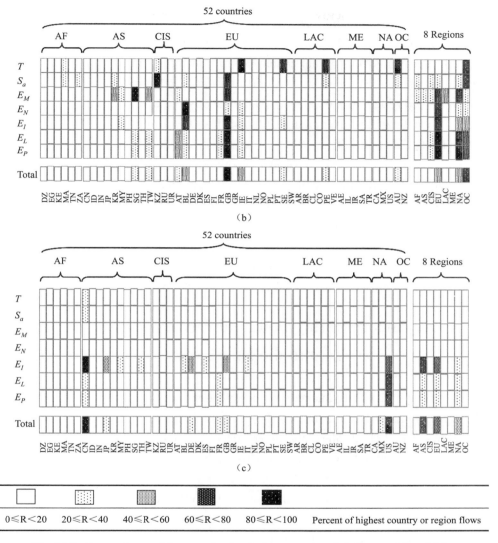

Fig. 15.5 Comparisons of the seven lead emission streams and their potential toxicities for the 52 countries and eight world regions in this study

(a) country-level emission streams black-degree-coded as the percentage of the emission in that stream in the highest magnitude country; (b) as for (a), but on a per-capita basis; (c) country-level potential toxicity streams, constructed by multiplying the absolute emission by the relative ranking of Table 17.1, and then black-degree-coding as the percentage of the highest potential toxicity for any of the emission streams in any country.

If we recast Fig. 15.5 (a) on the basis of emissions per capita, a different picture emerges[Fig. 15.5 (b)]. Here countries with large populations tend to rank lower, while those with modest populations but vigorous industrial activity are somewhat higher. So far as the individual emission streams are concerned, ①Tailings – Australia is highest, followed by Ireland, Sweden and Peru. ②Slag – Kazakstan is highest by far; followed by the United Kingdom. ③Fabrication & Manu-facturing Emissions – Singapore is largest, then the United Kingdom. ④Comatose stock – Belgium-Luxembourg is highest, then the United Kingdom and Ireland. ⑤Dissipation from Use – the United Kingdom is largest, then Belgium-Luxembourg. ⑥Landfill – The United Kingdom is largest, Austria next. ⑦Dispersal from WMR –The United Kingdom is largest, Austria next.

On a total basis, the per capita lead loss rankings have the United Kingdom highest, then Ireland and Belgium-. Several countries, including Singapore, Taiwan, Austria, and Sweden, are 20%–40% of the United Kingdom per capital level.

The regional per capita losses are shown at the right side of Fig. 15.5（b）. On this basis, different regions are most prominent for different loss steams: Oceania for tailings and slag, Europe for emissions from hibernation and dissipated losses, North America for other stages. Overall, Oceania is highest, Europe next, then North America.

15.4　Incorporating Potential Environmental Hazard into the Life Cycle

In section 15.2 of this chapter, we described the method developed to scale rates of lead loss by the risk of the forms of lead that they contain. If we apply this approach to the flows of Table 15.2, the global potential hazards of the flows are seen not to follow the absolute flow rates. In particular, dissipation from use（about 21% of the total on a mass basis, largely as "hibernating" cable sheathing）is responsible for about 41% of the global potential toxicity of lead emissions, while WMR emissions, including landfill （34% of the total on a mass basis）and dispersed losses（15% of the total on a mass basis）, are together responsible for around 41%. No other loss stream constitutes as much as 10% of the potential toxicity, by this characterization.

The risk methodology can also be applied at country and regional levels. The result is shown graphically on Fig. 15.5（c）. In this figure, the highest risk for any of the country-level or regional-level emissions is set to 100 and all others are scaled to that value. When this is done, it is seen that the only loss stream of significance is losses from Use – gasoline additives, ammunition, tire balancing weights, fishing weights, and other dissipative applications. China's rate of loss in Use is the highest of any country, and the United States next at 67% of China's rate, followed by Japan, Germany, and the United Kingdom. China is also the highest in total losses, while several other countries, including Japan, and France, are at 20%-40% of China's total loss value.

On a regional basis, Asia has the highest risk resulting from lead losses from products in use, then Europe. Overall, Europe and North America are at 74% and 59% of Asia value on our potential toxicity basis.

15.5 Discussion

To our knowledge, the results in this chapter constitute the first comprehensive study of emissions from the lead cycles of countries, regions, and the planet. We divided the lead emission into seven groups: tailings, slag, F&M emissions, dissipation from Use, hibernation, landfill, and dispersal from WMR, according to the life stage from which lead is emitted to the environment. In addition, we estimated the potential toxicity of the lead emissions for 52 countries and eight regions.

The comprehensive approach adopted to study the multi-level emissions of lead has demonstrated remarkable diversity and complexity in those flows. The lead losses are distributed across several environmental repositories, and losses of lead occur at each major life stage, the largest flows of which are from WMR. However, the highest potential toxicity occurs at the Use stage.

About 1/3 of all lead losses are uncontained solids on land (dispersal and dissipation from Use as lead additives), while 47% of the losses are to containment facilities on land.

With global lead flow into Use at about 6000 Gg Pb/year and losses of around 3000 Gg Pb/a, many opportunities exist for enhanced recovery and recycling efforts. In this vein, country and regional strategies to address lead loss should incorporate both risk and mass flow information. This generates different results, depending on the different life stage lead flows. For example, China could well focus on the early life stages, Australia on Mining, European countries on the Use stage, and the United States on the F&M and WMR stages.

We emphasize that the results of this research are indicative, not definitive. In many cases, data are not available on a country basis for the loss streams, and our calculations utilize regional or global average parameters such as processing efficiencies or rates of loss. Partly because somewhat more data were available on a multiple-country basis than on a national basis, and partly because the regional results benefit by summation, we regard the regional results as somewhat more reliable than some of the individual country parameters. Therefore, this work is properly seen as a guide to where further study would be of most value rather than a definitive quantification so far as lead losses are concerned.

A significant value of this work is its methodological approach. We have identified a number of loss streams for lead, developed analytical approaches to estimate their magnitudes at different spatial levels, originated a procedure to roughly rank losses on the basis of potential toxicity, and constructed coded displays to reveal patterns in large bodies of data. In each of these cases, the work can serve as guidance for similar studies dealing with other spatial levels, and with other metals, especially those with significant hazard potential.

Acknowledgements

This project was supported by a special fund of the State Key Laboratory of Water Environment Simulation (08ESPCT-Y) and the "National Basic Research (973) Program" Project (No. 2005CB 724204) of the Ministry of Science and Technology of China.

Appendix. Supplementary material

Supplementary data associated with this article can be found, in the online version, at doi: 10.1016/j.envpol.2009.05.003.

References

Beijing General Research Institute of Mining and Metallurgy. 2000. China Investigation Report on the Exploitation and Utilization of Lead-zinc Mineral Resource: Beijing.

Condor J M, Lanno R P, Basta N T. 2001. Assessment of metal availability in smelter soil using earthworms and chemical extractions. Journal of Environmental Quality, 30: 1231-1237.

Darling C T R, Thomas V G. 2003. The distribution of outdoor shooting ranges in Ontario and the potential for lead pollution of soil and water. Science of the Total Environment, 313: 235-243.

Editorial Staff of the Yearbook of the Nonferrous Metals Industry. 2002. China Nonferrous Metals Industry Yearbook, 2001. Beijing: China Nonferrous Metals Industry Yearbook Press. (in Chinese).

Giles J. 2004. Collateral damage. Nature, 427: 580-581.

Green Media Toolshe. 2005. Scorecard, the Pollution Information Site. http: //www.scorecard.org/chemical-profiles/summary.tcl?edf_substance _id¼7439%2d92%2d1 Accessed from Jul. 2005 to Jan. 2006.

Henstock M E. 1996. The Recycling of Non-Ferrous Metals. International Council on Metals and the Environment, Ottawa.

International Lead and Zinc Study Group (ILZSG). 2005. London: Principal Uses of Lead and Zinc.

International Lead and Zinc Study Group (ILZSG). 2007. Lead and Zinc in the World. Interactive Statistical Database. http://www.ilzsg.org/ (Accessed from Nov. to Dec. 2007)

Jiang S. 2000. The development and utilization of secondary nonferrous resource in China. China Resources Comprehensive Utilization, 1: 18-21 (in Chinese).

Jonsson A, Lindström M, Bo B. 2002. Phasing out cadmium and lead – emissions and sediment loads in an urban area. Science of the Total Environment, 292: 91-100.

Karlsson S. 1999. Closing the technospheric flows of toxic metals – modeling lead losses from a lead acid battery system for Sweden. Journal of Industrial Ecology, 3 (1): 23-40.

Klee R J, Graedel T E. 2004. Elemental cycles: a status report on human or natural dominance. Annual Review of Environment & Resources, 29: 69-107.

Leygraf C, Graedel T E. 2000. Atmospheric Corrosion. Wiley-Interscience, New York.

Li F Y, Li S S, Wang J. 1999. The present production status of recycled lead and its future in domestic and oversea. World Nonferrous Metals, 5: 26-30 (in Chinese).

Lohm U, Anderberg S, Bo B. 1994. Industrial metabolism at the national level: a case-study on chromium and lead pollution in Sweden, 1880-1980. In: Ayres, R.U., Semonis, U.E. (Eds.), Industrial Metabolism. United Nations University Press, Tokyo.

Ma Y G. 2000a. The present state of lead emissions, reasons and countermeasures. China Resources Comprehensive Utilization, 2: 26-27 (in Chinese).

Ma Y G. 2000b. Some suggestions on changing present state of secondary lead refining in China and its related policy and regulations. China Resources Comprehensive Utilization, 3 (1): 15-19 (in Chinese).

Mao J S, Lu Z W, Yang Z F. 2006. The eco-efficiency of lead in China's lead-acid battery system. Journal of Industrial Ecology, 10 (1/2): 185-197.

Mao J S, Yang Z F, Lu Z W. 2007. Industrial flow of lead in China. Transactions of Nonferrous Metals Society China, 17 (2): 400-411.

Mao J S, Dong J, Graedel T E. 2008a. The multilevel cycle of anthropogenic lead: Ⅰ. Methodology. Resources Conservation & Recycling, 52: 1058-1064.

Mao J S. Dong J, Graedel T E. 2008b. The multilevel cycle of anthropogenic lead: Ⅱ. Results and discussion. Resources Conservation & Recycling, 52: 1050-1057.

Nriagu J O, Pacyna J M. 1988. Quantitative assessment of worldwide contamination of air, water and soils with trace metals. Nature, 333: 134-139.

Ostergren J D, Brown G E, Parks G A, et al. 1999. Quantitative speciation of lead in selected mine tailings from Leadville, CO. Environmental Science & Technology, 33: 1627-1636.

Pacyna J M, Scholtz M T, Li Y F. 1995. Global budget of trace metal sources. Environmental Reviews, 3: 145-159.

Pechnikov A V, Guseva T V, Kemp R G. 1996. Environmental protection in the Russian Federation: a case study of lead contamination around a crystal production facility. Trans. Int. Chem. Eng, 74 (Part B), 189-196.

Rankin C W, Nriagu J O, Aggarwal J K, et al. 2005. Lead contamination in cocoa and cocoa products: isotopic evidence of global contamination. Environmental Health Perspectives, 113: 1344-1348.

Raw Materials Group. 2005. Raw Materials Data – the Mining Database. Stockholm. Sweden. Aug.http: //www.rmg.se/RMG2005/pages/database.htm.

Ritson P J, Esser B K, Niemeyer S, et al. 1994. Lead isotopic determination of historical sources of lead to Lake Erie, North America. Geochimica Et Cosmochimica Acta, 58: 3297-3305.

Rosman K J R, Chisholm W, Boutron C F, et al. 1994. Anthropogenic lead isotopes in Antarctica. Geophysical Research Letters, 21: 2669-2672.

Scheuhammar A M, Norris S L. 1996. The ecotoxicology of lead shot and lead fishing weights. Ecotoxicology, 5: 279-295.

Shotyk W, Weiss D, Appleby P G, et al.1998. History of atmospheric lead deposition since 12, 370^{14}C year BP from a peat bog, Jura mountains, Switzerland. Science, 281: 1635-1640.

Smith G R. 2000. Lead (2000). http: //minerals.usgs.gov/minerals/pubs/commodity/lead/380400.pdf. (Accessed Oct. 2005)

Soto-Jimenez M F, Hibdon S A, Rankin C W, et al. 2006. Chronicling a century of lead production in Mexico: stable lead isotopic composition analysis of dated sediment cores. Environmental Science & Technology, 40: 764-770.

Thomas V M. 1995. The elimination of lead in gasoline. Annual Review of Energy & the Environment, 20: 301-324.

Thornton I, Rautiu R, Brush S. 2001. Lead, the Facts. IC Consultants Ltd, London, UK. http: //www.ldaint.org/factbook/index.htm. (Accessed Oct. 2005)

Tukker A, Buist H, van Oers L, et al. 2006. Risks to health and environment of the use of lead in products in the EU. Resources Conservation & Recycling, 49: 89-109.

U.S. Environmental Protection Agency. 2005. Occupational Exposures and Environmental Releases of Lead Wheel-Balancing Weights Washington, DC.

US Geological Survey. 2000. Recycling-Metals. Minerals Yearbook, 2000 Washington, DC.

von Storch H, Hagner C, Costa-Cabral M, et al. 2002. Reassessing past European gasoline lead policies. EOS Transactions American Geophysical Union, 83: 393-399.

Walsh D C, Chillrud S N, Simpson H J, et al. 2001. Refuse incinerator particulate emissions and combustion residues for New York City during the 20th century. Environmental Science & Technology, 35: 2441-2447.

第三篇

外部效应与评估

第 16 章　铅酸电池中铅的生态效率
The Eco-efficiency of Lead in China's Lead-acid Battery System*

16.1　Introduction

16.1.1　Background

The consumption of resources and the environmental crisis that has developed in recent decades have led to a global focus on how to achieve sustainable development. To address this situation, the Factor 10 Club (Bleek, 1999) has proposed a tenfold leap in the utilization efficiency of energy and other resources within one generation. Because human economic systems are huge, complex systems, reaching such a goal would require action in every industrial field.

Lead (Pb) has been used in many industrial fields, such as mechanical, electrical, and chemical engineering, to take advantage of its properties, such as high density and tenacity, low rigidity and melting point, ease of machining and smelting in a foundry, sound resistance, and radiation attenuation (Dong, 2000). About 6.4 million metric tonnes1 of metallic lead was consumed annually around the world in the last years of the twentieth century, and 60 to 70 percent of this total was used to manufacture lead-acid batteries (LABs) (Ma, 2000). In addition, both total lead consumption and the proportion of lead consumed in LABs have been increasing in recent years. Because lead is produced from mineral ore, it is a nonrenewable resource. Moreover, the lead emissions from lead mining, smelting, the manufacture of LABs, and other production processes are harmful or poisonous to the ecosystem and particularly to humans. Thus, lead-related industries have obvious impacts on the lead mineral resource, on human health, and on the environment.

In China, the conflict between rapid economic development and environmental

---

* Mao J S, Lu Z W, Yang Z F. 2006. Eco-efficiency of Lead in China's Lead-acid Battery System. Journal of Industrial Ecology, 10 (1/2): 185-197.

deterioration caused by lead-using industries has become quite serious in recent years. On the one hand, annual production of both metallic lead and LABs has increased rapidly (Table 16.1). On the other hand, both the consumption of lead ore and the resulting anthropogenic lead flows have gone far beyond the environmental carrying capacity, as suggested by the increasing proportion of lead ore that is imported and the magnitude of the anthropogenic lead flow, which is about 13 times the estimated natural flow. Therefore, studying the LAB system and using lead as the representative material for this system, we can study the eco-efficiency (defined below) of lead as a means of protection both lead ore resources and the environment.

Table 16.1 Annual production of lead-acid batteries and refined lead in China

Year	1990	1991	1992	1993	1994	1995	1996	1997	1998	1999	2000
Batteries (GWh)[a]	6.980	5.146	6.837	7.773	—	7.080	9.487	—	—	10.394	11.881
Lead (kt)[b]	296.5	319.7	366	411.9	467.9	607.9	706.2	707.5	756.9	918.4	1099.9

Data source: a. China Machinery industries Yearbook, China Power and Electrical Equipment Yearbook China Foreign Trade Yearbook.

b. China Nonferrous Metals Industry Yearbook (1990-2001).

Note: The quantification of batteries in GWh represents the total energy available from one charging-recharging cycle.

Studies of lead ore consumption in the Swedish LAB system (Karlsson, 1999) have revealed that the recovery rate of discarded batteries is the most important factor in determining the consumption of lead ore in the system. The higher the recovery rate, the lower the lead ore consumption. However, Karlsson's conclusion only applies to steady-state systems in advanced countries, because the study was based on the actual state of Sweden with constant annual production of LABs, advanced technology, and a high recovery rate. In many developing countries, such as China, the annual production of LABs is increasing rapidly, but technology is lagging behind, and under these conditions, Karlsson's study does not adequately describe the true situation. Thus a more general conclusion is required for the LAB system and a theoretical study of the eco-efficiency of lead in the LAB system is necessary. In addition, studying the present status of China's LAB system and analyzing existing problems in the LAB system would suggest how to improve the eco-efficiency of lead in LAB systems everywhere. This study would thus have very significant implications for the protection of lead ore resources and the environment.

16.1.2 The Present Study

In this article, the LAB system was chosen for study, with an emphasis on lead flow within the system, by means of life-cycle assessment. Lead mining, concentration, smelting, and refining, as well as the manufacturing of LABs and their use and recovery, are the main components of the LAB system.

"Eco-efficiency" is generally defined as the so-cietal service provided per unit of environmental impact (OECD 1998). In the present study, the societal service provided by the concerned system and the environmental impacts are defined somewhat differently.

Because the main function of LABs is to store and deliver electrical energy (Shi, 1983), we have defined the societal service provided by LABs as the total estimated electrical energy delivered by the LABs produced in a single fiscal year (Mao and Lu, 2003a). We have assumed that

(1) The capacity of an LAB in one charge-discharge cycle is expressed as E_i, where E represents the energy in kilowatt-hours (kWh), and the subscript i represents a specific LAB. The total capacity of all LABs produced in a given year represents the annual output of the LAB systems in energy units; that is, $E = \sum E_i$, with units of kilowatt-hours per year (kWh/year).

(2) The amount of energy delivered depends on the movement of electrons between the two electrodes of an LAB, and lead is the main material in both electrodes. The greater the lead content in an LAB, the more energy can be delivered. Thus, the energy delivery capacity should be directly proportional to the lead content of the LAB. In practice, the ratio of the total output of LABs in energy units (i.e., E) to the total output of LAB in lead content (expressed as P, with units of tonnes per year [tonne/year]) is defined as the average specific energy of the LABs, which is rep-resented by F (i.e., $F = E/P$, with units of kilowatt-hours per tonne [kWh/tonne]).

(3) An LAB usually has a life-span of several years, and this life-span is represented by $\Delta\tau$, with units of years (year). An LAB can be reused for hundreds of charge-discharge cycles during its life span. In the present study, the total estimated electrical energy delivered by the LABs produced in one year is represented by the annual output of the LABs in energy units multiplied by their life span (i.e., $S = E \cdot \Delta\tau$, with units of kilowatt-hours [kWh]).

To express the relationship between S and P more clearly, we can use the equation:

$$S = F \cdot P \cdot \Delta\tau \qquad (16.1)$$

where $\Delta\tau$, the life-span of the LAB, is represented by an average value to simplify the data collection and calculations.

In the present study, the environmental impacts have been defined as the impact of the LAB system on the lead ore resource and the impact of lead emissions on the environment. These two parameters are called the "lead ore resource load" and the "lead emission load" respectively. The former is defined as the annual lead ore consumption by the LAB system and is represented by R; the latter is defined as the annual lead emissions into the environment and is represented by Q. Both kinds of load have units of tonnes of lead content per year (t/year).

Based on these two kinds of environmental impact, the eco-efficiency of lead in the LAB system can be divided into resource efficiency (RE) and environmental efficiency (EE). RE is represented by r and is expressed as

$$r = \frac{S}{R} \tag{16.2}$$

EE is represented by q and is expressed as:

$$q = \frac{S}{Q} \tag{16.3}$$

Both RE and EE use kilowatt-hour-years per tonne (kWh · year/t) as their units.

From equation (16.2) and equation (16.3), it can be deduced that a higher eco-efficiency means reduced consumption of lead ore or reduced lead emissions for the same level of societal service provided by the LAB system. Alternatively, these equations can be interpreted as providing more societal services with the same environmental impacts.

Our study is composed of two parts: discussion of the theory and its application. In the theoretical study, we have emphasized the analysis of lead flow in the LAB system. Based on that analysis, we derive a quantitative relationship between the societal service provided by the LAB system and the corresponding environmental impacts. To conclude the theoretical discussion, we discuss the changing factors that determine the eco-efficiency of lead. In the application section, we will study the present status of the LAB system in China based on statistical data. Problems that exist within the system will be analyzed, and ways to improve the situation will be proposed.

16.2　Primary Regulations

16.2.1　Methodology: the Lead-flow Diagram in the LAB System

In the LAB system, lead will flow through every stage of the life cycle of a battery, and the law of conservation of mass (in which in-puts equal outputs) will be obeyed (Kleijn, 2000). A lead-flow diagram can be drawn to reflect the directions of lead flow and the distribution of lead based on the application of the conservation law at every stage.

We assume that:

(1) The time spent in various production processes can be ignored, because it usually lasts only a few weeks and is thus very short compared to the years of an LAB's life span.

(2) The average life-span of an LAB remains constant during the study period.

(3) Each LAB becomes obsolete $\Delta \tau$ years after its production, and some of the LAB will become old lead scraps that can be recycled as a secondary source of lead.

(4) The trade in LABs can be ignored because the few kilotonnes (kt) of net import of lead scraps (Table 16.2) remains small compared with the hundreds of kilotonnes of lead that are consumed (351 kt in 1999) and the lead content (323 kt in 1999) in LABs.

Table 16.2 The trade in lead scraps in China

Year	1990	1991	1992	1993	1994	1995	1996	1997	1998	1999	2000	Average
Exports/ (kt)	6.132	5.250	1.500	0.621	1.510	0.589	0.152	0.061	1.751	1.060	0.037	
Imports/ (kt)	1.690	0.140	5.712	7.322	5.793	5.690	0.820	0.204	0.007	—	0.050	
Net imports/ (kt)	−4.442	−5.110	4.212	6.701	4.283	5.101	0.668	0.143	−1.744	—	0.013	0.983

Data source: China Foreign Trade Yearbook.

Based on these assumptions, the lead-flow diagram for an LAB's life cycle in reference year τ is illustrated in Fig.16.1, in which the production of LABs in reference year τ is represented by P_τ.

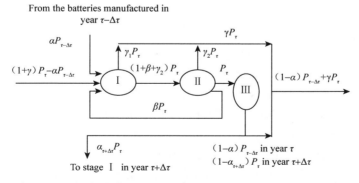

Fig. 16.1 The lead-flow diagram for the lead-acid battery life cycle

Stage I : lead mining, concentration, smelting and refining

Stage II: battery manufacturing

Stage III: use of LAB

Some explanations for Fig. 16.1 are as follows:

(1) The production of primary and secondary lead have been treated as a single process, and are expressed by stage Ⅰ.

(2) The annual production of LABs changes yearly, and the production in year $\tau - \Delta\tau$ is $P_{\tau-\Delta\tau}$.

(3) The average life-span of an LAB is $\Delta\tau$. The LABs manufactured in year τ will become obsolete in year $\tau + \Delta\tau$, and the old lead scraps that become production inputs in year τ come from the batteries produced in year $\tau - \Delta\tau$.

(4) The recycling rate, represented by α [with unit of tonne/tonne (t/t)], has been defined as the ratio of the old lead scraps reused in lead production in year τ to the total LAB production $\Delta\tau$ years ago. Under these conditions, $\alpha_{\tau+\Delta\tau} P_\tau$ of old lead scraps will become inputs for lead production in year $\tau + \Delta\tau$ and $\alpha_\tau P_{\tau-\Delta\tau}$ will become inputs in year τ. The subscript τ for the recycling rate in year τ has been omitted in Fig. 16.1 for simplicity.

(5) The moderate recycling rate, represented by β [with units of (t/t)], has been defined as the ratio of the lead scrap produced in LAB manufacturing that become inputs for lead production to the total production of batteries in the same year.

(6) The lead emission ratios in stages Ⅰ and Ⅱ, represented by γ_1 and γ_2 [with units of (t/t)], have been defined as the ratio of the lead emissions in the corresponding stage to the lead content of batteries produced in the same year. The sum of the two ratios is defined as the "overall lead emission ratio" and is represented by γ (i.e., $\gamma = \gamma_1 + \gamma_2$).

Fig. 16.1 shows that the lead ore resource load in year τ can be expressed mathematically as:

$$R = (1+\gamma)P_\tau - \alpha P_{\tau-\Delta\tau} \qquad (16.4)$$

and the lead emission load in year τ as:

$$Q = \gamma P_\tau + (1-\alpha)P_{\tau-\Delta\tau} \qquad (16.5)$$

This method has been called element flow analysis (EFA), and is a subset of substance flow analysis (Kleijn, 2000), a methodology that is used to analyze the relationships between an industrial system and its environment (Hansen and Lassen, 2003). Some characteristics of EFA are as follows:

(1) Only one element in the product being studied (lead in the present study) is traced and used to represent the overall product;

(2) The time interval between manufacturing and disposal of the products (LAB in the present study) is considered;

(3) Changes in the annual production of the product (LAB) are permitted.

Lu (2000) first proposed adapting this method by incorporation the time difference between production and disposal and changes in production. The method had been successfully used to study steel scrap (Lu, 2000) and iron emissions (Lu, 2002). The method was improved by emphasizing the final product to create a more convenient link with the societal service provided by the product and by taking the fiscal year as the statistical period to facilitate data gathering (Mao and Lu, 2003a).

16.2.2　Results and Discussion

1. Resource Efficiency

Based on the definition of resource efficiency given earlier, we can substitute equation (16.1) and equation (16.4) into equation (16.2), with the following result (resource efficiency will be abbreviated by RE in the text and represented by r in equations):

$$r = \frac{F \cdot \Delta\tau}{1+\gamma-\alpha p} \tag{16.6a}$$

where p represents the ratio between the level of production one life span prior to the current year and the level of production in the current year and is expressed as $p = P_{\tau-\Delta\tau}/P_{\tau}$. This value is always positive.

If we assume that the production of LABs increases linearly with time at an annual growth rate ρ:

$$\frac{P_{\tau-\Delta\tau}}{P_{\tau}} = 1 - \rho\Delta\tau \tag{16.6b}$$

where $1-\rho\Delta\tau \geqslant 0$ when p is positive.

If we substitute equation (16.6b) into equation (16.6a), we get:

$$r = \frac{F \cdot \Delta\tau}{1+\gamma-\alpha+\alpha\rho\Delta\tau} \tag{16.7}$$

Equation (16.7) shows that the RE of lead in the LAB system is a function of the recycling rate (α), the lead emission rate (γ), the annual growth rate (ρ), the LAB life-span ($\Delta\tau$), and the specific energy (F). Discussion of some qualitative details concerning the influences of the above factors on RE follows:

(1) The influence of recycling rate (α) on RE (r): r will increase with increasing α independently of how the other factors change. This rule can be treated as the foundation for changes in RE along with changes in other factors.

(2) The influence of annual growth rate (ρ) on the relationship r-α: in general, ρ affects both the rate of change in r with changing α and the maximum

value of r. RE increases more slowly and reaches a smaller maximum value with increasing production of LABs than with constant production. The faster the production increases the more obvious the effect. The opposite situation occurs with decreasing production of LABs, which suggests that it is easier to obtain a high RE with decreasing production.

(3) The influence of lead emission rate (γ) on the relationship between r and α: γ affects the rate of change r with changing α and the maximum and minimum values of r. The lower the value of γ, the faster r increases and the higher the maximum and minimum values that it attains. With higher values of γ, the opposite occurs, which suggests that high RE can be more easily attained at a low value of γ.

(4) The influence of LAB life-span ($\Delta\tau$) on the relationship between r and α: $\Delta\tau$ only affects the value of r, but do not affects the rate of change r with changing α. The longer the life-span, the greater the value of r. With shorter life-spans, the opposite is true.

(5) RE will increases linearly with increasing specific energy of LAB (F), which can be easily concluded from equation (16.6a) and equation (16.7).

2. Environmental Efficiency

If we substitute equation (16.1) and equation (16.5) into equation (16.3), we get the following equation for environmental efficiency:

$$q = \frac{F \cdot \Delta\tau}{\gamma + (1-\alpha)p} \quad (16.8a)$$

If we assume that the production of LABs increases linearly with time, we can substitute equation (16.6b) into equation (16.8a) to get:

$$q = \frac{F \cdot \Delta\tau}{\gamma + (1-\alpha)(1-\rho\Delta\tau)} \quad (16.8b)$$

Equation (16.8b) shows that the environmental efficiency (EE) of lead in the LAB system is a function of the recycling rate (α), the lead emission rate (γ), the annual growth rate (ρ), the LAB life-span ($\Delta\tau$), and the LAB specific energy (F). This reveals certain inevitable relationships between EE and RE. Combining equation (16.6a) and equation (16.8a) provides:

$$\frac{1}{q} = \frac{1}{r} - \frac{1-p}{F \cdot \Delta\tau} \quad (16.9a)$$

If we assume that the production of LABs increases linearly with time, substituting equation (16.7) into equation (16.9a) produces:

$$q = \left[\frac{1}{r} - \frac{\rho}{F}\right]^{-1} \qquad (16.9b)$$

Equation (16.9b) shows that EE will always increase with increasing RE. This is the basic relationship between EE and RE. Further analysis can show that the annual growth rate (ρ) and the specific energy (F) will affect the rate of change EE with respect to RE. In brief, the value of r/q will equal 1 with constant production of LABs, whereas values of this ratio (i.e., r/q) will be less than and greeter than 1, respectively, with increasing and decreasing production of LAB. The influence of F on the rate of change of EE with respect to RE is thus related to changes in production trends. With increasing production, higher values of F lead to slow increase of EE with respect to RE, and lower values of EE for a given value of RE. Conversely, with decreasing production of LABs, higher values of F lead to faster increase in EE with respect to RE, and higher values of EE for a given value of RE.

Given the influence of specific energy (F) on RE, we can speculate that increasing specific energy while decreasing the production of LABs will increase EE more effectively than that other solution.

16.3 A Case Study: the Eco-efficiency of Lead in China's LAB System

16.3.1 Brief Description of Lead Flow in the LAB System

The case study described in this section has been carried out based on statistical data for those portions of China's lead-using industry that were related to the LAB system in 1999.

It is known that about 351 kt of metallic lead were used in the manufacturing of LABs in 1999 (Li et al., 1999). Of these lead inputs, 92% entered the LAB system as lead content of batteries, 3.56% was recycled as new lead scraps, and the remaining 4.44% was released into the environment as lead wastes or lead emissions (Mao and Lu, 2003a). The average life span of an LAB was estimated to be 3 year (Lan and Yin, 2000).

Based on the quantities and sources of the lead scrap recycled in 1999 (Yang and Ma, 2000), about 90.9 kt of lead in obsolete LABs and 12.5 kt of lead in new scrap were estimated as the lead input into lead production; the overall recovery rate in secondary lead smelting and refining was estimated at between 80 and 88 percent, and a value of 86.4 percent was used for the calculations in this case study (based on a lead flow balance sheet for a secondary refinery). Thus, 89.3 kt of secondary refined lead was obtained in total. The other lead input into the manufacture of LABs would be primary refined lead, which was estimated as 261 kt.

During the production of primary lead, many processes are involved, including lead mining, concentration, smelting, and refining. The recovery rate in lead mining and concentration was 83.8% in 1999, versus 92.8% for lead smelting and refining. Therefore, to obtain 264 kt of primary refined lead would require the consumption of ore containing 336 kt of lead.

For an LAB life span of 3 years, obsolete LABs re-cycled in 1999 would have been manufactured in 1996. Using the data in table 1, which represents about 77% of actual 1996 production and about 78% of actual 1999 production, the production of LABs in 1996 can be estimated at 292 kt of lead content, assuming that both the LAB life span and the LAB specific energy remained constant.

Based on this analysis, the lead-flow diagram for the life cycle of LABs in China in 1999 is illustrated in Fig. 16.2 (with units of kt).

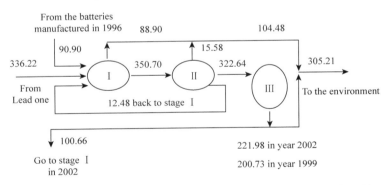

Fig. 16.2 The lead flow diagram for the LAB life cycle for China (1999)

16.3.2 Data Sources

The sources of the data related to lead flows in China's LAB system in 1999 are listed in Table 16.3.

Table 16.3 Sources of data for the case study

Data type or name	Data source	Agency responsible for compiling the data
Recovery rate in lead mining	China Investigation Report on the Exploitation and Utilization of Lead-zinc Mineral Resource 2000	Beijing General Research Institute of Mining and Metallurgy
Recovery rate in lead mining, concentration, smelting, and refining	China Nonferrous Metals Industry Yearbook (1990-2001)	Editorial staff of the yearbook of the nonferrous metals industry
Data related to lead scraps and lead recycling	Published literature or actual manufacturing data	Partly provided by the China association for metals recycling
Data related to battery manufacturing	Report on the environmental impacts for some lead-acid battery companies	Research Institute of Environment Science
Battery performance and its profile	China Statistic Report on Lead-acid Batteries	Shenyang Research Institute of Storage Battery
Annual production of lead-acid batteries	China Machinery industries Yearbook China Power and Electrical Equipment Yearbook	Editorial staff of the yearbook of machinery industries Editorial staff of the yearbook of nonferrous metals industry, power and electrical equipment
Export of LAB and lead scraps	China Foreign Trade Yearbook	Editorial staff of China Foreign Trade Yearbook

16.3.3 Results and Discussion

Based on Fig. 16.2, and using the previously defined symbols from the theoretical discussion, the production of LABs in 1999 can be expressed as $P = 323$ kt/year, and the resource and environmental loads can be expressed as $R = 336$ kt/year, $Q = 305$ kt/year, respectively. Based on the data in table 1 and the assumptions described earlier, the expected societal service value provided by the LAB produced in 1999 would be $S = 40$ gigawatt-hours (GWh). Thereby, the RE and EE of lead in China's LAB system in 1999 can be expressed as $r = 119$ (kWh·year/t) and $q = 131$ (kWh·year/t), respectively. The specific energy of the LABs can be expressed as $F = 41.3$ (kWh/t) based on equation (16.1). The values of the other factors that influence eco-efficiency can be expressed as follows: $\alpha = 0.312(t/t)$, $\beta = 0.039(t/t)$, $\gamma = 0.324(t/t)$, $\rho = 0.032$, and $p = 0.904(t/t)$ based on Fig. 16.2 and the previous calculations.

Similarly, we can estimate the RE, the EE, and the values of the factors that influence lead flow in Sweden's LAB system based on the data provided by Karlsson (1999). To simplify this comparison, the values of the two eco-efficiencies and their underlying factors for China and Sweden are summarized in Table 16.4. For this comparison, the specific energy of LABs in Sweden was taken as 40 (kWh/t). The results in Table 16.4 show that both the RE and the EE of lead in Sweden's LAB system had reached 15.8 (MWh·a/t), which represents 133 and 121 times the corresponding values for China. The main reasons for this difference are as follows:

(1) The lead recycling rate for Sweden has reached 0.99 (t/t), which means that nearly all of the obsolete LABs were recycled. In contrast, available data indicate that the corresponding rate for China is only 0.312 (t/t), which would mean that nearly 70% of the old lead scrap is not recycled.

(2) The lead emission rate for Sweden's LAB system is only 0.002655 (t/t), which means that there are almost no lead emitted from the system. In contrast, China's emission rate is 0.324 (t/t), which means that nearly 33% of the lead inputs used in LAB system were lost into the environment.

(3) The production of LABs in Sweden has remained constant for at least 5 years, whereas production in China has increased rapidly during the same period.

(4) The life-span of LABs in Sweden is about 5 years, versus only 3 years for China.

Table 16.4 Comparison of eco-efficiency and relevant parameters between China and Sweden

	Resource efficiency kWh·a/t	Environmental efficiency kWh·a/t	Recycling rate / (t/t)	Moderate recycling rate / (t/t)	Lead emission rate / (t/t)	Annual growth rate / (t/t)	Life-span /a
China	118.90	130.98	0.312	0.039	0.324	0.032	3
Sweden	15 804.00	15 804.00	0.990	0.1236	0.002655	0	5
Sweden/China	132.92	120.66	3.173	3.169	0.0082	0	1.667

It is easy to see that to improve the eco-efficiency of lead in China's LAB system, attention should focus on increasing the lead recycling rate and reducing the lead emission rate. Because the production of LABs is likely to continue to increase in the long term and the LAB life span will change only slowly in the short term, further discussion of the reasons for China's low recycling rate and high emission rate is necessary. This discussion will help identify potential countermeasures that will help improve the eco-efficiency of lead in China's LAB system.

1) Reasons for the Low Recycling Rate

The recycling rate relates mainly to the domestic consumption of LABs, collection of obsolete LABs, trade in lead scraps, the integrity of the data gathered, and so on (Mao and Lu, 2003b).

The annual production and domestic consumption of LABs for a recent 10-year period are summarized in Table 16.5. In general, about 7.62% of the obsolete LABs cannot be recycled back into China's LAB system because of the export of LABs.

Table 16.5 The production and domestic consumption of lead-acid batteries in China

Year	1986	1987	1988	1989	1990	1991	1992	1993	1994	1995	1996	average
Production/GWh[a]	3.220	5.072	4.550	—	6.980	5.146	6.837	7.773	—	7.080	9.487	
Export/GWh[b]	0.158	0.477	0.007	—	0	1.968	0.020	0.014	0.022	0.011	0	
export/production/%	4.91	9.40	0.15	—	0	38.24	0.29	0.18	—	0.16	0	7.62

Data source: a. The data for production is from China Machinery industries Yearbook, b. The data for export is from China Foreign Trade Yearbook..

Table 16.2 shows that the trade in lead scraps is nearly balanced during this period compared with the lead consumption in the LAB system and lead content in LABs. lead content in LABs. Because lead scrap exports nearly equal lead imports for the LAB system, we have ignored the trade in LABs in this study.

Ma (2000) and Yang and Ma (2000) have reported that there were about 300 secondary lead refineries in China in 1999 and that most of these facilities were privately owned and operated on a small scale. About half of the obsolete LABs are collected and recovered by these private refineries and these data may not be counted in the statistics. This situation results in a potential understatement of the recycling rate equivalent by 0.312 (t/t).

Consequently we can estimate that about 29.9% of the obsolete LABs in 1999 were not recovered and were thus lost into the environment.

Here we need to mention that if we consider a reasonable actual value of lead recycling rate, that is, if both the official statistics and the private estimated values were considered, the lead recycling rate would be 0.624 (t/t). In this case the RE and EE would become $r = 163$ (kWh·year/t) and $q = 180$ (kWh·year/t), respectively. But, such a result is uncertain because it was calculated based on estimated data.

2) Reasons for the High Emission Rate

We conducted further study of the lead loss in various production-related processes in China's LAB system in 1999. The results are summarized in Table 16.6, which shows that most of the lead loss occurred during lead concentration, followed by refining, then manufacturing.

Table 16.6 Lead losses during the production processes of the China's lead-acid battery system

	Lead concentration	Primary lead refining	Secondary lead refining	Manufacturing of LAB	Total
Lead loss/kt	54.47	20.34	14.09	15.58	104.48
Percentage of total/%	52.13	19.47	13.49	14.91	100

The recovery rate during lead concentration was only 81% to 86%, which is 5% to 15% lower than recovery levels in other countries (BGRIMM, 2000). This results mainly from the poor quality of the lead ore resources in China, which have a low lead-to-zinc ratio (1∶2.6 in the lead-containing material) and contain more than 50 kinds of metals (e.g., copper, silver, gold) in a complex ore that makes the concentration process unusually difficult (BGRIMM, 2000).

In the smelting and refining processes, Ma and Yang (2001) and Yang and Ma (2000) have reported that about half of the lead (including both primary and secondary lead) was smelted and refined by small-scale enterprises using outdated technology. Most of these enterprises still utilize a traditional sintering and blast-furnace approach that resembles the process used in more developed countries during the 1980s and that has since been replaced by more advanced processes.

3) Possible Measures to Improve Eco-efficiency

To improve the lead recycling rate, several measures should be adopted (Mao and Lu, 2003b; 2003c):

(1) Learn from the management experience of more developed countries. Implement laws, regulations, and mechanisms for lead recycling that would encourage more effective lead recycling paths (Sakuragi, 2002).

(2) Extend the responsibility of LAB companies so that they must sell "services" instead of only "products," thereby facilitating the recycling of LABs (Stahel, 1994; 1997).

(3) Enhance public awareness of environmental protection and develop an attitude that wastes are also resources (David and Nasrin, 1994). Levy a tax upon the consumers of LABs and charge a tax for lead emissions to the environment (Turner et al., 1998). These steps would greatly improve the re-cycling of lead scrap.

To reduce lead emissions, the following tasks must be accomplished:

(1) Implement a special license for companies involved in lead production that strictly stipulates the production scale, technology used, and measures required for environmental protection. Without exception, close companies with unacceptably inefficient operations in lead mining, concentration, smelting, and refining, and ban the creation of such companies in the future.

(2) Develop new technologies for lead recovery. Eliminate the use of outdated technology by small-scale companies and promote the spread of clean production technologies for lead. Establish a favorable economic policy that encourages the application of new advanced technologies.

（3）Improve public awareness of the harm caused by lead emissions and teach methods to prevent harm to humans. Teach good habits that can reduce lead emissions and encourage people to devote their best efforts to this project.

We also suggest that the statistic work should be improved so that the official statistics can represent the actual status.

4）Forecasts for China's LAB System

Ma（2004）reported that a technical policy for the prevention of pollution by obsolete lead-acid batteries has been promulgated. The situation for China's LAB system is thus expected to improve in the coming years.

If the eco-efficiency of lead in China's LAB system can be improved to 50 times the value in 1999, producing the same quantity of LABs as in 1999 would only require 6.72 kt of lead in ore, and lead emissions into the environment would be reduced to about 6.10 kt. The status of China's lead ore resources and environment would improve greatly. Given the fact that Sweden's eco-efficiency for lead is more than 100 times the value in China, an increase of only 50 times is a feasible goal.

16.4　Conclusions

（1）The concept of societal service for the LAB system was defined quantitatively. This represents a very significant insight for the transition to a service-oriented society.

（2）The concept of eco-efficiency was applied to the LAB system. Based on the lead-flow diagram for the life cycle of LABs, the principal factors underlying the resource efficiency and environmental efficiency of lead in the LAB system were obtained. This improved understanding of the system provides a robust theoretical basis for improving lead flows within China's LAB system and thus reducing its environmental impact.

（3）The current status of the eco-efficiency of lead in China's LAB system in 1999 was studied. The results show that the re-source efficiency and the environmental efficiency are only 119 and 131 kWh·a/t, respectively, which are less than 1% of the levels achieved by Sweden. The main reasons for this difference were China's lower lead recycling rate, higher lead emissions, increasing production of LABs, and shorter LAB life span.

（4）The primary reasons for the low eco-efficiency of lead in China's LAB system are the inefficient management of lead scrap, poor-quality lead ore, and abundance of inefficient small-scale lead-related plants. Several measures to improve this situation were proposed.

Notes

Tonne (t) indicates metric ton. One tonne = 1 megagram (Mg, SI) ≈1.1 short tons.

References

Bleek F S. 1999. A report by the factor 10 club. Factor 10 Institute. http：//www.Factor 10 .org/

BGRIMM (Beijing General Research Institute of Mining and Metallurgy). 2000. China Investigation Report on the Exploitation and Utilization of Lead-zinc Mineral Resource. (in Chinese)

David T A, B Nasrin. 1994. Wastes as Raw Materials. The greening of industrial ecosystems. Washington DC: National Academy Press, 69-89.

Dong J G. 2000. The manual of application materials. Beijing: China Machine industry press, 1122-1128. (in Chinese)

Halada K, R. Yamamoto. 2001. A review of the development of resource productivity concept and eco-efficient technology. Northeastern University. International symposium on industry and ecology 2001. P.R. China. Shenyang: Northeastern University, 14-36.

Hansen E, C Lassen. 2003. Experience with the use of substance flow analysis in denmark. Journal of Industrial Ecology, 6 (3/4): 201-219

Karlsson, S. 1999. Closing the technospheric flows of toxic metals — Modeling lead losses from a LAB system for Sweden. Journal of Industrial Ecology, 3 (1): 23-40.

Kleijn R. 2000. IN=OUT: the trivial central paradigm of MFA? Journal of Industrial Ecology 3 (2/3): 8-10.

Lan X H, J H Yin. 2000. The lead-recycling industry in developing. China Resources Comprehensive Utilization. 8: 9-21. (in Chinese)

Li F Y, S S Li, J Wang. 1999.The present production status of recycled lead and its future in domestic and oversea. World Nonferrous Metals, 5: 26-30. (in Chinese)

Lu Z W. 2000. A study on the steel scrap resources for steel industry. Acta Metallurgica Sinica, 36 (7): 728-734. (in Chinese)

Lu Z W. 2002. Iron-flow analysis for the life cycle of steel products: A study on the source index for iron emission. Acta Metallurgica Sinica, 38 (1): 58-68. (in Chinese)

Ma Y G. 2000. The present state of lead emissions, reasons and countermeasures. China Resources Comprehensive Utilization, (2): 26-27. (in Chinese)

Ma Y G. 2004. The promulgation of the technical policy for the prevention of pollution by obsolete lead-acid batteries. Chinese Journal of Power Sources, 28 (2): 100-100. (in Chinese)

Ma Y G, H Y Yang. 2001. The recovery of the obsolete lead-acid battery and the secondary lead refining. Environmental Herald, (1): 52-53. (in Chinese)

Mao J S, Z W Lu. 2003a. Resource-service efficiency of lead in lead-acid battery. Journal of Northeastern University (nature science), 24 (12): 1173-1176 (in Chinese).

Mao J S, Z W Lu. 2003b. A Study on Causes of Low Recovery of Scrap Lead. World Nonferrous Metals, 11: 24-32. (in Chinese)

Mao J S, Z W Lu. 2003c. Study on Resource Efficiency of Lead for China. Research of Environmental Science, 17 (3): 78-80. (in Chinese)

Sakuragi Y. 2002. A new partnership model for Japan: promoting a circular flow society. Corporate Environmental Strategy, 9 (3): 292-296

Shi X L. 1983. Lead acid storage batteries. Beijing: People's Mail & Telecommunication Press, 5-6. (in Chinese)

Stahel W R. 1994.The utilization-focused service economy: resource efficiency and product life extension. The greening of industrial ecosystems. Washington DC: National Academy Press, 178-190.

Stahel W R. 1997. The functional economy: cultural and organizational change. The industrial green game. Washington DC: National Academy Press, 91-100.

Turner R K., R Salmons, J Powell, et al. 1998. Green taxes, waste management and political economy. Journal of Environmental Management, 53: 121-136.

Yang C M, Y G Ma. 2000. The recovery and recycling of discarded lead-acid batteries in China. China electrician technique association LAB council. Thesis collection for the 7th national annual meeting on lead-acid batteries. Guangdong Nanhai: Shenyang storage battery institute, 198-202. (in Chinese)

第 17 章　关于中国铅的资源效率的研究
Study on the Resource Efficiency of Lead for China*

铅的资源效率，是指投入单位铅矿石所能产出的精铅量（Halada and Yamamoto，2001；Eco-Efficient 2002 Fair，2002）。因此，铅的资源效率越高，获得同等数量的精铅所消耗的铅矿资源越少。

近年来，中国精铅产量快速增长，而铅矿资源明显短缺，需要进口大量的铅精矿来满足国内需求。如能大幅提高铅的资源效率，则会大大缓解中国铅量增长与资源短缺之间的矛盾。为此，首先要研究铅的资源效率的变化规律；然后，在估算中国铅的资源效率现状及其各影响参数数值的基础上，分析中国铅业中存在的问题，并依此提出改进建议。

17.1　铅的资源效率及其变化规律

17.1.1　铅的资源效率的定义

根据定义，若以"年"作为统计期，则某年铅的资源效率为

$$r = \frac{P}{R} \tag{17.1}$$

式中，r 为铅的资源效率（t/t）；P 为精铅年产量（t）；R 为生产 P 吨精铅投入的铅矿石数量，按含铅量（t）计算。

17.1.2　铅的资源效率变化规律

在精铅生产中，根据铅的流入量等于流出量（Kleign，1999），列出铅的平衡方程，则

$$R + R_z + R_j = P + Q \tag{17.2}$$

式中，R、R_z、R_j 分别为铅矿石、折旧废铅、加工废铅（陆钟武，2000；陆钟武，

---

* 毛建素，陆钟武. 2004. 关于中国铅的资源效率研究. 环境科学研究，17（3）：78-80.

2002）的投入量，按含铅量（t）计算；Q 为废铅排放量（t）。

由于铅制品通常具有几年的使用寿命，因而，第 τ 年回收的折旧废铅 $R_{z\tau}$ 来自第 $\tau-\Delta\tau$ 年生产的精铅，数量为

$$R_z = \alpha P_{\tau-\Delta\tau} \tag{17.3}$$

式中，α 为铅的大循环率，是某年生产出来的精铅经过一个生命周期（Graedel and Allenby，1995）后，形成的折旧废铅量所占的比率（t/t）；$P_{\tau-\Delta\tau}$ 为第 $\tau-\Delta\tau$ 年的精铅产量（t）；$\Delta\tau$ 为从精铅生产到铅制品报废之间的"时间差"（Kleign，1999；陆钟武，2000），近似取铅制品的平均使用寿命（a）。

第 τ 年回收的加工废铅，来自第 τ 年生产的精铅，数量为

$$R_j = \beta P_\tau \tag{17.4}$$

式中，β 为铅的中循环率，是加工废铅量与精铅产量的比值（t/t）。

废铅排放量 Q 是在精铅的生产过程中发生的，在第 τ 年，其数量为

$$Q = \gamma P_\tau \tag{17.5}$$

式中，γ 为铅的排放率，是废铅排放量与精铅产量的比值（t/t）。

另外，根据式（17.1），为获得精铅产量 P_τ，第 τ 年需要投入铅矿石

$$R = \frac{P_\tau}{r} \tag{17.6}$$

将式（17.3）~式（17.6）代入式（17.2），整理后得到

$$r = \frac{1}{1+\gamma-\beta-\alpha p} \tag{17.7}$$

式中，$p = P_{\tau-\Delta\tau}/P_\tau$ 为第 wygd 年与第 τ 年的精铅产量比。$p=1$ 表示该期间精铅产量保持不变；$p<1$ 表示精铅产量持续上升；$p>1$ 表示精铅产量持续下降。

由式（17.7）可知：当其他条件不变时，铅的循环率（包括 α、β）越高，资源效率 r 越高；铅的排放率 γ 越低，资源效率 r 越高；铅的资源效率与产量比有关，且产量比的数值越高，资源效率随大循环率的提高速度越快。相反，产量比的数值越低，资源效率随大循环率的提高速度越慢。

17.1.3 资源效率估算式

应用中，可根据某一国家（地区）的铅业统计数据，估算铅的资源效率。方法如下：

$$r = \eta_x \eta_y \frac{P}{P_1} \tag{17.8}$$

式中，P_1 为原生铅年产量（t）；η_x、η_y 分别为采/选、冶炼过程中的铅收率（%）。

17.2 实例应用——中国铅的资源效率分析

17.2.1 中国铅的资源效率现状

根据中国有色工业统计数据，按照式（17.8）估算各年铅的资源效率，如表 17.1 所示。可见，近 10 年来，中国铅的资源效率只有 0.85~1.06（t/t），平均为 0.91（t/t）。

表 17.1　中国铅的资源效率估算表（1990~2000）
Table 17.1　Estimated values of RE for China（1990-2000）

年份	1990	1991	1992	1993	1994	1995	1996	1997	1998	1999	2000	平均值
精铅产量/kt	296.5	319.7	366	411.9	467.9	607.9	706.2	707.5	756.9	918.4	1099.9	—
原生铅产量/kt	268.3	273.4	317.7	367.7	408.0	432.5	562.6	583.8	664.5	821.0	997.9	—
选矿收率/%	81.00	82.73	83.49	82.95	83.33	81.46	82.93	83.07	84.36	83.82	86.26	83.22
冶炼收率/%	94.85	94.39	93.91	94.61	94.17	92.44	92.49	92.75	93.69	93.49	93.52	93.66
资源效率/(t/t)	0.85	0.91	0.90	0.88	0.90	1.06	0.96	0.93	0.90	0.88	0.89	0.91

注：表中各年精铅、原生铅产量以及选矿、冶炼收率的数据取自各年《中国有色金属工业年鉴》。

在瑞典铅酸电池系统中，投入 0.253 kt 铅矿石可获得 22.472 kt 铅金属（Karlsson，1999），铅的资源效率高达 88.82（t/t）。相比之下，中国铅的资源效率十分低下。

17.2.2 原因分析

如果估算出中国铅的循环率 α 和 β、产量比 p 和排放率 γ 的数值，并与瑞典铅酸电池系统的相应数据进行对比，那么，便可找出中国资源效率低下的原因。

1. 资源效率影响参数的估算

1）估算方法

（1）铅的循环率：由式（17.3）、式（17.4）可知，在得到各年的精铅产量、折旧废铅量和加工废铅量的情况下，可以折算出相应年份铅的循环率 α、β 的数值。

考虑废铅经过再生过程，形成再生铅，并且，目前中国铅的再生收率为 80%~95%（杨春明和马永刚，2000），取 88%，因此，用各年的再生铅量除以 0.88，即得到相应年份的废铅量。在废铅中，折旧废铅约占 86%（姜松，2000），其余为加工废铅，占 14%，用各年的废铅量分别乘以这两个百分数，得到各年的折旧废铅量和加工废铅量。按照这种方法，由式（17.3）、式（17.4）整理得到

$$a_\tau = 0.977 \frac{P_{2\tau}}{P_{\tau-\Delta\tau}} \quad (17.9)$$

$$\beta_\tau = 0.159 \frac{P_{2\tau}}{P_\tau} \quad (17.10)$$

式中，$P_{2\tau}$ 为第 τ 年的再生铅产量（t）。

（2）产量比：与各年的精铅产量 P_τ、铅制品的使用寿命 $\Delta\tau$ 两个因素有关。研究中采用铅酸电池、含铅电缆和铅材的加权平均寿命，加权系数取各种铅制品在铅消费中的比例。其中，铅酸电池的使用寿命按 3 年（兰兴华和殷建华，2000）计算，电缆和铅材的使用寿命按 15 年（姜松，2000）计算，相应的消费比例依次取 65%和 7%（蔡显弟等，1996）。由此估算出 $\Delta\tau = 4a$。这样一来，某年的产量比 p 应等于 $P_{\tau-4}/P_\tau$。

（3）铅的排放率：在估算出各年 r、α、β 和 p 的数值的基础上，按照式（17.7），可折算出各年的铅排放率 γ 的数值。

2）估算结果

按照前文中的估算方法，结合中国铅业统计数据，可估算出铅的循环率、产量比和排放率的数值。估算结果汇总到表 17.2 中。

表 17.2 中国铅的资源效率影响参数估算表（1990～2000）
Table 17.2 Estimated values of the influencing factors of RE for China（1990-2000）

年份	1990	1991	1992	1993	1994	1995	1996	1997	1998	1999	2000	平均值
精铅产量/kt	296.5	319.7	366.0	411.9	467.9	607.9	706.2	707.5	756.9	918.4	1099.9	
再生铅产量/kt	28.2	46.3	48.3	44.3	59.9	175.3	143.6	123.7	92.3	97.4	102.0	
大循环率/(t/t)	0.115	0.184	0.196	0.143	0.197	0.536	0.383	0.293	0.193	0.157	0.141	0.231
中循环率/(t/t)	0.015	0.023	0.021	0.017	0.020	0.046	0.032	0.028	0.019	0.017	0.015	0.023
产量比/(t/t)	0.808	0.771	0.660	0.733	0.634	0.526	0.518	0.582	0.618	0.662	0.642	0.650
铅排放率/(t/t)	0.284	0.264	0.261	0.258	0.256	0.271	0.272	0.274	0.249	0.257	0.229	0.261

注：表中各年精铅、再生铅产量的数据取自各年《中国有色金属工业年鉴》。

由表 17.2 可见，在 1990～2000 年，中国铅的大循环率为 0.115～0.536（t/t），平均为 0.231（t/t）；中循环率为 0.015～0.046（t/t），平均为 0.023（t/t）；铅的排放率为 0.249～0.284（t/t），平均为 0.261（t/t）；精铅产量比为 0.518～0.808（t/t），平均为 0.650，远小于 1，表明中国精铅产量持续高速增长。

2. 与瑞典铅酸电池系统的对比

1992～1995 年，瑞典铅酸电池系统步入生产稳定期。在该系统的精铅生产阶段，由于利用大量废铅作为生产原料，每年仅投入 0.253 kt 铅矿石，便可获得 22.472 kt 的铅金属（Karlsson，1999），铅的资源效率高达 88.82（t/t），是目前中国铅业中铅的资源效率的近百倍。这既表明了中国铅的资源效率十分低下，又暗示出中国铅的资源效率的改善潜力十分巨大。

为了更清楚地了解中国铅业与瑞典铅酸电池系统在铅的资源效率方面的差距及其产生的原因，及其产生的原因，根据 Karlsson 研究中的数据，计算得到瑞典铅酸电池系统中铅的资源效率及其影响参数的数值。通过对比，发现在铅的循环率方面，瑞典铅酸电池系统中铅的大、中循环率分别达 0.881（t/t）、0.110（t/t）；而我国铅业的大、中循环率分别仅为 0.231（t/t）、0.023（t/t）（取近 10 年的平均数据）。铅的循环状况很差，是造成中国铅的资源效率低下的重要原因之一；在废铅排放方面，瑞典的铅排放率为 2.360（kg/t），而我国却高达 0.261（t/t），中国废铅排放严重，是造成中国铅的资源效率低下的又一重要原因；在精铅产量变化方面，瑞典至少稳定了一个铅酸电池的生命周期——5 年，而中国仍在以产量比 0.650 的速度高速增长。中国铅产量增长较快，是造成中国铅的资源效率低下的第三个重要原因。

17.2.3　提高铅的资源效率的途径及建议

1. 提高铅的资源效率的途径

根据上面的分析，原则上可通过提高中国铅业中铅的循环率、降低铅的排放率、提高产量比这三种途径来提高中国铅的资源效率。

根据表 17.2 中国铅业中 α、β、γ、p 的估算结果，若假设 β、γ、p 保持不变，将 α 的数值提高到 1（t/t），则按式（17.7）可计算出 r =1.70（t/t），表明，目前情况下，单纯依靠提高 α 来提高 r，最大只能达到 1.70（t/t）。若假设 α、β、p 的数值保持不变，将铅的排放率 γ 降低到 0，则铅的资源效率也只能达到 1.21（t/t）。精铅产量的变化主要受市场需求的拉动，估计近期内精铅产量还将持续增长，因而，较难通过提高产量比来提高资源效率。不止如此，即使将 p 提高到 1，铅的资源效率也只能达到 0.99（t/t）。可见，在目前情况下，无论单独采用哪一种途径，都不可能大幅度地提高中国铅的资源效率。

但是，如果从以上三个方面同时抓起，那么，将较容易收到良好的效果。比如，在将 α 提高到 0.990（t/t），γ 降低到 0.050（t/t），p 提高到 1 的情况下，铅的资源效率将提高到 27.03（t/t）。

2. 提高铅的资源效率的建议

（1）在铅业生产过程中，加强管理，淘汰落后工艺、技术和设备，推行"清洁生产"，提高铅的利用率，减少废铅的排放。

（2）借鉴国外先进的管理经验，建立废铅回收法规，完善回收机制，促进废铅的回收，提高循环率。例如，延伸产品生产企业的责任，实行生产企业废铅回收责任制；制作铅制品回收标签；向消费者收取产品消费税等。

（3）宣传可持续发展战略，加强宏观调控力度，引导消费心理，寻找铅的替代品，降低铅的消费需求，从而提高精铅产量比。

17.3 结　　论

（1）引入了铅的资源效率这一概念。并获得了铅的资源效率的变化规律。

（2）近10年来，中国铅的资源效率只有0.85～1.06（t/t），平均为0.91（t/t），与瑞典铅酸电池系统相比，铅的资源效率十分低下。

（3）造成中国铅的资源效率低下的原因，一是近10年来，中国铅产量高速增长；二是铅的循环状况较差；三是废铅排放严重。

（4）可以预料，如果加强宏观调控力度，从以上三个方面着手改善中国铅业，那么，中国铅的资源效率可望得到较大的提高。

参 考 文 献

蔡显弟，冯君丛，江达，等. 1996. 中国铅的消费及回收. 世界有色金属，(6)：37-40.

姜松. 2000. 中国再生有色金属资源的开发利用. 中国资源综合利用，(1)：18-21.

兰兴华，殷建华. 2000. 发展中的再生铅工业. 中国资源综合利用，(8)：19-21.

陆钟武. 2000. 关于钢铁工业废钢资源的基础研究. 金属学报，36（7）：728-734.

陆钟武. 2002. 钢铁产品生命周期的铁流分析——关于铁排放量源头指标等问题的基础研究. 金属学报，38（1）：58-68.

杨春明，马永刚. 2000. 中国废铅蓄电池回收和再生铅生产. 第七届全国铅酸蓄电池学术年会论文全集. 广东南海，198-202.

Eco-Efficient 2002 Fair. 2002. Eco-efficiency—Creating Well-being and Business by Increasing Resource Productivity[EB/OL]. Http://www.eco-efficient.net/teh.html.

Graedel T E，Allenby B R. 1995. Industrial Ecology. New Jersey：Prentice Hall：5-8.

Halada K，Yamamoto R. 2001. A Review of the Development of Resource Productivity Concept and Eco-efficient Technology. International Sysposium on Industry and Ecology 2001. Shenyang：Northeastern University. PRC，14-36.

Karlsson S. 1999. Closing the technospheric flow of toxic metals—modeling lead losses from a lead-acid battery system for Sweden. Journal of Industrial Ecology，3（1）：23-40.

Kleign R. 1999. In=out：the trivial central paradigm of MFA?. Journal of Industrial Ecology，3（2，3）：11-12.

第18章 铅元素人为循环释放物的风险评价
Risk Assessment of Lead Emissions from Anthropogenic Cycle[*]

18.1 Introduction

Lead is one of the most abundant and toxic heavy metals in the environment (Niisoe et al., 2011). There are both natural and anthropogenic sources for lead emissions, and the anthropogenic sources dominate the emissions (Klee and Graedel, 2004). According to study, more than 95% of the lead within the biosphere is of anthropogenic origin (Smith and Flegal, 1995). Although lead abatement programs are provided in many developed countries, lead risk is still an important concern in the developing countries such as China (He et al., 2009). Threaten imposed by lead pollution to human health and the ecosystem still deserves our intensive attention. In this context, the risk assessment can be used to support the decision making in lead pollution management.

A wide variety of studies are already done on lead risk assessment, which can roughly be classified into the risk assessment of human health and the ecological. Human health risk assessment is defined as the process which estimates the likelihood of adverse health effects on humans, who may be exposed to chemicals in contaminated environmental media (NAS/NRC, 1983). Up to date, the human health risk assessment are generally determined by gauging the concentration in the environment, applying the method recommended by U.S. EPA (Li et al., 2014; Liu et al., 2014). On the other hand, the ecological risk assessment evaluates the likelihood that adverse ecological effects may occur or are occurring as a result of exposure to one or more stressors (EPA, 2014). For the ecological risk assessment, the method of hazard quotient is most widely applied (Liu et al., 2014; EL-Said and Youssef, 2013; Luo et al., 2010). All the present studies provide a clear way to measure lead risk. However, they fail to take the pollution sources into account, which are an integral part of risk formation. A study

---

[*] Liang J, Mao J S. 2016. Risk assessment of lead emissions from anthropogenic cycle. Transaction of Nonferrous Metals Society of China, 26 (1): 248-255.

attempted to assess the risk of lead losses, but it lacks a detailed speciation as well as the analysis of environmental fate and exposure (Mao et al., 2009). Thus, information on which process (or life cycle stage) should be paid attention to, and which chemical forms should be taken priority is still missing. This kind of knowledge is significant in helping direct limitation of social consumption and governmental resources management.

In this study, a model to assess lead risk was established. We estimated the source emissions and analyzed the environmental fate and exposure. Factors were applied to estimate the environmental decrease and exposure probability. Finally, the risk scores showing the levels of risk were calculated.

18.2 Methodology

18.2.1 Framework for Lead Risk Assessment

The framework for lead risk assessment consists of four components: source estimation, environmental fate analysis, exposure analysis and risk assessment (Fig. 18.1). The source estimation refers to the quantification and speciation of emissions from anthropogenic cycle. Environmental fate analysis refers to the analysis of the move and transformation of chemicals in the environment. Exposure analysis estimates the likelihood of exposure to chemicals in contaminated environmental media. And risk assessment includes three types of risk: the human health risk, the ecological risk and the total risk.

The risk of lead emissions (R) can be influenced by four factors:

$$R = Q \cdot f_e \cdot f_x \cdot E \quad (18.1)$$

where Q (t) is the emission quantity from the source; f_e is the fate coefficient, indicating the quantity decrease in the environment; f_x is the exposure coefficient, indicating the likelihood of lead exposure; E is the effect factor, showing the toxicity of lead emissions, represented by the unit risk score in the Indiana Relative Chemical Hazard (IRCH) ranking system (Green Media Toolshed, 2013). IRCH ranking system provides the unit human health risk score, unit ecological risk score and unit total risk score for various lead chemicals. As the scores in IRCH ranking system are obtained from the sum of the points assigned, therefore they do not have any units. In this study, we calculated the risk by multiplying the unit score by the emission quantities with a unified unit of t (ton). For easy comparison, the outcomes of the calculation were defined as scores without units.

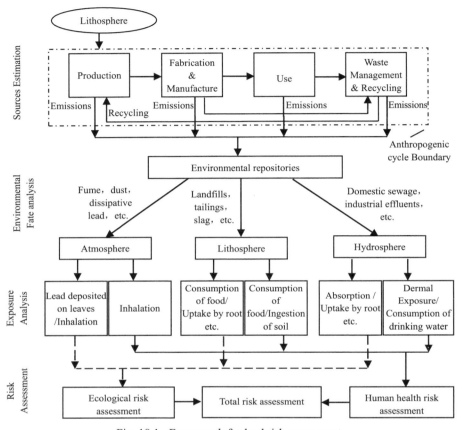

Fig. 18.1　Framework for lead risk assessment

The total risk for lead emissions integrates both the human health risk and the ecological risk. According to the definition of total risk in IRCH ranking system, the total risk R_T is defined as:

$$R_T = [(1.15 \cdot R_H) + (R_E / 3.5)] / 2 \qquad (18.2)$$

where R_H is the human health risk and R_E is the ecological risk. The coefficients of 1.15, 3.5 and 2 were applied by the IRCH ranking system based on how much the human health risk or the ecological risk contributes to the total risk.

18.2.2　Model for Lead Risk Assessment

1. Source estimations

The anthropogenic cycle includes all the stages in lead life cycle: Production,

Fabrication & Manufacture, Use and Waste Management & Recycling. At Production stage, the primary lead is refined from lead ores, and secondary lead is refined from lead waste. At F&M stage, lead product is manufactured. The Use stage is the phase where lead product provides service to human and satisfies the demands. At the Waste Management & Recycling stage, lead discards are recycled, disposed or landfilled. Due to the differences between the four stages, the lead chemical forms emitted vary from stage to stage. In our previous research, we have studied the quantities and chemical forms emitted from lead life cycle in 2010 for China, detailed calculation can be referred to the reference (Liang and Mao, 2014a; 2014b). We show the chemical forms and their quantities in Table 18.1.

Table 18.1 Chemical forms and their quantities emitted from anthropogenic lead cycle in 2010 in China

Life-cycle stage	PbO	$PbSO_4$	Pb	$PbCO_3$	PbO_2	PbS	$PbCl_2$	Others	Total
Production/t	66	160	0	110	6	89	0	18	450
Fabrication &Manufacture/t	47	5	19	0	0	0	0	1	72
Use/t	214	96	218	41	72	0	7	91	740
Waste Management & Recycling/t	195	176	70	45	69	2	41	27	625
Total/t	523	437	307	196	147	91	48	137	1887

2. Environmental fate analysis

In this section, we firstly analyzed what media the chemicals enter, and then we considered the technological treatment and natural decrease, which together result in a decrease in quantity of emissions.

The environmental fate depends on the receiving media as well as the chemicals themselves. The types of lead wastes discharged from the life cycle stages have already been analyzed in previous study (Liang and Mao, 2014b). In this study, we estimated the fate of chemical forms according to the possible fate of the wastes. For example, the tailings are emitted and mainly go to soil at Production stage. As $PbSO_4$ mainly exists in the tailings, we assume that for $PbSO_4$ emissions at production stage, 0.8 of them go into soil, and the rest go into water and air with an equal proportion, respectively. The fate of all the other chemicals can be obtained with similar analysis.

Due to human treatment methods, some emissions are removed technologically when introduced into air, water and soil. According to China Statistical Report on Environment in 2010 (Luo, 2011), the removal rate from effluents by standard waste water treatment methods is 0.9, with 0.1 still left in water; The removal rate of lead in soil is estimated to be 0.3 based on the treatment rate of multiple solid wastes recorded in the China Statistical Yearbook on Environment (National Bureau of Statistics of PRC, 2011). For the removal rate in the air, 0.95 of the total is assumed to be removed because the collection efficiency of dust collector in China is about 95% (Deng et al., 2013).

For lead emissions that stay in the environment, some have little mobility, and therefore they have minimal exposure potential and pose little or no danger to human or ecosystem. Decrease factors were used here to measure how much emissions exactly have the risk potential. For lead in the water, the precipitation reactions, along with strong absorption by suspended particle, causes a sharp decrease in quantity. The soluble lead was deduced from the solubility product constant (Beneš, 1985). For lead in the soil, lead is immobilized through sorption or precipitation processes, and usually remains on the surface. The soluble lead in soil was estimated from the absorption rate in soil (Ma et al., 2010). For lead in the air, the atmospheric deposition, a transport process from air to water or soil, causes a quantitative decrease in the atmosphere. This decrease factor was estimated based on the information that atmospheric deposition is about 1mg/d in polluted areas (Renberg et al., 2000).

3. Exposure analysis

Exposure analysis includes the analysis of exposure endpoint, routes and likelihood. The endpoint of health risk is human body. The main routes for human exposure are consumption of the contaminated food and drinking water, ingestion of soil and dust, inhalation absorption and dermal absorption. Lead in the soil can be exposed by food consumption or soil ingestion. For the food consumption, the exposure likelihood is represented by dietary absorption rate, which is 50% (UNEP, 2010). For the soil ingestion, as the bioavailability of lead in soil is about 60% of that in water and food, therefore, the absorption rate of soil lead is 0.3 (EPA, 2013a). The inhalation intake and dermal intake was estimated based on absorption coefficient (Chen et al., 2012) and skin permeability (EPA, 2013b), respectively. As for lead in the water, human can be exposed by water consumption or having bath. As a study on the drinking water in China shows that there is no health risk caused by lead (Gu et al., 2011), we consider the risk from drinking water is insignificant.

The endpoints of ecological risk include the animals and plants in ecosystem. Due to the richness of biodiversity and the complexity of ecosystem, the existing studies on ecological risk assessment just started. The ecological exposure risk is related to many factors, such as body weight, contact route, population dynamics. Some detailed exposure factors are given for the some specific animal species (EPA, 2013c), however, a general factor has not been given for all the animals and plants. In this study, we will not give the exposure factors for the ecological risk assessment, and just assume that all the lead emissions remaining in environment are to pose an ecological risk in the long run.

All the environmental decrease factors and exposure factors in the risk assessment model can be seen in Table 18.2.

Table 18.2　Fate factors and exposure factors for lead emissions in risk assessment

Type	Subdivision	Decrease factors	Reference
Decrease factors in the environment	Decrease rate caused by water treatment	0.1	Luo, 2011
	Decrease rate caused by solid waste treatment	0.3	National Bureau of Statistics of PRC, 2011
	Decrease rate caused by dust collection	0.05	Deng, et al., 2013
	Natural decrease factor in water	10^{-5}	Ma, et al., 2010
	Natural decrease factor in soil	10^{-2}	Renberg, et al., 2000
	Natural decrease factor in air	10^{-2}	UNEP, 2010
Exposure factors	Exposure by food and water consumption	0.5	U.S. EPA, 2013a
	Exposure by soil or dust ingestion	0.3	Chen, et al., 2012
	Exposure from inhalation	0.5	U.S. EPA, 2013b
	Exposure by dermal absorption	10^{-5}	Gu et al., 2011

18.2.3　Risk Assessment

In this study, we calculated the risk scores for all three kinds of risks based on the unit risk scores from the IRCH ranking system (Green Media Toolshed, 2013). The unit scores for chemicals such as $PbSO_4$, PbS and Pb are already available from this ranking system, however, the scores for $PbCO_3$, PbO_2 and PbO cannot be obtained. Estimations were made by applying the assignment principles of this ranking system, which indicate that the risk score of a chemical relies on its presence on the regulatory and target lists of the government.

According to the IRCH ranking system, the lowest ecological risk score is 75 for lead thiocyanate. Lead thiocyanate exists in the government regulation list of Dangerous for the Environment (Nordic Council of Ministers, 2013) while $PbCO_3$ does not. This indicates that $PbCO_3$ has a lower risk than lead thiocyanate according to the principals of this system. We assumed $PbCO_3$'s risk score to be 74. As PbO and PbO_2 are on this very list, the score 76 was assigned to them.

Then, the unit human health risk scores for PbO, PbO_2 and $PbCO_3$ were estimated. We obtained some information regarding the chemical's health effects from International Chemical Safety Card (National Institute for Occupational Safety and Health, 2013), using Pb as a reference. Pb, PbO and PbO_2 are similar in many properties except that PbO_2 is more active than Pb with strong oxidability. The unit risk score of PbO and PbO_2 was estimated to be 22 and 23. As $PbCO_3$ has a low capacity to impact human health, its score was set to be 11, using the scores for other metals in this system for reference. All the unit risk scores for lead chemicals are shown in Table 18.3.

Table 18.3　Unit risk scores for lead chemicals according to the IRCH ranking system

Chemicals	Unit human health risk score	Unit ecological risk score	Unit total risk score
Pb	22	140	33
$PbCl_2$	22	130	32
PbS	22	86	25
$PbSO_4$	31	100	33
$PbCO_3$	11	74	17
PbO_2	23	76	24
PbO	22	76	23
Others	22	103	29

18.3　Results and Discussions

18.3.1　Human Health Risk Assessment

Based on the model of lead risk assessment, the lead intake from food consumption was the most while the dermal intake was the least in 2010. It is known from previous study that lead emissions were 1887 t from the source in 2010 for China, mainly from the Use stage (39%) (Liang and Mao, 2014a). However, after the processes of the environmental fate and risk exposure, lead emissions that reach human bodies mainly were from the Waste Management & Recycling stage (42% of the total) and the Production stage (33% of the total). The health risk also mainly came from the Waste Management & Recycling stage and Production stage.

In terms of chemical forms, PbO (523 t) and $PbSO_4$ (437 t) were the most from the source, occupying 28% and 23%, respectively (Liang and Mao, 2014a).

After the processes of environmental fate and risk exposure, the quantity that reached human bodies was 1.3 t, which equals 0.9 mg on per capital level, with $PbSO_4$ being 0.29 mg/ca, PbO 0.22 mg/ca and $PbCO_3$ 0.10 mg/ca (population in China set as 1.36 billion). Based on equation (18.1) and the unit risk scores, the total human health risk score was calculated to be 30. For PbO and $PbSO_4$, their health risk scores were 12 and 7, respectively. Altogether, these two chemical forms occupied 63% of the total. This percentage was higher than their quantitative percentage in the source emissions (51%). The distribution of human health risk among different life cycle stages and chemicals is shown in Fig. 18.2.

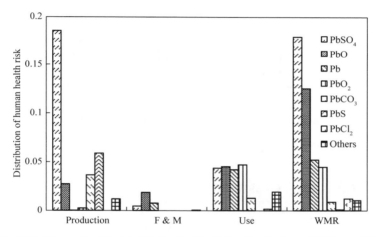

Fig.18.2 Human health risk distribution for the lead emissions of China in 2010

Note: F&M, Fabrication & Manufacture; WMR, Waste Management & Recycling.

18.3.2 Ecological Risk Assessment

Following the assessment model, the lead emissions that reach the ecosystem were quantified as 2.7 t. For the source emissions, they mainly came from the Use stage (as mentioned above), while the emissions posing an ecological risk mainly came from the Waste Management & Recycling stage and the Production stage (altogether 65% of the total). Based on the unit ecological risk scores and equation (18.1), the ecological risk score was calculated to be 255, with the stages of Waste Management & Recycling and stage contributing most.

In terms of chemical forms, the most chemicals emitted were PbO (523 t, 28%) and $PbSO_4$ (437 t, 23%). The most chemical quantitatively reaching the ecosystem was PbSO$_4$. For the total risk score, PbSO$_4$ (87, 38%), PbO (48, 28%) and Pb (42, 22%) contributed most. The distribution of this risk among the stages and chemicals is shown in Fig. 18.3.

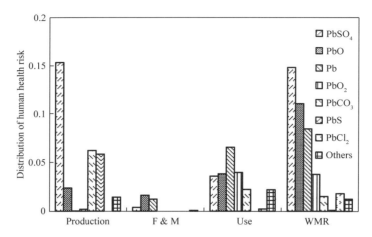

Fig. 18.3　Ecological risk distribution for the lead emissions of China in 2010

Note: F&M, Fabrication & Manufacture; WMR, Waste Management & Recycling.

18.3.3　Total Risk Assessment

With equation (18.2), the total risk was calculated as 54. The total risk mainly came from the Waste Management & Recycling stage (43%) and the Production stage (32%). As for the chemical forms, the total risk was mainly caused by PbSO$_4$ (36%), PbO (20%) and Pb (14%). The ecological risk was larger than the human health risk, and the total risk score was between them (Table 18.4). The total lead risk score is significant in that the risk score of lead can be compared with the scores of other metals such as copper and zinc to measure their overall risks.

Table 18.4　Human health risk, ecological risk and the total risk score in 2010 in China based on IRCH ranking system

Life cycle stage	Human health risk score	Ecological risk score	Total risk score
Production	10	80	17
Fabrication & Manufacture	1	8	2
Use	6	58	12
Waste Management & Recycling	13	109	23
Total	30	255	54

The quantity of emissions that caused an ecological risk was twice the quantity of the health risk, but the total ecological risk was over eight times the health risk, which probably resulted from the higher unit ecological risk scores (Fig. 18.4). For both risks, $PbSO_4$ and PbO were the primary chemicals that contributed most. Although the emission quantity of $PbCO_3$ causing risk was no less than Pb, its human and ecological risks were less. By applying equation (18.2), the chemicals' contribution to the total risk can be obtained from the chemicals' contribution to the human health risk and ecological risk. Pb contributed more than $PbCO_3$ to the total risk. To conclude, $PbSO_4$, PbO and Pb altogether occupied 71% of the total risk, which implies that these three chemicals should be taken priority for pollution control.

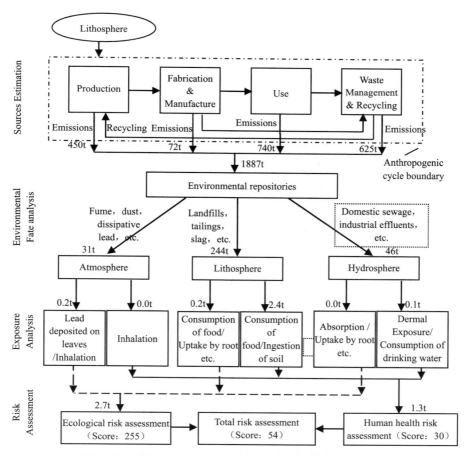

Fig. 18.4 Risk assessment of lead emissions in China in 2010

18.3.4　Uncertainty Analysis

In this section, we identified the main sources of uncertainty in the risk estimates presented in this study:

(1) The uncertainty of the hypothesis in the environmental process: When lead emissions enter the environment, their chemical forms will definitely change due to the interactions between them. In this study, we applied the chemical forms from the source to evaluate the risk, without considering the specific changes of chemicals in the environment, which may lead to some imprecision. In addition, the environmental fate of chemicals was estimated by assigning factors, which only showed a possible scenario in the environment.

(2) Uncertainty in the exposure process: As the technology differs greatly from one area to another, the lead pollution is at different levels. In addition, some people are more vulnerable to lead pollution and thus are exposed to more risk.

(3) Uncertainty in source estimation: The quantities of chemicals were quoted from the previous work, in which we considered the chemical forms being emitted perhaps after one transformation.

(4) Uncertainty in factor estimation: Because we tried to estimate the lead risk in a new method, the studies that can be referred to are limited. The determination of factors such as the decrease factors and exposure factors were attempted, and their precision remained to be checked in the future.

(5) Uncertainly in the unit risk scores: The scores in the IRCH ranking system does not cover all the chemicals, and the estimation for the unlisted ones cannot be as precise as the listed ones.

18.4　Conclusions

In this study, a model for lead risk assessment was established, and the human health risk, ecological risk and total risk were assessed. The source emissions were 1887 t in 2010 for China. The lead that reached human body was quantified to be 1.3 t (0.9 mg/ca) while the lead that reached ecosystem was 2.7 t. Based on the unit risk score obtained from the IRCH ranking system, the human health risk was calculated to be 30, the ecological risk was 255 and the total risk was 54.

In terms of life cycle stages, lead emissions mainly came from the Use stage while lead emissions reaching human and ecosystem mainly came from Waste Management & Recycling and Production stage. These two stages also contributed most to the total risk.

In terms of chemical forms, the most chemicals discharged from the source were PbO and $PbSO_4$. The most significant chemical for the health risk was PbO while the major chemical for ecological risk was $PbSO_4$. For the total risk assessment, $PbSO_4$, PbO and Pb altogether contributed 71%, which implies that those three forms should be taken priority for lead pollution control.

References

Beneš P, Čejchanová M, Havlík B. 1985. Migration and speciation of lead in a river system heavily polluted from a smelter. Water Research, 19 (1): 1-6

Chen L G, Xu Z C, Liu M, et al. 2012. Lead exposure assessment from study near a lead-acid battery factory in China. Science of the Total Environment, 429: 191-198.

Deng S, Zhang F, Liu Y, et al. Lead emission and speciation of coal-fired power plants in China. China Environmental Science, 33 (7): 1199-1206. (in Chinese)

El-said G F, Youssef D H. 2013. Ecotoxicological impact assessment of some heavy metals and their distribution in some fractions of mangrove sediments from Red Sea, Egypt. Environmental Monitoring and Assessment, 185: 393-404.

Green Media Toolshed. 2013. Scorecard: the pollution information site. 2013-12-11.http: //scorecard.goodguide.com/chemical-profiles/

Gu C L, Wang J P, Wang S. 2011. Exposure assessment for heavy metals in drinking water. Science and Technology of Food Industry, 32 (11): 374-376. (in Chinese)

He K M, Wang S Q, Zhang J L. 2009. Blood lead levels of children and its trend in China. Science of the Total Environment, 407 (13): 3986-3993.

Klee R J, Graedel T E. 2004. Elemental cycles: A status report on human or natural dominance. Annual Review of Environment and Resources, 29: 69-107.

Li Z Y, Ma Z W, Kuijp T J, et al. A review of soil heavy metal pollution from mines in China: Pollution and health risk assessment. Science of the Total Environment, 468-469: 843-853.

Liang J, Mao J S. 2014a. Environmental losses from anthropogenic lead flow and their accumulation in China: A dynamic analysis. Transactions of Nonferrous Metals Society of China, (24): 1125-1133.

Liang J, Mao J S. 2014b. Speciation analysis of lead losses from anthropogenic flow in China. Environmental Science, 5 (3): 1191-1197. (in Chinese)

Liu G N, Yu Y J, Hou J, et al. 2014. An ecological risk assessment of heavy metal pollution of the agricultural ecosystem near a lead-acid battery factory. Ecological Indicators, 47: 210-218.

Liu J L, Wu H, Feng J X, et al. 2014. Heavy metal contamination and ecological risk assessments in the sediments and zoobenthos of selected mangrove ecosystems, South China. CATENA, 119: 136-142.

Luo W, Lu Y L, Wang T Y, et al. 2010. Ecological risk assessment of arsenic and metals in sediments of coastal areas of northern Bohai and Yellow Seas, China. AMBIO, 39: 367-375.

Luo Y. 2011. China statistical report on environment in 2010 (1st Ed.). Beijing: China Environmental Science Press. (in Chinese)

Ma L, Xu R, Jiang J. 2010. Adsorption and desorption of Cu (II) and Pb (II) in paddy soils cultivated for various years in the subtropical China, Journal of Environmental Science (China), 22 (5): 689-695.

Mao J S, Cao J, Graedel T E. 2009. Losses to the environment from the multilevel cycle of anthropogenic lead. Environmental Pollution, 157: 2670-2677.

NAS/NRC (National Academy of Sciences/National Research Council). 1983. Risk assessment in the federal government: Managing the process. Washington, DC: National Academy Press.

National Bureau of Statistics of PRC, Ministry of Environmental Protection of PRC. 2011. China statistical yearbook on environment. Beijing: China Statistics Press.

National Institute for Occupational Safety and Health. 2013. International chemical safety card. 2013-12-26. http://www.cdc.gov/niosh/ipcsneng/neng1001.html.

Niisoe T, Harada K H, Hitomi T, et al. 2011. Environmental ecological modeling of human blood lead levels in East Asia. Environmental Science and Technology, 45 (7): 2856-2862.

Nordic Council of Ministers. 2013. Dangerous for the Environment: list of chemicals. 2013-11-09. http://scorecard.goodguide.com/chemical-groups/one-list.tcl?short_list_name=dfe.

Renberg I, Brannvall M, Bindler R, et al. Atmospheric lead pollution history during four millennia (2000 BC to 2000 AD) in Sweden. AMBIO, 29 (3): 150-156.

Smith D R, Flegal A R. 1995. Lead in the biosphere: Recent trends. AMBIO, 24 (1): 21-23.

U.S. EPA. 2013a. Estimation of relative bioavailability of lead in soil and soil-like materials using in vivo and in vitro methods. 2013-12-11. http://www.epa.gov/superfund/bioavailability/lead_tsd_main.pdf.

U.S. EPA. 2013b. EPA/600/8-91/011B. Dermal Exposure Assessment: Principles and applications. 2013-12-11. http://rais.ornl.gov/documents/DERM_EXP.PDF.

U.S. EPA. 2013c. Wildlife Exposure Factors Handbook. 2013-12-11. http://ofmpub.epa.gov/eims/eimscomm.getfile?p_download_id=489605.

U.S. EPA. 2014. Framework for ecological risk assessment. 2014-03-03. http://www.epa.gov/raf/publications/pdfs/FRM-WRK_ERA.PDF.

UNEP (United Nations Environment Programme). 2010. Final review of scientific information on lead. 2013-12-11. http://www.unep.org/chemicalsandwaste/Portals/9/Lead_Cadmium/docs/Interim_reviews/UNEP_GC26_INF_11_Add_1_Final_UNEP_Lead_review_and_apppendix_Dec_2010.pdf.

第 19 章　物质的循环流动与价值循环流动
The Material Circular Flow and Value Circular Flow*

循环经济是保护资源与环境、实现可持续发展的重要途径。循环经济的核心是经济系统中物质的循环流动（Robert，1994），比如废物的资源化（David and Nasrin，1994）、工业生态园内的资源共享和废物交换利用（Henning，1997）等。物质作为价值的载体，伴随物质的循环流动，将存在价值的循环流动。这一流动势必对经济系统产生重大影响。因此，建立价值循环流动对分析有关的经济问题具有重大的现实意义。然而，由于循环经济刚刚起步，较为深入的理论研究为数不多，至今尚未见到价值循环流动的报道。

陆钟武（陆钟武，2002a；陆钟武，2002b）曾针对产品中的某一种元素 M，深入地研究了工业物质的循环流动，并获得了相应的规律。如果以此为基础，结合经济学的价值理论，把单位质量的元素 M 具有的价值定义为元素 M 的"价位"，就能得知物质循环流动下的价值流动。本章将沿着这个思路，分析产品生命周期中元素 M 的循环流动和价位变化，从而建立产品生命周期中价值的循环流动。

19.1　元素 M 的循环流动

工业产品的生命周期通常包括工业材料的生产、产品的制造、使用、以及到达使用寿命报废后的废物再生等几个阶段（Graedel and Allenby，1995），是一个漫长的物质流动过程。假设元素 M 是产品中的一个主要组分，那么针对元素 M，可绘制产品生命周期的元素 M 流图，如图 19.1 所示。图中 I、II、III 分别代表产品生命周期的三个不同阶段。

图 19.1 中各变量含义如下：

R_k——天然资源投入量，t；

R_z、R_j——分别为折旧废物、加工废物的投入量，t；

---

* 毛建素，陆钟武. 2003. 物质的循环流动与价值循环流动. 材料与冶金学报，2（2）：157-160.

P_m——产品制造阶段生产原料的投入量,同时也是工业材料生产阶段的产出量,t;

P——产品产量,t;

W_1、W_2、W_3——分别为工业材料生产阶段、产品制造阶段,以及产品使用报废后,排放的废物和污染物的数量,t;

W——元素 M 的排放总量,$W = W_1 + W_2 + W_3$,t。

请注意,以上各种物料的数量均按元素 M 的含量计算。

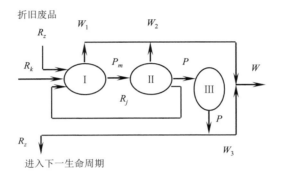

图 19.1 产品生命周期元素 M 流图

Fig. 19.1 Element flow diagram for product life cycle

第Ⅰ阶段——工业材料的生产;第Ⅱ阶段——产品制造;第Ⅲ阶段——产品使用

从图 19.1 不难看出,在元素 M 的流动过程中,不仅有从天然资源流向环境的单向流动,而且还有循环流动:

(1)大循环——工业产品经使用报废后,部分元素 M 返回工业材料的生产阶段,作为原料,重新利用。如图 19.1 中折旧废物 R_z 的流动。

(2)中循环——产品制造阶段产生的废物中,部分元素 M 返回工业材料的生产阶段,作为原料,重新利用。如图 19.1 中加工废物 R_j 的流动。

此外,还有小循环——某一生命周期阶段产生的废物中,部分元素 M 返回生命周期的同一阶段重新利用。主要表现为企业内部下游工序的废物,返回上游工序,作为原料,重新利用。由于小循环是物质在生命周期内的同一阶段内部进行的循环,因此图 19.1 中没有标出。

就整个产品系统而言,由于这三种循环流动的存在,使得投入单位天然资源所产出的工业产品的数量显著增加,而排向环境的废物、污染物的数量明显减少,换句话说,资源效率和环境效率(Eco-Efficient 2002 Fair,2002)大大提高,从而起到了节约天然资源、保护环境的作用。

19.2 元素 M 的价位变化

如果将产品的价值分摊给他的各组成元素，那么产品中的每种元素都将成为价值的载体。若用元素 M 代表产品，则产品的价值就简化成元素 M 的价值。在产品生命周期中，伴随产品价值的形成和消退，元素 M 的价位将不断变化。

19.2.1 元素 M 的价位

定义某股物流中，单位质量的元素 M 离开生命周期某一阶段时的价值，为该股物流在这一阶段元素 M 的价位。他等于相应物流的价值除以物流中元素 M 的质量，单位是元/t。比如，某钢铁厂生产出 1 万 t 钢材，价值 1500 万元，由此算得钢铁生产阶段铁元素的平均价位为 1500 元/t。

元素 M 的价位反映工业系统对元素 M 的加工程度。元素 M 的价位越高，表示其加工程度越高，工业附加值也越高。较为明显的例子是，金属产品（如钢铁、铝、铜等）中元素 M 的价位，普遍高于相应金属矿粉中元素 M 的价位，但却低于金属制品（如钢瓶、铝罐、铜管等）中金属元素 M 的价位。

另外，元素 M 的价位还受产品类型、生命周期阶段、地域、技术装备、管理水平等的影响。例如，汽车部件中铁元素的价位可高达几千元，甚至上万元，而铸铁下水管道中的铁元素的价位通常只有一千元左右。

工业系统对外部环境造成的经济影响，是经济系统外部性问题（Deacon，1998）的研究内容。当某一工业部门向环境索取元素 M 时，如果使资源环境受损，即外部不经济性，那么该工业部门应向资源管理部门缴纳资源使用费（Annegrete，1998），这时资源流中元素 M 的价位按正值计算；反之，若向环境索取元素 M 使外部环境受益，即外部经济性，则该工业部门可从受益部门回收部分资金，这时资源流中元素 M 的价位按负值计算。当某一工业部门向环境排放元素 M 时，如果使外部环境受损，即外部不经济性，那么该工业部门应向环境治理部门缴纳排污费（Turner et al.，1998），这时，排放物流中元素 M 的价位应按负值计算；反之，若向环境排放元素 M 使外部环境受益，即外部经济性，则该工业部门可通过销售废物资源的方式，从受益部门回收部分资金或通过免缴排污费的方式，节省部分资金，这时废物流中元素 M 的价位按正值计算。

19.2.2 元素 M 的价位变化

在产品生命周期中，伴随着工业生产活动，生产原料发生一系列物理化学变化，并转化成不同的工业产品，元素 M 的价位得到不同程度的提升；在产品的使用阶段，伴随产品的使用，组成产品的物质不断磨损，产品的服务能力逐渐削减，元素 M 的价位逐渐降低。元素 M 的价位随产品生命周期阶段不断变化的现象，称为元素 M 的价位变化。按照这种思路，绘制出元素 M 的价位随产品生命周期的变化过程，得到图 19.2，称为元素 M 的生命周期价位变化图。不难看出，元素 M 的价位仅在经过生命周期各阶段时才会发生变化。

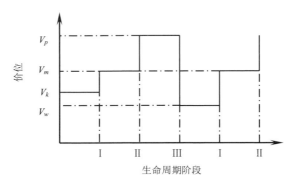

图 19.2 元素 M 生命周期价位变化

Fig. 19.2 The changing of value level for element M life cycle

通常情况下，在产品的生命周期中，元素 M 主要以天然资源、工业材料、产品和排放的废物 4 种形式存在，相应地，元素 M 的价位大致分为以下 4 种：

（1）天然资源中元素 M 的价位，用 V_k 表示；

（2）工业材料中元素 M 的价位，用 V_m 表示；

（3）产品中元素 M 的价位，用 V_p 表示；

（4）排放的废物，看做废物资源，其中的元素 M 的价位，用 V_w 表示。

例如，在铅酸电池生命周期中，铅的提炼、铅酸电池制造、使用等过程构成其生命周期的不同阶段。根据 1999 年中国对外贸易部的统计数据，可计算得到不同阶段铅元素的价位：V_k =150.62 美元/t；V_m =468.19 美元/t；V_p =2381.45 美元/t；V_w =145.94 美元/t。由此可看出，铅元素的价位随着生命周期阶段的升降变化过程。

19.2.3 价值的循环流动

不难理解，在产品生命周期中，伴随每股元素 M 的循环流动，存在着价值的循环流动，而且，每股价值流的流量，应等于元素 M 的质量流量乘以元素 M 在该生命周期阶段相应物流中的价位。根据图 19.1 中的元素 M 的流量，以及图 19.2 中元素 M 的价位，可以绘制出产品生命周期元素 M 的价值流动图，如图 19.3 所示。该图也可称为产品生命周期的**价值循环流动图**。

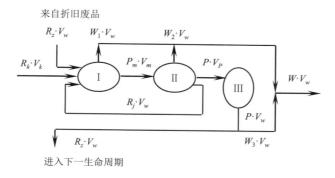

图 19.3 产品生命周期价值循环流动图

Fig. 19.3 Value circular flow diagram for product life

由图 19.3 可见，在第Ⅰ阶段，伴随物料的投入，共有 R_kV_k、R_zV_w、R_jV_w 三股价值流入，其中 R_zV_w、R_jV_w 是伴随元素 M 的循环流动产生的，它的存在将替代一定数量的与天然资源有关的价值流入。比如，当天然资源与废物资源具有相同的元素 M 的收得率时，这两股价值循环流将替代数量为 $(R_z+R_j)V_k$ 的与天然资源有关的价值流，从而，将可节省数量为 $(R_z+R_j)\cdot(V_k-V_w)$ 的价值。在第Ⅱ阶段，价值流 R_jV_w 是伴随元素 M 的中循环产生的，它的存在不仅减少了排污费 R_jV_w，而且还增加了数量为 R_jV_w 价值产出，共增加收入 $2R_jV_w$。在第Ⅲ阶段，价值流 R_zV_w（与第Ⅰ阶段的价值流入 R_zV_w 不一定相同）是伴随元素 M 的大循环产生的，它的存在将减少消费者的排污费 R_zV_w，还由于售出废物资源使消费者回收资金 R_zV_w，这两项合计，共提高经济收入 $2R_zV_w$。综上所述，对于生产者和消费者来说，价值的循环流动，都将起到"节支增收"的作用。

按照上述方法，可分析价值循环流动对各经济部门造成的影响。例如，对于铅酸电池系统，根据中国 1999 年铅业统计数据，并考虑前文中得到的铅元素的价位，可整理得到铅酸电池生命周期价值循环流动图，如图 19.4 所示。图中单位为百万美元。由此可算出，在第Ⅰ阶段，由于铅元素的循环流动，将"节支"48.39 万美元。在第Ⅱ阶段，由于循环价值流 182.4 万美元的存在，可"增收"364.8 万美元。在第Ⅲ阶段，若能够回收 31.2%的废旧电池，则将形成 1470 万美元的价值循环流动，从而可帮助消费者回收资金 2940 千万美元。在此再次看到，价值的循环流动，可为生产者和消费者带来可观的经济效益。

总体看来，所得的价值循环流动具有以下几个特点：①建立在物质循环流动的基础上，可结合物质的循环流动规律，特别是物质的动态循环规律（陆钟武，

2002a；陆钟武，2002b），来研究经济的动态活动规律。②将经济系统以外的环境，看作有机的一个整体，因此无论是作为天然资源中的元素 M，还是作为废物中的元素 M，都具有一定的价值，体现了工业系统与自然环境之间的经济关系。③天然资源与废物资源中元素 M 的价位的不同，说明这两种资源在元素 M 的使用性能上的差别，这种差别是分析废物资源利用与经济收益之间关系的基础。④由于资源（包括天然资源和废物资源）中元素 M 具有一定的价位，因此可以在价值循环流动基础上，运用经济手段，进行资源与环境的管理。

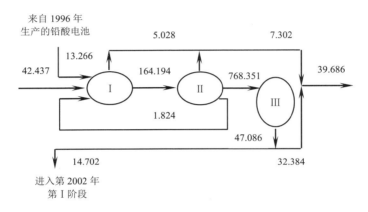

图 19.4 1999 年中国铅酸电池生命周期价值流动

Fig. 19.4 The value circular flow for lead-acid battery life cycle in 1999 of China

19.3 结　　语

本文选用产品中的某一种元素 M 作为产品的代表性物质，分析了产品生命周期中物质的循环流动。提出了产品生命周期"元素 M 的价位"这一概念。在此基础上，分析了产品生命周期中元素 M 的价位变化过程。把物质看作价值的载体，得到了工业物质循环流动下价值的循环流动。价值的循环流动将对经济、社会等产生深远的影响。

参 考 文 献

陆钟武. 2002a. 钢铁产品生命周期的铁流分析——关于铁排放量源头指标等问题的基础研究. 金属学报，38（1）：58-68.

陆钟武. 2002b. 论钢铁工业的废钢资源. 钢铁，37（4）：66-70.

Annegrete B. 1998. Taxing virgin materials：an approach to waste problems. Resources，Conservation and Recycling，22：15-29.

David T A, Nasrin B. 1994. Wastes as Raw Materials. The Greening of industrial Ecosystems. Washington DC: National Academy Press: 69-89.

Deacon R T. 1998. Research Trend and Opportunities in Environmental and Nature Resource Economics. Environmental and Resource Economics, 11 (3, 4): 383-397.

Eco-Efficient 2002 Fair. 2002. Eco-efficiency—Creating Well-being and Business by Increasing Resource Productivity. Http://www.eco-efficient.net/teh.html.

Graedel T E, Allenby B R. 1995. Industrial ecology. New Jersey: Prentice Hall: 5-8.

Henning G. 1997. The Industrial Symbiosis at Kalundborg, Denmark. The Industrial Green Game. Washington DC: National Academy Press: 117-123.

Robert A, Frosch. 1994. Closing the Loop on Waste Materials. The Greening of industrial Ecosystems. Washington DC: National Academy Press: 37-47.

Turner R K, Salmons R, Powell J, et al. 1998. Green taxes, waste management and political economy. Journal of Environmental Management, 53: 121-136.

第20章 物质循环流动对价值源强的影响
The Influence of Recycling of Materials on Value Source Intensity*

20.1 Introduction

20.1.1 Background

"Circular economy" approach aims to regulate anthropogenic flows of materials according to the natural flow of materials in a closed (recycling) loop so that the impact of human activities on an ecosystem can be greatly reduced in terms of resource consumption, energy use, waste emissions, labor costs, etc. As a result, human development can be harmonized with the natural ecosystem(Frosch, 1994). Therefore, many related practices have been carried out around the world, including in Japan (Sakuragi, 2002) and the Netherlands (Hekkert and Joosten, 2000; Joosten et al., 1999) and the circular economy has been chosen as China's preferred strategy for the 21st century (Yu and Wang, 2001).

In general, the circular economy focuses on the recycling of materials in human economic systems, in which a series of product life cycle stages are included, such as the production of the material, manufacture of products from this material, consumption of the products, and discarding or recycling of the products after they have been abandoned (Graedel and Allenby, 1995). At present, the circular economy mainly emphasizes the recycling of the residues generated in every stage of the life cycle by means of utilization of the wastes(David and Nasrin, 1994), or exchange and utilization of the wastes in "eco-industry" parks (Grann, 1997). Whether recycling of the residues can generate economic benefits will be the key point in determining whether it is possible to successfully implement the circular economy, especially for developing countries. Ayres (1994) divided the industrial uses of materials into three classes according to their

---

* Mao J S, Ying Z W, Lu Z W. 2008. The Influence of Recycling of Material on Value Source Intensity. 2008 Proceedings of Information Technology and Environmental System Sciences, ITESS'2008 (Part 2): 1078-1084.

technological and economic feasibility under present prices and regulations. However, how can the recycling of materials generate economic benefits? What kind of conditions should be met for this to occur? To answer these questions, it would be necessary to conduct a deep theoretical study of the influence of recycling of materials on the economy. Because the circular economy is still a new economic pattern, there have been few studies to answer these questions. We conducted the present study to determine how the recycling of materials influences an economic system and to validate this understanding through a case study. This will not only answer these questions, but will also help to promote and further develop the use of the circular economy.

20.1.2 This Study

1. Evaluation index

In economic systems, a producer always acts as a value amplifier by transforming various inputs(such as labor, materials, energy, and capital)into a product that possesses special functions that meet social demand. During this process, the value of the material inputs is increased. Therefore, we can treat a producer as a source of and creator of value. It is easy to understand that for producers, their net income is usually the most important consideration, and this parameter is often related to the product output (quantity), the value increase per product, and the inputs of labor and facilities. Since the latter inputs are often directly proportional to the scale of production, we can filter out the influence of production scale on net income per unit of production; in that case, the value increase per product will be the key factor that determines the net income of producers. For this reason, we chose the value increase per product as an indirect index for evaluating the economic benefits for a producer. This increase is calculated as the value source intensity (VSI):

$$v_i = \frac{\Delta V_i}{P_i} \qquad (20.1)$$

where v_i is the VSI at stage i in a product's life cycle($/t); ΔV_i is the value increase at stage i ($), which equals the output value minus the input value; and P_i is the output of the principal product at stage i (t).

It is easy to understand that VSI is always proportional to the producer's net income under a certain production scale. In the present study, we have used VSI as the index for evaluating the producer's economic income.

2. General idea

Since the core content of a circular economy is the recycling of materials in economic system, studying the influence of recycling of materials on an economic system must begin with material flow analysis and an analysis of the recycling of materials. In this aspect, Lu (2000) developed a model for "element flow analysis" in product's life cycle and investigated the steel scrap for steel industry, and then Lu (2002) progressed to a deeper study of source index for iron emission in steel product system. All the studies validate the model and the method used. We have used Lu's method in the present study.

Beyond question, material substances in a product are the carriers of the product's value. Thus a value circular flow (VCF) must exist corresponding with the recycling of material in product's life cycle, and thereby we can analyze the VCF based on the analysis of the recycling of materials. Mao and Lu (2003a) researched this specific relationship, and the method used and the results obtained thereby will be referenced in the present study. Based on the concept of VCF, we will also analyze the value flow for every stage of a product's life cycle to identify the rules that govern changes in VSI.

To demonstrate how this approach can be applied, we will use the lead-acid battery system as an example. Based on data in the literature on this system, we will study the circular flow of lead (Pb) and its influence on VSI for China and Sweden to validate our predicted theoretical outcomes.

3. Contents

This study comprises two parts: the theoretical study and a case study of its application. In the former part, we will analyze the recycling of materials and VCF, and then study the rules that govern changes in VSI for all production stages in a product's life cycle. In the latter part, we will use the lead-acid battery system as an example of this approach by calculating the recycling of lead metal and VSI for all production stages in China and Sweden. In this way, the results of the theoretical study will be illustrated and made more concrete.

20.2 Primary Rules (Theoretical Study)

20.2.1 Methodology

1. Recycling of materials

In general, a material flow is pulled by a product flow in circular economy and the recycling of materials is one of the main flow in product life cycle. Thus, we can study the recycling of materials according to the product flow in its life cycle.

For a product composed of many kinds of substance, we can select one primary element of the product to represent the product itself. Then, based on the law of material conservation, i.e., inputs of that element equal outputs (Kleijn, 2000) for every stage, and we can draw an element-flow diagram for that element to indicate its distribution and direction of flow throughout the product's life cycle (Fig. 20.1). The numbers I, II, and III represent the three stages of the product's life cycle.

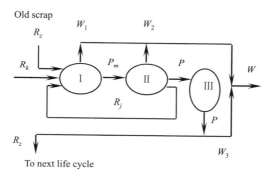

Fig. 20.1 The element-flow diagram for the element chosen to represent the product throughout its life cycle

stage I: material production; stage II: product manufacture; stageIII: product use

In Fig. 20.1, R_k represents the input of natural resources to stage I; R_z represents old scrap (i.e., post-consumer products) that serves as an input to stage I; R_j represents new scrap (i.e., scrap generated during production processes) that serves as an input to stage I; P_m represents the output of materials from stage I that becomes the material input to stage II; P represents the output of products from stage II; W_1, W_2, and W_3 represent the wastes generated at stages I, II, and III, respectively; and W represents the sum of the wastes generated during all three stages (i.e., $W = W_1 + W_2 + W_3$). The above-mentioned parameters will be calculated in tons of element content.

Fig. 20.1 shows that there are three kinds of recycling of materials in addition to the material flow in the form of natural resource linearly into the environment (Lu, 2003):

(1) a large recycling flow: some of the element returns to stage I as a material input in the form of old scrap (R_z in Fig. 20.1);

(2) an intermediate recycling flow: some of the element in the form of new scrap generated at stage II returns to stage I as an input (R_j in Fig. 20.1);

(3) A small recycling flow: some of the element in the form of new scrap generated in a certain stage returns to the same stage as an input. Although this is part of the material flow for a product's life cycle, we have not presented this in Fig. 20.1 because this kind of material flow happens only within a stage (not between stages).

To evaluate the recycling of materials, we have defined resource efficiency (r_1) and an element emission ratio (γ_1) for the representative element at stage I:

$$r_1 = \frac{P_m}{R} \tag{20.2}$$

$$\gamma_1 = \frac{W_1}{P_m} \tag{20.3}$$

where R represents the total quantity of the element used as inputs (i.e., Rz + Rk).

We have also defined utilization efficiency (r_2) and an element emission ratio (γ_2) for the representative element at stage II:

$$r_2 = \frac{P}{P_m} \tag{20.4}$$

$$\gamma_2 = \frac{W_2}{P} \tag{20.5}$$

The units of these four factors [equation (20.2) –equation (20.5)] have been chosen as (t/t) for compatibility with the case study presented later in this paper.

Fig. 20.1 shows that the greater the large recycling flow (R_z), the greater the resource efficiency. This means that the output of the product per unit of natural resource input will increase. In addition, the greater the recycling of materials, the lower the emissions of the element at various stages for a given level of output; this means that there will be less waste (W) generated per product. For the overall product system, the three kinds of recycling of materials can help to increase the product output per unit of natural resource input and reduce the waste or emissions per unit of product. Thus, a recycling of materials can be beneficial in terms of conservation of natural resources and in terms of environmental improvement.

2. Value circular flow

(1) The value level (VL) of the representative element.

If we apportion the value of a product among its components, every element in the product becomes the carrier of value. If we choose one element to represent the overall product, the value of the product can be simplified by focusing our analysis on the value of that element. We define the value carried by a unit mass of this element as it moves from one stage to the next as the element's value level (VL; $/t in our case study) at this stage. We can calculate VL by dividing the value of the material outflow by the mass of the element.

The VL of the element reflects the degree of value-added as a result of manufacturing for the element in the economic system, and will be influenced by many factors, such as the production pattern, stages in the product's life cycle, location of the manufacturing plant, technology and equipment employed in the manufacturing process, and so on. Therefore, we must choose a benchmark to allow us to compare the values of a given product at given stages. To meet this need, we have defined the VL of the representative element in terms of the average price in international trade for a given year as the reference value level (RVL) in that year. We have also calculated a VL based on the actual price, and defined this as the actual value level (AVL) of the element in the same year.

The economic system will also affect the exterior environment. In this context, the VL of the element in scrap that is reused will have a positive value because of its effect on saving material inputs. In contrast, the VL of the element in wastes or emissions into the environment will have a negative value because of the resulting environmental impact.

(2) Changes in VL during a product's life cycle.

It is easy to see that during the production activities in a product's life cycle, the material input will be changed physically or chemically and this material may thereby be transformed into different products. During this process, the element's VL will progressively increase. During use of the products, however, the VL of the element will gradually decrease. These changes in VL during the product's life cycle can be expressed as shown in Fig. 20.2, which shows how the element's VL changes between the various life cycle stages.

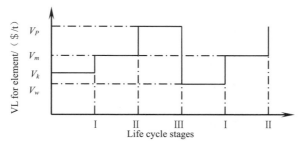

Fig. 20.2 The changes in value level (VL) for the element chosen to represent the product during the product's life cycle

V_k, V_m, V_p and V_w represent the VL of the element in natural resources, in industrial materials, in products, and in wastes and emissions, respectively.

(3) The value circular flow.

It is easy to see that a value circular flow (VCF) ill accompany the circular flow of the element in the product's life cycle and that the flux of every value flow should equal the quantity (e.g., mass) of the representative element multiplied by the VL of the element in the material flow. Based on the quantity in Fig. 20.1 and the VL in Fig. 20.2, we can draw a VCF diagram of the element throughout the product's life cycle (Fig. 20.3), which indicates that the flux of every value flow is tightly related to the corresponding mass flow and the VL of the element in these flows.

(4) The value source intensity (VSI).

Based on Fig. 20.1 and Fig. 20.2, the VSI for stages I and II in a product's life cycle can be derived using equation (20.1), where the output of stages I and II are $P_1 = P_m$ and $P_2 = P$, respectively.

Because the VL of the element can be defined in terms of RVL and AVL, and its value will become positive or negative depending upon whether the scrap is recycled, the VSI should be defined accordingly. Let us assume that the VL of the element equals the RVL and that all of the element in scrap will be utilized, partly by the same system and partly by other systems. This state is termed the "reference state", and the corresponding VSI for a certain stage in the product's life cycle is called the reference value source intensity (RVSI). Similarly, the VSI for a certain stage in a product's life cycle calculated based on the actual state is called the actual value source intensity (AVSI). RVSI represents the maximum value of AVSI. AVSI reflects the actual level for a certain region or producer in a product's life cycle stage. The higher the AVSI for a producer, the greater the producer's competitiveness will be. AVSI can thus be treated as an index for evaluating the management and operation of companies.

20.2.2 Results and Discussion

Based on the above discussion, we can derive the VSI for the production stages in a product's life cycle based on Fig. 20.3. The resulting equations are summarized in Table 20.1.

Table 20.1 shows that RVSI will always be greater than AVSI for every production stage. This means that the utilization of scrap as a resource can provide greater benefits than releasing these materials into the environment. The value of RVSI minus AVSI equals $2\gamma_i V_w$, which means that the lower the element emission ratio, the smaller the distinction between RVSI and AVSI. AVSI will equal RVSI at zero emissions of the representative element.

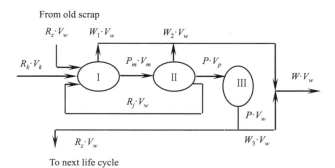

Fig. 20.3 The value circular flow (VCF) diagram for a product's life cycle

Table 20.1 The formulas for calculating value source intensity (VSI) during the production stages of a product's life cycle

	Material production stage	Product-manufacturing stage
RVSI	$(V_m - V_w) - \frac{1}{r_1}(V_k - V_w)$	$(V_p - V_w) - \frac{1}{r_2}(V_m - V_w)$
AVSI	$(V_m - V_w) - \frac{1}{r_1}(V_k - V_w) - 2\gamma_1 V_w$	$(V_p - V_w) - \frac{1}{r_2}(V_m - V_w) - 2\gamma_2 V_w$
RVSI-AVSI	$2\gamma_1 V_w$	$2\gamma_2 V_w$

The RVSI for stage I is related to the resource efficiency (i.e., r_1) and the difference in VL, including the difference in VL between industrial material and scrap ($V_m - V_w$) and between natural resources and scrap ($V_k - V_w$). RVSI will be related only to the resource efficiency at a constant difference in VL; thus, the greater the resource efficiency, the greater the RVSI. This results from the fact that higher resource efficiency means both an effective use of materials and a greater replacement of natural resources (with high VL) with scrap (with lower VL). This in turn results in a greater economic return per unit of product. It is important to note that the VSI for stage I will decrease with increasing resource efficiency if the opposite difference in VL exists between natural resources and scraps (i.e., $V_k < V_w$). In this situation, the use of scrap offers lower economic returns but still saves natural resources and reduces emissions into the environment. This situation mainly occurs when the scrap resource is limited, as is the case in most developing countries (Mao and Lu, 2003b) and some administrations with an irrational resource policy, such as countries that offer subsidies for the exploitation of scarce natural resources. These situations require an improvement in the relevant resource policies (Turner et al., 1998; Annegrete and Bruvoll, 1998).

The RVSI for stage II is related to the utilization efficiency (i.e., r_2) and the difference in VL, including the differences between industrial material and scrap ($V_m - V_w$) and between products and scrap ($V_p - V_w$). Because the VL of the representative element in the product is always higher than that in scrap, increasing the utilization efficiency can help increase RVSI, and the RVSI for this stage will be related only to the utilization efficiency at a constant difference in VL. In this case, the greater the utilization efficiency, the greater the RVSI will be.

Based on our theoretical approach, the value circular flow has the following characteristics:

(1) The VCF is built upon the recycling of materials. As a result, the dynamic rules that govern this economic system can be researched by building upon previously studied dynamic rules for recycling of materials.

(2) The economic system and its external environment were treated as an integration. The representative element has a value whether it is present in a natural resource or in scrap.

(3) The difference in the VL of this element between natural resource and scrap indicates the difference in the practicality of obtaining the element from the two kinds of resource. This is the basis for analyzing the relationship between the utilization of scrap and its economic benefit.

(4) Because the representative element has a VL, irrespective of its source, both resources and the environment can be managed through an economic instrument based on the VCF.

20.3 A Case Study: the VSI for the Production Stages in the Lead-acid Battery System

20.3.1 Methodology

1. General introduction

Lead (Pb) is the most important element in the lead-acid battery system because:
(1) It accounts for 80% of the weight of these batteries (Shi, 1983).
(2) It comes from a nonrenewable resource (lead ore).

(3) Lead emissions into the environment are hazardous to living organisms in the ecosystem.

For this reason, we chose lead as the representative element for lead-acid batteries. Accordingly, discussion of this system will focus on the production of lead, and on the manufacture, use, discarding, and recycling of the batteries.

We chose this system in 1999 for China and the corresponding system from 1992 to 1995 for Sweden in average for our comparison, and studied the circular flow of lead and its influence on the VSI for two production stages: the lead production stage (stage I) and the manufacturing stage (stage II).

Because the mineral resource and the amount of waste entering the environment have not been properly evaluated in China, we cannot calculate the VL of lead in lead ore and waste. In order to validate the outcomes from our theoretical study, we have thus chosen to use concentrated lead instead of lead ore and collected scrap lead instead of waste. We have used this data to calculate the VL of lead in lead ore and in scrap lead.

2. Data sources

We used two types of data in this case study: data on lead fluxes and data on the VL of lead. The first type of data was taken from published studies for China (Mao and Lu, 2003c) and for Sweden (Karlsson, 1999). The second type of data will be computed based on statistical data from the China Foreign Trade Yearbook (2000). The trade value of concentrated lead divided by the lead content equals the VL of lead in concentrated lead, which represents the VL of lead in the lead mineral resource; the trade value of refined lead, batteries, and scrap lead divided by the corresponding lead content equals the VL of lead in these materials. Each value represents the average value based on import and export data.

20.3.2 Results and Discussion

Based on the data presented by Mao and Lu (2003c), we have drawn the lead flow diagram for China's lead-acid battery system in 1999 as Fig. 20.4. Based on these data, and on the resource efficiency and lead emission ratio for the lead production stage, the utilization efficiency and lead emission ratio for the manufacturing stage can be calculated. In the same way, based on the data of Karlsson (1999), we have calculated the corresponding parameters for Sweden. The results are summarized in Table 20.2.

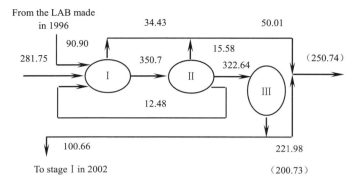

Fig. 20.4 The lead flow diagram for the life cycle of lead-acid batteries (LAB) in China in 1999
All quantities are kt.

Table 20.2 Comparison of the circular flow of lead in the lead-acid battery systems of China and Sweden

	$r_1/(t/t)$	$r_2/(t/t)$	$\gamma_1/(t/t)$	$\gamma_2/(t/t)$
China	1.245	0.920	0.098	0.048
Sweden	96.034	0.890	0.002	0
Ratio: Sweden/China	77.136	0.967	0.020	0

The VL of lead for different life cycle stages in 1999 was calculated as:

$$V_P = 2381.45(\$/t)$$

$$V_k = 150.62(\$/t)$$

$$V_m = 468.19(\$/t)$$

$$V_w = 145.94(\$/t)$$

By substituting these results in the formulas in Table 20.1, the RVSI and AVSI for the various production stages can be obtained (Table 20.3).

Table 20.3 Comparison of the value source intensity (VSI, $/t) in the production stages of the lead-acid battery systems of China and Sweden

Stage	RVSI		AVSI	
	I	II	I	II
China	318.49	1885.24	289.89	1871.23
Sweden	322.20	1873.43	321.62	1873.43
Difference: Sweden − China	3.71	−11.81	31.73	2.20

Table 20.3 shows that AVSI during the lead production stage in Sweden is 31.73 $/t higher than that in China, which means that Swedish lead refineries can earn 31.73 $ more than those in China for each tonne of lead output. This difference results from the greater resource efficiency in Sweden, which is 77.136 times the value in China. As well, Swedish emissions of lead into the environment equal only about 2% of Chinese emissions.

For the manufacture of lead-acid batteries, the AVSI for Sweden is 2.20 $/t higher than that for China, which means that Swedish manufacturers earn 2.20 $ more than those in China per tonne of lead content. That result from near-zero lead emissions into the environment and near-zero loss of scrap lead in Sweden. However, because of the lower utilization efficiency in Swedish manufacturing, the RVSI for Sweden is actually lower than that in China, though other reasons such as erroneous or inaccurate data may also be responsible for this difference.

20.4 Conclusions

(1) We have proposed the VSI concept and used it as an important index for evaluating the economic returns for producers.

(2) By using a primary component element to represent an overall product, we can model an element-flow diagram based on the product's life cycle and use the diagram to analyze the recycling of materials.

(3) Based on the recycling of materials, in which we treat the material as the carrier of value, and by developing the concept of VL for the representative element, we can derive the VCF, which provides a basis for analyzing the influence of recycling of materials on an economic system.

(4) The VSI for a producer is related to two main factors: the recycling of materials and the VL of the representative element in the product's life cycle stages. The more efficient the recycling of materials, the greater the resource efficiency; the lower the element emission ratio, the greater the VSI.

(5) We used the lead-acid battery system for a case study, in which lead was the representative element, and calculated the resulting VSI for producers in China and Sweden. The outcomes illustrate and support our theoretical models, and show that VSI is higher in Sweden than in China.

Acknowledgments

I am very thankful to various people for explaining and contributing data on specific industrial processes, but especially SUN Chuan-yao, YANG ling, FENG Jun-cong, and YAO Hong-fei. I gratefully acknowledge the economic support provided by the "National Basic Research (973) Program" Project (No.2005CB724204) of the Ministry of Science and Technology of China.

References

Annegrete B, 1998. Taxing virgin materials: an approach to waste problems, Resources, Conservation and Recycling, 22: 15-19.

Ayres R U. 1994. Industrial Metabolism: Theory and Policy. The Greening of Industrial Eco-systems. Washington D.C: National Academy Press, 23-37.

David T A, Nasrin B. 1994. Wastes as Raw Materials. The Greening of Industrial Ecosystems, Washington DC: National Academy Press, 69-89.

Frosch R A. 1994. Closing the Loop on Waste Materials. The Greening of Industrial Ecosystems, Washington DC: National Academy Press, 37-47.

Graedel T E. Allenby B R. 1995. Industrial Ecology. Prentice Hall, New Jersey, 110.

Grann H. 1997. The Industrial Symbiosis at Kalundborg, Denmark. The Industrial Green Game, Washington DC: National Academy Press, 117-123.

Hekkert M P, Joosten L A J. 2000. Analysis of the paper and wood flow in the Netherlands. Resources Conservation and Recycling, 30 (1): 29-48.

Joosten L A J. Hekkert M P, Worrell E, et al. 1999. STREAMS: a new method for analyzing material flows through society. Resources Conservation and Recycling, 27 (3): 249-266.

Karlsson, S. 1999. Closing the technospheric flows of toxic metals—Modeling lead losses from a lead-acid battery system for Sweden. Journal of Industrial Ecology, 3 (1): 23-40.

Kleijn R. 2000. IN=OUT: the trivial central paradigm of MFA? Journal of Industrial Ecology, 3 (2/3): 8-10.

Lu Z W. 2000. A study on the steel scrap resources for steel industry. Acta Metallurgica Sinica. 36 (7): 728-734. (in Chinese, with English abstract)

Lu Z W. 2002. Iron-flow analysis for the life cycle of steel products: A study on the source index for iron emission. Acta Metallurgica Sinica, 38 (1): 58-68. (in Chinese, with English abstract)

Lu Z W. 2003. Study on some aspects of circular economy, Research of Environmental Sciences. 16(5): 1-10(in Chinese, with English abstract).

Mao J S, Lu Z W. 2003a. Value circular flow and recycling of materials, Journal of Materials and Metallurgy, 2 (2):

157-160（in Chinese，with English abstract）.

Mao J S，Lu Z W. 2003b. On lead scrap resources for lead industry，World Nonferrous Metals，（7）：10-14.（in Chinese，with English abstract）.

Mao J S，Lu Z W. 2003c. Resource-service efficiency of lead in lead-acid battery，Journal of Northeastern University（Nature Science version），24（12）：1173-1176.（in Chinese，with English abstract）.

Sakuragi Y. 2002. A new partnership model for Japan: promoting a circular flow society. Corporate Environmental Strategy，9（3）：292-296（in Japanese，with English title）.

Shi X L. 1983. Lead Acid Storage Batteries. Beijing: People's Mail & Telecommunication Press，5-6（in Chinese）

Turner R K，Salmons R，Powell J，et al. 1998. Green taxes，waste management and political economy. Journal of Environmental Management，53：121-136.

Yu D H，Wang J N. 2001. Circular economy: the strategy for the 21st century. Research on Recycling Resource，5：2-5（in Chinese）.

第 21 章　工业物质循环的若干收益
Several Benefits From Recycling of Industrial Materials*

21.1　Introduction

21.1.1　Background

Circular economy (CE) regulates the anthropogenic flow of materials according to the natural flow of materials in a closed loop, so that the impact of human activities on ecosystem could be reduced to a level that the human economic system harmonizes and evolves with the natural ecosystem (Frosch, 1994). Recently, CE has been promoted in many countries such as Japan (Sakuragi, 2002), Netherlands (Joosten et al., 1999; Hekkert and Joosten, 2000), and has been chosen as the strategy for sustainable development of China in the 21 century (Yu and Wang, 2001).

The core of CE is the circular flow of materials in human economic system. This flow can be treated as the results from the circular flow of various materials for product's life cycle, which series of stages such as the production of materials, the manufacturing of products, and the use, disposal and recovery is mainly constituted (Graedel and Allenby, 1995). Since the circular flow of materials in human economic system often involves many industrial processes and thereby the materials are usually called industrial materials, it is usually termed as the circular flow of industrial material (CFIM). At present, the practice of CE emphases on the recycling of scraps produced at various of stages in product's life cycle, such as the waste utilization (David and Nasrin, 1994), or waste exchange and utilization in eco-industry parks (Grann, 1997), and so on. Ayres sorted the industrial material uses into 3 classes according to their technological and economical feasibility under present prices and regulations (Ayres, 1994). However, what, how, and how much can we benefit

---

* Mao J S, Lu Z W, Yang Z F. 2004. Several benefits from recycling of industrial materials. Proceedings of the Fifth Annual Conference for Young Scientists of China Association for Science and Technology. Nov.2-5. Shanghai, China. 489-498.

from CFIM? The study on these aspects is still very little and insufficient in publications. But the above questions must be answered first because they are the key factors that influence the promotion of CE, particularly for developing countries. Therefore, to study the impact of CFIM on the environment and its benefits theoretically and to validate it through case study will help us not only to answer the above questions but also to promote CE to a further development.

21.1.2　The Present Study

Since the core of CE is the CFIM, that is essentially different from the liner flow of industrial materials (LFIM) in traditional economy, where the material flows from natural resource through industrial products to environment. In this article, we started from the analysis of CFIM and learned from the methods of element flow analysis (EFA) (Lu, 2002), which considers changes in production and the time interval between production and disposal of products. By tracing a certain element M in product of its flow in product's life cycle, we established the relationships between a product system and its environment.

Considering that the elements in products are the carriers of product's economic value and thereby a circular flow of value (CFV) must exist accompanying CFIM, thus we analyze the CFV based on the model of CFIM. We learn and reference the method and some results obtained by Mao and Lu (Mao and Lu, 2003a) to establish a model of CFV. Based on the models of CFIM and CFV, we compared the difference between CFIM and LFIM, and the differences of natural resource consumption, environmental emissions, social service, and economic incomes between CFIM and LFIM, and thus we derived the benefits from CFIM on natural resource, the environment, the social service, and economic incomes.

As a case study, by taking a lead-acid battery (LAB) system as an example, we study the circular flow of lead in the system and its benefits on lead ore resource, lead emissions, production of LABs, and economic incomes of producers based on literature data. Through the case study, we validate the theory and show its application.

The present article is composed of three parts: in the first part, we introduce the method used, which includes the models of CFIM and CFV, and the definition of the various of benefits; In the second part, we present and discuss the theoretical results of the various of benefits; In the third part, we carry out a case study and show the application of the theory.

21.2 Methodology

21.2.1 CFIM and Its Material Benefits

As mentioned above, the CFIM can be treated as the results of the circular flow of various materials in product's life cycle. Because a product is composed of various of material elements, if we assume that element M is the main component and is the representative of the product, the CFIM will be simplified as the M-flow for product's life cycle.

1. The M-flow Diagram for product's life cycle

A M-flow diagram for a product's life cycle means a diagram reflecting the directions of M-flow and the distribution of M based on the application of the "Conservation Law" (in which inputs equal outputs) (Kleijn, 2000) at every stage of a product's life cycle. Fig. 21.1 is a M-flow diagram for a product's life cycle in reference year τ, in which we have assumed that: ①The average life span of the product was $\Delta\tau$ years, which remained constant during the study period. ②The time spent in various production processes was ignored, since it was usually very short compared to the years of a product's life span. ③The products became obsolete $\Delta\tau$ years after its production and some of the products would be recycled as a secondary source of M in the same year as its obsolete.

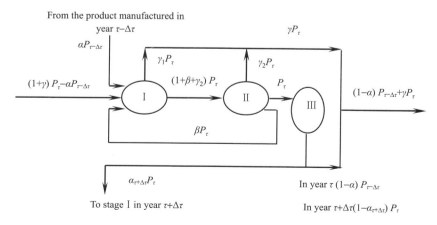

Fig. 21.1 The M-flow diagram for a product's life cycle

stage I: Production of industrial materials; stage II: Manufacture of products; stage III: Use of products

Some explanations for Fig. 21.1 are as follows:

(1) The production of primary and secondary materials have been treated as a single process, and are expressed by stage Ⅰ.

(2) The annual production of products changes yearly, and the production in reference years τ and $\tau-\Delta\tau$ are P_τ and $P_{\tau-\Delta\tau}$ (with unit of t/y, in M content) respectively.

(3) The recycling of scrap products was termed as large recycling of industrial materials (Lu, 2003), and thus the large recycling rate, represented by α with unit of (t/t), has been defined as the ratio of the scrap products (in M content) reused in material production in year τ to the total products (in M content) produced $\Delta\tau$ years ago. Under these conditions, $\alpha_{\tau+\Delta\tau}P_\tau$ of M in scrap products will become inputs for material production in year $\tau+\Delta\tau$ and $\alpha_\tau P_{\tau-\Delta\tau}$ would become inputs in year τ. The subscript τ of α in year τ has been omitted in Fig. 21.1 for simplicity.

(4) The recycling of scraps produced in product manufacturing was termed moderate recycling of industrial materials (Lu, 2003) and thus the moderate recycling rate, represented by β with units of (t/t), has been defined as the ratio of the scraps produced in product manufacturing that become inputs for material production to the total products produced in the same year.

(5) The emission ratios in stages Ⅰ and Ⅱ, represented by γ_1 and γ_2 (with units of (t/t)), have been defined as the ratio of the emissions (in M content) in the corresponding stage to the output of products in the same year. The sum of the two ratios has been defined as the "overall emission ratio" and is represented by γ (i.e., $\gamma=\gamma_1+\gamma_2$).

Besides, it exists another recycling of scraps in every stage of product's life cycle, that the scraps are produced and reused in the same stage. This recycling of scraps has been termed as small recycling of industrial materials (Lu, 2003). The level of this recycling of scraps could be represented indirectly by the emission ratios. Under the same production scale, the lower the emission ratio, the more materials will be recycled in the small recycling of scraps. Because the small recycling of scraps usually happens in the same stage, it is not illustrated in Fig. 21.1.

This method has been called "element flow analysis" (EFA), which is a subset of "substance flow analysis" (Joosten, 1999; Ayres, 1994). Lu first proposed adapting this method by incorporation the time difference between production and disposal and the changes in production (Lu, 2002a). The method had been successfully used to study iron emissions (Lu, 2002a) and steel scraps (Lu, 2002b). The method had been improved by emphasizing the final product (the product in the present paper) so as to create a more convenient link with the social service provided by the product and by taking the fiscal year as the statistical period to facilitate data gathering (Mao and Lu, 2003b).

2. The LFIM

If all the scraps produced in product's life cycle were not recycled, all the scraps would be discharged into the environment as emissions. In this case, the flow of industrial materials in product's life cycle would behave the liner flow of materials starting from natural resource to the environment through a production of products, and was named the liner flow of industrial materials (LFIM), the M-flow diagram for product's life cycle would be simplified as Fig. 21.2.

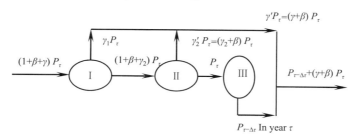

Fig. 21.2 The M-liner flow diagram for a product's life cycle

Fig. 21.2 was illustrated under an assumption of the same annual production of products (i.e., P_τ) and the same utilization rates of M in various materials (such as the natural resources, scraps, and so on) and in various production processes as those in Fig. 21.1.

3. The Material Benefits from CFIM

Fig. 21.1 and Fig. 21.2 show that the relationship between a product system and its environment exhibit as follows:

(1) To provide products (represented by P_τ, with unit of t/y) to the society to meet certain social services;

(2) To consume natural resource and form the resource load, which is represented by R in M content, with unit of t/y;

(3) To discharge scraps and pollutants and form the emission load, which is represented by Q in M content with unit of t/y.

On one side, Because of the difference between CFIM and LFIM, the resource load under CFIM is different from that under LFIM to obtain the same quantity of products. The difference of resource load resulted from the difference between the two kinds of material flow is the saved natural resource brought by CFIM, and is termed as the resource benefit, which is represented by ΔR (t/y). Similarly, The difference of emission load resulted from the difference between CFIM and LFIM under the same output of products is the reduced environmental emissions caused from CFIM, and is termed as the environmental benefit, which is represented by ΔQ (t/y).

On the other side, in the case of the same resource load or/and emission load, the output of products under CFIM is different from that under LFIM. The difference in output of products resulted from this case is termed as the service benefit, which is represented by ΔP with unit of t/y. The service benefit is usually sorted into the resource-related service benefit and the emission-related service benefit, that the former is the service benefit calculated under the same resource load (represented by $\Delta P_{\tau R}$) and the latter is the service benefit calculated under the same emission load (represented by $\Delta P_{\tau Q}$).

The resource benefit, the environmental benefit, the service benefit (including the resource-related service benefit and the emission-related service benefit) are the material benefits from CFIM. The values discussed above belong to absolute value.

In application, in order to reflect the essential of the benefits from CFIM, we need to sieve out the influence of production scale on the benefits and thereby definite the ratio of absolute benefits to the output of products (P_τ) as the corresponding relative value of benefits.

21.2.2 The CFV and Its Economic Benefits

If we apportion the value of a product among its components, every element in the product will become a carrier of economic value. If we represent the product by an element M, the value of a product will be simplified as the value of element M. We thus deduce that a circular flow of value (CFV) exist in a product system accompanying the CFIM. The CFV would result in an economic impact.

1. The CFV for Product's Life Cycle

It is easy to understand that the CFV is accompanying every flow of element M in product life cycle and the flux of every flow of value should equal to the mass of element M multiplied by the economic value per unit mass of element M in the corresponding M-flow. Thereby, if we obtain the economic value for per unit mass of element M at every stage of a product's life cycle, we will be able to illustrate a circular flow of value

(CFV) diagram for product's life cycle based on Fig. 21.1.

Mao and Lu (2003a) have studied the circular flow of value (CFV) caused by CFIM and defined the economic value carried by the material outflow departing from a certain stage divided by the mass of element M in the flow as the value level (VL) of element M at the corresponding stage, which was represent by V with unit of dollar/t. In a product's life cycle, the VL of element M changes along with the stages, i.e., the VL of element M is upgraded successively accompanying various of production activities, however, the VL of element M is whittled gradually during the use of products. The changing of VL of element along with the stages in product life cycle can be illustrated in Fig. 21.3, which shows that the VL of element M rise or fall only during the passing of life cycle stages. V_N, V_M, V_P, and V_Q in Fig. 21.3 represent the VL of element M in natural resource, industrial materials, products, and the scraps, respectively.

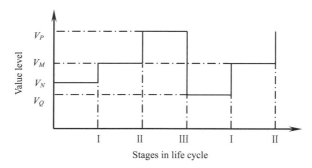

Fig. 21.3 The changing of value level for element M in its life cycle

Based on Fig. 21.1 and Fig. 21.3, we can calculate every value flow caused by CFIM and obtain the CFV diagram under CFIM, which is illustrated as Fig. 21.4.

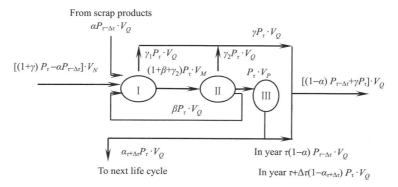

Fig. 21.4 Circular flow of value diagram for a product's life cycle

2. The Value Flow under LFIM

Following the same thought as in section 2.1.1, we obtain the value flow for a product's life cycle under LFIM based on Fig. 21.2 and Fig. 21.3, which is illustrated as Fig. 21.5.

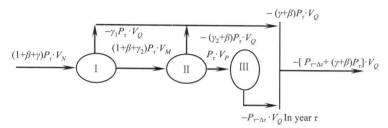

Fig. 21.5 The value flow for a product's life cycle under M-liner flow

In this case, we need supplement that the value of wastes takes negative value. Because all the wastes produced in product system would be emitted to the environment as environmental emissions and be assumed to blight the environment, thus to be charged by the environmental protection agency for environmental improvement.

3. The Economic Benefits from CFV

On economic viewpoint, producers always act as value amplifiers and are the source of economic value.

In order to analyze the economic benefits from CFIM, we define the difference between the outflow and the inflow of value for a certain stage in product's life cycle as the value increment of the corresponding stage, which is represented by ΔV_i with units of dollar/y. Here, the subscript i represents the stage of product's life cycle. It is easy to understand that, for producers, the net income should equal to that the value increment (i.e., ΔV_i) minus the corresponding values of various inputs, such as labors, facilities, and so on. The more the value increment, the more the producer will obtain the net income.

By comparing the value flow in Fig. 21.4 with that in Fig. 21.5, we deduce that the value increments for producers under CFIM are different from that under LFIM, because of the different pattern of material flow and the different value of element M in scraps (or wastes, emissions). Here we defined the difference of value increment for producers between CFIM and LFIM as the economic benefit of the producer from CFIM, which is represented by $\Delta(\Delta V)_i$, with unit of dollar/y. The value of $\Delta(\Delta V)_i$ belongs to absolute value of economic benefits. In application, we define the ratio of the absolute economic benefit to the output of products, i.e., $\Delta(\Delta V)_i / P_\tau$, as the relative value of economic benefits to reflect the essential economic benefits from CFIM.

In this section, we need note that we had assumed that in the case of CFIM, all the wastes (scraps, emissions) produced in product system would be reused, some by the components in product system itself, the others by producers outside of the product system, and therefore the value of wastes were positive. On the contrary, in the case of LFIM, all the wastes (scraps, emissions) produced in product system would emit to the environment as emissions, and thus the value of wastes were negative. Besides, we also assumed that the VL of element M in various wastes such as emissions and scraps remained constant.

21.3　Results and Discussion

21.3.1　Results on Various Benefits

The above analysis indicates that we can obtain the following four kinds of benefits from CFIM: the resource benefit, the environmental benefit, the service benefit, and the economic benefit. Based on the models for CFIM, LFIM, CFV, and the value flow under LFIM, we calculate the values of the above four kinds of benefits and summarize them in Table 21.1.

Table 21.1　Several benefits from circular flow of material industrial

	Resource benefit	Environmental benefit	Service benefit		Economic benefit	
			Resource-related	Emission-related	Material Production stage	Product manufacture stage
Absolute indices	$\beta P_\tau + \alpha P_{\tau-\Delta\tau}$	$\beta P_\tau + \alpha P_{\tau-\Delta\tau}$	$(r-r')R$	$(q-q') \cdot Q$	$(1+\beta-\gamma_1-\frac{1}{r}) \cdot P_\tau \cdot (V_N - V_Q)$	$(\beta + 3\gamma_2) \cdot P_\tau \cdot V_Q$
Relative indices	$\beta + \alpha p$	$\beta + \alpha p$	$1 - \frac{r'}{r}$	$1 - \frac{q'}{q}$	$(1+\beta-\gamma_1-\frac{1}{r}) \cdot (V_N - V_Q)$	$(\beta + 3\gamma_2) \cdot V_Q$

Note: $r = \frac{1}{1+\gamma-\alpha p}$, $r' = \frac{1}{1+\beta+\gamma}$, $q = \frac{1}{\gamma+(1-\alpha)p}$, $q' = \frac{1}{p+\beta+\gamma}$, $p = \frac{P_{\tau-\Delta\tau}}{P_\tau}$, the relative indices are calculated from the absolute indices divided by the output of products (P_τ).

21.3.2　Discussion on Various Benefits

1. The Resource Benefit and Environmental Benefit

The table 21.1 shows that the resource benefit or the environmental benefit from

CFIM is the sum of quantities of large recycling (i.e., $\alpha P_{\tau-\Delta\tau}$) and moderate recycling (i.e., βP_τ). The more the recycling of industrial materials, the more will be the resource benefit or the environmental benefit.

The relative values of resource benefit and environmental benefit is a function of the large recycling rate, the moderate recycling rate, and the ratio of production in product's life cycle (i.e., $p = P_{\tau-\Delta\tau}/P_\tau$). The higher the recycling rate, the more will be the resource benefit and environmental benefit. A decreasing production will be helpful to more resource benefit and environmental benefit.

2. The Service Benefit

The Table 21.1 also shows that the resource-related service benefit is a function of the resource efficiencies under CFIM and LFIM, i.e., r and r', and the resource load under CFIM, i.e., R. Here the resource efficiency means the resource load per unit of products (Halada and Yamamoto, 2001), which is represented by r with units of (t/t). The resource efficiency is usually a function of the large recycling rate, emission rate, and the ratio of production (see the notes for Table 21.1). Increased recycling rate, reduced emission rate, and decreasing of production will help to improve the resource efficiency (Mao and Lu, 2003b). The relative value of resource-related service benefit is only related to the ratio of the two resource efficiencies, i.e., r'/r, the lower the ratio of resource efficiency, the higher will be the relative value of the resource-related service benefit which means the more resource-related service benefit from CFIM.

Similar results could be obtained for the emission-related service benefit if we instead the resource efficiency and resource load by the environmental efficiency (Annegrete, 1998) (represented by q with unit of (t/t)) and emission load, respectively. The only difference is that the increasing of production will help to improve the environmental efficiency.

The above analysis also indicates that the following factors such as resource efficiency, the environmental efficiency, the recycling rate, the emission rata, and the ratio of production, reflect the relationships of material flow between components in a product system and that between product system and its environment. We thereby call them as the factors of CFIM. The higher the resource and environmental efficiency, or the higher the recycling rate, or the lower the emission rate, means the better recycling of industrial materials.

3. The Economic Benefit

The economic benefit from CFIM for material production stage is a function of the factors of CFIM and difference of VL between the natural resource and wastes (i.e., $V_N - V_Q$). In the case of $V_N > V_Q$, the higher the resource efficiency, or the higher the moderate recycling rate, or the less the emissions, the higher will be the economic benefit for the stage. This is because that a higher resource efficiency, or higher moderate recycling rate, or lower emission rate, means not only an effective use of materials in the stage but also more natural resource in higher VL replaced by scraps in lower VL. Moreover, the higher the value of $V_N - V_Q$, the higher will be the economic benefit for this stage. What we need to point out here is that a negative economic benefit will occur under converse VL difference between natural resource and scraps, i.e. $V_N < V_Q$. In that situation, the utilization of scraps can bring no economic returns although it still save natural resource and reduce emissions to the environment. That situation mainly occurs at some districts that lack of scraps resource such as most developing countries (Lu, 2002b) or some districts with irrational resource policy such as the countries under subsidy policy for scarce resources. This situations need to be changed by an improvement of resource policy (Annegrete, 1998; Turner et al., 1998).

The economic benefit from CFIM for product manufacturing stage is tightly related to the VL of element M in scraps, the moderate recycling rate, and the emission rate at this stage. The higher the VL of element M in scraps, or the higher the moderate recycling rate and emission rate at this stage, the higher will be the economic benefit for this stage. This is mainly caused by that all the emissions from a product system under CFIM are assumed to be reused by other industrial producers.

21.4 A Case Study: the Benefit From the Circular Flow of Lead in Lead-acid Battery System

21.4.1 Methodology

1. General introduction

The lead (Pb) is the most important element in lead-acid battery (LAB) because that it holds 80 percent of the weight (Shi, 1983), it comes from nonrenewable resource lead ore, and the lead emissions in environment may be hazardous and poisonous to the ecosystem. Therefore, we choose lead as the representative of LAB. Accordingly, the product system in this study will briefly include the production of lead, the manufacture of LAB, the use, discard and recycling of LAB.

In the case study, we choose the LAB system of China in 1999 as an example and analyzed the various benefits from the circular flow of lead in LAB system base on the obtained factors of CFIM and the obtained VL of lead. Also we forecast the future benefits by replace the factors of CFIM for China in 1999 with those for Sweden in 1992 to 1995 which represent the international advanced level for LAB system at present.

Considering that the mineral resource and the waste to environment have not been evaluated properly at present, the VL of lead in lead ore and waste cannot be calculated. In order to show the application, we use concentrated lead instead of lead ore, the collected lead scraps instead of lead wastes (scraps and emissions). Based on that, the VL of lead in lead ore and that in lead scraps will be calculated respectively.

2. Data source

Two types of data are involved in the case study for LAB systems: the data for lead flux at various life cycle stage, or the date for the factors of CFIM, and the data for VL of lead.

The first type of data for China in 1999 is illustrated in Fig. 21.6 (with unit of kt/y), which was taken from the literature (Mao and Lu, 2003b). The data for Sweden was calculated based on the literature (Karlsson, 1999). The corresponding factors for CFIM in China and Sweden were listed in Table 21.2.

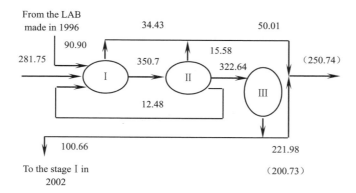

Fig. 21.6 The lead flow diagram for lead-acid battery system in 1999 of China

Table 21.2 The compare of the factors for CFIM of the LAB system between China and Sweden

	Resource efficiency r	Environmental efficiency q	Large recycling rate α	Moderate recycling rate β	Emission rate γ_1	Emission rate γ_2	Gross emission rate γ	Ratio of production p
	/ (t/t)	/ (t/t)	/ (t/t)	/ (t/t)	/ (t/t)	/ (t/t)	/ (t/t)	/ (t/t)
China	1.145	1.287	0.312	0.039	0.107	0.048	0.155	0.904
Sweden	85.47	85.47	0.990	0.1236	0.002 655	0.000	0.002 655	1

The second type of data will be estimated based on the statistic data in China Foreign Trade Yearbook. In details, the trade value of concentrated lead divide by the lead content is the VL of lead in concentrated lead, which represents the VL of lead in lead mineral resource; and the trade value of refined lead, LAB and lead scraps divide by each lead content is the VL of lead in refined lead, LAB, and lead scraps, respectively. Each value takes the average value of that from importing and exporting. The outcomes for V_P, V_N, V_M, and V_Q in 1999 were estimated as 2381.45, 150.62, 68.19, and 145.94 USD/t, respectively.

21.4.2 Results and Discussion

1. Result

Fig. 21.6 shows that the overall lead flow for LAB system of China in 1999 exhibits as follows:

$$P = 322.64 \text{ kt}; \quad R = 281.75 \text{ kt}; \quad Q = 250.74 \text{ kt}$$

Table 21.3 Several benefits from the circular flow of lead in LAB system in the case study

		Resource benefit	Environmental benefit	Service benefit		Economic benefit	
				Resource-related	Emission-related	Material production	Product manufacture
Relative indices	Present	0.321	0.321	0.268	0.292	0.274	26.707
	Future	1.114	1.114	0.990	0.990	5.191	18.038
Absolute indices	Unit	kt/y	kt/y	kt/y	kt/y	Thousand USD/y	Thousand USD/y
	Present	103.57	103.57	86.47	94.21	88.40	8 616.75
	Future	359.29	359.29	319.41	319.41	1 674.82	5 819.78

Note:

(1) The present relative benefits were estimated on the status of lead flow of China in 1999, i.e., $r = 1.145$, $r' = 0.838$, $q = 1.287$, $q' = 0.911$, $p = 0.904$;

(2) The future relative benefits were estimated on the status of lead flow of Sweden in 1992-1995, i.e., $r = q = 85.47$, $r' = q' = 0.888$, $p = 1$;

(3) The absolute benefits were estimated on the output of LAB of China in 1999, i.e., $P_r = 322.64$ kt/y;

(4) The value level of lead were taken as $V_N - V_Q = 4.68$ USD/t, $V_Q = 145.94$ USD/t.

Substituting the values for the various factors of CFIM and the value levels of lead at different stages into the formulae in Table 21.1, we can obtained the values for various of benefits. The outcomes were listed in Table 21.3, in which the outcomes for future were estimated based on the same overall lead flow in 1999.

2. Discussion

Table 21.3 shows that the circular flow of lead in LAB system can save lead ore resource, reduce lead emissions, raise the output of LAB and the economic incomes for lead refineries and LAB manufactories. For instance, to produce 1 ton of LAB, compared to the LFIM, the CFIM in 1999 will save 0.321 tons of concentrated lead, reduce 0.321 tons of lead emissions, raise 0.274 and 26.707 USD of incomes for lead refineries and LAB manufactories, respectively, or, under the same resource load and/or emission load, raise about 1/4 to 1/3 of LABs. If we can improve the circular flow of lead in LAB system to the current level of Sweden (in 1992-1995), to produce the same output of LAB as in 1999, we would save 359.29 kt of concentrated lead and reduce 359.29 kt of lead emissions, raise 1.67 and 5.82 millions USD of incomes for lead refineries and LAB manufactories per year, respectively. Or, under the same lead ore resource load and/or lead emission load, raise about 320 kt of lead-acid batteries per year. For the manufacturing of lead-acid batteries, the economic benefit in the future case is lower than that at present case, the reason for that is mainly caused by the lower utilization efficiency of lead in manufacture of LAB in Sweden than that in China, although some other reasons, like false data, may exist in the investigate data for China.

21.5 Conclusions

(1) By taking an element M in product as the representative and analyzing the M-flow in product life cycle, we developed a model of CFIM, which is the foundation of the study on material benefits from CFIM.

(2) By treating the elements in product as the carriers of economic value, and introducing the concept of value level of element M at stages in product life cycle, we developed a model of CFV, which is the foundation of the study on economic benefits from CFIM.

(3) We defined several important concepts such as the resource benefit, environmental benefit, the service benefit, and the economic benefit. By comparing the difference of material flow between CFIM and LFIM and the difference of value flow between the

two kinds of material flow, we derived the theoretic values for the above four kinds of benefits. The result shows that the above four aspects of benefits were tightly related to the recycling of materials, and the better the recycling of materials, the more will be the benefits. Besides, the economic benefit is still related to the VL of element M in natural resource and scrap.

(4) By taking the LAB system as the background and taking lead as the representative of LAB, we estimated the resource benefit, the environmental benefit, the service benefit, and the economic benefit caused by the industrial circular flow of lead in LAB system. By taking the present level of lead circular flow in Sweden as the future level of China for LAB system, we forecasted the future of the above four aspects of benefit.

In one word, the outcomes showed that the CFIM can save natural resource, reduce environmental emissions, raise the output of products and raise the economic incomes for producers of industrial materials and products.

References

Annegrete B. 1998. Taxing virgin materials: an approach to waste problems, Resources Conservation and Recycling, 22: 15-19.

Ayres R U. 1994. Industrial Metabolism: Theory and Policy. The Greening of Industrial Eco-systems. Washington D.C: National Academy Press, 23-37.

David T A, Nasrin B. 1994. Wastes as Raw Materials, The Greening of Industrial Ecosystems. Washington DC: National Academy Press, 69-89.

Frosch R A. 1994. Closing the Loop on Waste Materials. The Greening of industrial Ecosystems. Washington DC: National Academy Press, 37-47.

Graedel T E, Allenby B R. 1995. Industrial ecology. New Jersey: Prentice Hall, 110.

Grann H. 1997. The Industrial Symbiosis at Kalundborg, Denmark. The Industrial Green Game. Washington DC: National Academy Press, 117-123.

Halada K, Yamamoto R. 2001. A review of the development of resource productivity concept and eco-efficient technology. Northeastern University. Shenyang. Northeastern University, P.R.China: International Symposium on Industry and Ecology, 14-36.

Hekkert M P, Joosten L A J. 2000. Analysis of the paper and wood flow in the Netherlands. Resources Conservation and Recycling, 30 (1): 29-48.

Joosten L A J, Hekkert M P, Worrell E. 1999. STREAMS: a new method for analysis material flows through society. Resources Conservation and Recycling. 27 (3): 249-266.

Karlsson S. 1999. Closing the technospheric flows of toxic metals—Modeling lead losses from a LAB system for Sweden. Journal of Industrial Ecology, 3 (1): 23-40.

Kleijn R. 2000. IN=OUT: the trivial central paradigm of MFA?. Journal of Industrial Ecology, 3 (2/3): 8-10.

Lu Z W. 2002a. Iron-flow analysis for the life cycle of steel products: A study on the source index for iron emission. Acta Metallurgica Sinica, 38 (1): 58-68.

Lu Z W. 2002b. On the steel scrap resources for steel industry. Iron & Stell, 37 (4): 66-70.

Lu Z W. 2003b. Study on some aspects of circular economy. Research of Environmental Sciences, 16 (5): 1-10.

Mao J S, Lu Z W. 2003a. Circular flow of value and material circular flow, Journal of Materials and Metallurgy, 2 (2): 157-160 (in Chinese, with English abstract).

Mao J S, Lu Z W. 2003b. Resource-service efficiency of lead in lead-acid battery, Journal of Northeastern University (nature science), 24 (12): 1173-1176.

Sakuragi Y. 2002. A new partnership model for Japan: promoting a circular flow society. Corporate Environmental Strategy, 9 (3): 292-296.

Shi X L. 1983. Lead Acid Storage Batteries. Beijing: People's Mail & Telecommunication Press, 5-6.

Turner R K., Salmons R, Powell J, et al. 1998. Green taxes, waste management and political economy. Journal of Environmental Management, 53: 121-136.

Yu D H, Wang, J M. 2001. Circular economy: the strategy for the 21st century. Research on Recycling Resource, 5: 2-5.

第四篇

废铅管理与政策

第 22 章 论工业中的废金属资源
On Metal Scrap Resource for Industry*

22.1 引　　言

　　工业中金属的来源有两个：一是金属矿石，二是废金属。前者是天然资源，后者是回收的再生资源。如果工业中多用废金属，少用金属矿石，那么将不仅有利于保存金属矿产资源，而且还有利于减少废金属的环境排放，起到改善环境的作用（David and Nasrin，1994；Frosch，1994）。近些年来，在欧洲各国，掀起了废金属资源利用的热潮（Bertram et al.，2002；Spatari et al.，2003；Sagar and Frosch，2004），并收到了很好的环境效益、社会效益和经济效益。在中国，伴随循环经济战略的实施（余德辉和王金南，2001），废金属物质的循环利用也将成为工业发展的重要内容（陆钟武，2003）。但是，要做到这一点，首先需要有充足的废金属资源。

　　目前，世界各国的废金属资源的实际情况差别很大，有的国家，比如美国（U.S. Department of the Interior，2002a），废金属资源较充足，可以大力发展再生金属业。而有的国家，比如中国，废金属资源不足，再生金属业难以为继（中国有色金属工业年鉴编辑委员会，1990～2001）。由此可见，废金属资源还决定着一个国家冶炼业的总体结构。

　　为什么有的国家废金属资源比较充足，而有的国家却相对短缺？如何衡量一个国家的废金属资源状况？废金属资源的状况与哪些因素有关？这些因素如何影响废金属资源的状况？废金属资源短缺的国家，会不会永远短缺？什么时候会发生变化？只有清楚地回答以上问题，才能把一个国家的废金属资源问题弄清楚。

　　陆钟武（陆钟武，2000；陆钟武，2002）曾采用"废钢指数"作为衡量一个国家钢铁工业废钢资源状况的指标，并把废钢指数与钢产量的变化联系起来，同时注意到钢铁生产到报废之间的"时间差"，分析了钢产量变化对废钢资源状况的影响。毛建素和陆钟武（毛建素和陆钟武，2003）进一步把废金属的循环状况

---

* 毛建素，陆钟武. 2003. 论铅业的废铅资源. 世界有色金属，7：10-14.

考虑进来，研究了废铅资源问题。然而，由于以上研究中，是在逐次考虑某一种影响因素的情况下，分别展开研究的，因此在系统性、规律性方面稍显不足。本研究将综合考虑各种影响因素，通过分析参考年份废金属的来源和数量，获得废金属指数的数学表达式。在讨论废金属指数与其各种影响因素的变化关系的基础上，深入理解废金属资源的变化规律。

除此之外，本章还将以中国有色金属业为例，分析铜、铝、铅、锌4种金属的废金属资源状况。并应用所得废金属指数的变化规律，分析中国废金属资源中的问题，并预测其发展前景。

22.2 理论研究

22.2.1 若干基本概念

1. 废金属的分类

通常，一个国家的废金属，按其来源可划分为以下4种（陆钟武，2002；姜松，2000）：

（1）内部废金属。这是产生于生产企业内部，同时又作为企业自身的生产原料重新利用的废金属。通常，这部分废金属不进入市场流通，因此不作为本章研究的废金属资源。

（2）加工废金属。这是来自国内金属制造业，并返回金属再生厂，作为生产原料重新利用的废金属。通常，这部分废金属在其产生后几周内就能返回金属再生厂，所以又称"短期废金属"。

（3）折旧废金属。这是金属制品完成其使用寿命后报废时形成的废金属。由于折旧废金属的形成，需要经过金属的生产、金属制品的制造、使用、报废和回收等若干生命周期阶段（Graecled and Allenby，1995），一般要经历几年甚至十几年的时间，所以折旧废金属又称"长期废金属"。

（4）进口废金属。这是从其他国家进口的废金属。由于通常情况下，只有废金属资源短缺的国家才进口废金属，因此本章中不把它看做是进口国的废金属资源。

2. 废金属的数量

定义某一国家在某年内回收的废金属的数量（按其中的金属含量计算），为这一国家的废金属资源量，用符号 R_w 表示，单位是 t/a。根据前文中的分析，废金属资源量应等于加工废金属与折旧废金属的数量之和。

假设：

（1）金属制品的使用寿命是 $\Delta\tau(a)$，并在研究涉及的时间尺度内保持不变。

（2）忽略金属制品生命周期中各生产过程所消耗的时间。这种情况下，所有废金属都在其形成当年返回到金属再生企业，同时 $\Delta\tau$ 可代表金属的一个工业生命周期。

（3）假设金属产量以年均变化率 ρ 线性变化，即

$$P_\tau = P_0(1+\rho\tau) \tag{22.1}$$

式中，P_0、P_τ 分别为起始年份和第 τ 年的金属产量（t/a）。

同时定义：

（1）某年回收的加工废金属量与该年金属产量的比值为加工废金属循环率，并用符号 β [单位是（t/t）]表示；

（2）金属产量中，经过金属制品的使用过程，在金属制品报废后形成的折旧废金属数量所占的比例，为折旧废金属循环率，并用符号 α [单位是（t/t）]表示。

在上述条件下，不难设想，第 τ 年的加工废金属来自当年生产的金属，数量为 βP_τ，折旧废金属来自第 $\tau-\Delta\tau$ 年生产的金属，数量为 $\alpha P_{\tau-\Delta\tau}$。两种废金属合计后，并根据式（22.1）把金属产量随年份的变化考虑进来，可以整理得到第 τ 年的废金属资源量为

$$R_{w\tau} = (\alpha+\beta)P_0 + (\alpha+\beta)P_0\rho\tau - \alpha P_0\rho\Delta\tau \tag{22.2}$$

为方便起见，式（22.2）省略了表示第 τ 年的 α、β 的下标符号 τ。

应用中，加工废金属循环率（α）和折旧废金属循环率（β）统称金属循环率。

3. 评价指标

一个国家废金属资源是否充足，是相对于这个国家在该年度的金属产量而言的。为此，定义一个国家某年的废金属资源量与该年的金属产量之比，为这一国家在该年的废金属指数，代表符号为 S [单位（t/t）]，即

$$S = \frac{R_{w\tau}}{P_\tau} \tag{22.3}$$

废金属指数（S）是衡量一个国家废金属资源充足程度的判据。S 值愈大，废金属资源愈充足；反之，S 值愈小，废金属资源愈短缺。

22.2.2 废金属指数基本规律

将式（22.1）、式（22.2）代入式（22.3），整理后得

$$S = \alpha + \beta - \frac{\alpha\rho\Delta\tau}{1+\rho\tau} \tag{22.4}$$

可见，废金属指数是金属循环率、金属产量年均变化率、金属生命周期以及

年份的函数。为更清楚地理解废金属指数的变化规律,讨论如下:

(1) 废金属指数与金属产量变化之间的关系。

假设:金属产量按以下三种情况变化:①保持不变,$\rho=0$;②持续增长,$\rho=0.1$;③逐渐下降,$\rho=-0.1$。同时假设,$\Delta\tau=5a$,$\alpha=0.5(t/t)$,$\beta=0.05(t/t)$。

在以上条件下,金属年产量变化和废金属指数变化的情况如图 22.1。图中的金属年产量采用无因次值,即

$$\overline{P_\tau}=1+\rho\tau \tag{22.5}$$

$\overline{P_\tau}$ 是无因次金属年产量,它是第 τ 年的金属产量与起始年份金属产量的比值。

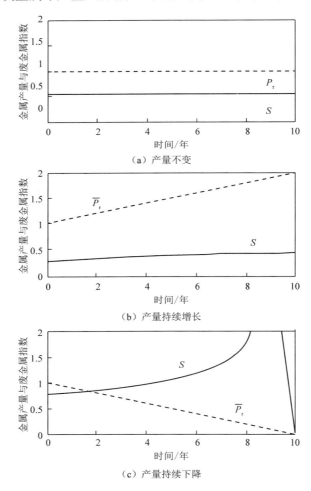

图 22.1 不同产量变化情况下的废金属指数随年份的变化

Fig. 22.1 The metal scrap index changing to years under different variations of metal production

由图 22.1 可见，由于金属产量变化的情况不同，导致废金属指数截然不同。在金属产量保持不变的情况下[图 22.1（a）]，废金属指数表现为与金属产量平行的水平线，表示废金属指数不变，图 22.1（a）中的数值为 0.55。在金属产量持续增长情况下[图 22.1（b）]，废金属指数在起始年份的数值较低，图 22.1（b）中的数值为 0.30，并以较金属产量的增长率更慢的速率缓慢增长。表示废金属资源相对短缺；而且不难理解，产量增长愈快，S 值愈低，废金属资源愈短缺。在金属产量持续下降情况下[图 22.1（c）]，废金属指数在起始年份的数值较高，图 22.1（c）中的数值为 0.80，并以不断增长的速率提高，表示废金属资源相对充裕，甚至一定时期以后，出现废金属指数远大于 1，表示废金属资源过剩的局面。而且可以推测，产量下降愈快，S 值愈高。

为了掌握废金属资源充足状况出现的时间，定义废金属指数为 1 的年份为临界年份，用 τ_{crit} 表示。令式（22.4）中 $S=1$，可以解得

$$\tau_{crit} = \frac{\alpha \Delta \tau}{\alpha + \beta - 1} - \frac{1}{\rho} \tag{22.6}$$

不难验证，在其他条件不变的情况下，金属制品的使用寿命越长，或者产量下降的速度越快，或者循环率越高，临界年份出现的时间也越早。

（2）废金属指数随其他参数的变化。

由式（22.4）可见，废金属指数（S）与加工废金属循环率（β）呈线性增长关系，较为简单，不予讨论。而与折旧废金属循环率（α）、使用寿命（$\Delta\tau$）的关系较为复杂。为此假定，$\beta = 0.05(t/t)$，分别考察 $\Delta\tau$ 为 5、10 年情况下，第 5 年的 S 值。在上述条件下，根据式（22.4）绘制 S 随 α 的变化曲线，得到图 22.2。

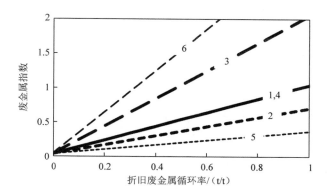

图 22.2 废金属指数随金属循环率的变化

Fig. 22.2 The metal scrap index changing to recycling rate

线 1 $\rho=0$，$\Delta\tau=5a$；线 2 $\rho=0.1$，$\Delta\tau=5a$；线 3 $\rho=-0.1$，$\Delta\tau=5a$；
线 4 $\rho=0$，$\Delta\tau=10a$；线 5 $\rho=0.1$，$\Delta\tau=10a$；线 6 $\rho=-0.1$，$\Delta\tau=10a$。

由图 22.2 可见，(a) 所有曲线均呈上升趋势，表示 α 越高，S 也越高。(b) S 与金属制品使用寿命（$\Delta \tau$）的关系受金属产量的变化状况的影响：图 22.2 中线 1 与线 4 重叠，表示产量不变情况下，S 与 $\Delta \tau$ 无关；图 22.2 中线 5（代表 $\Delta \tau$=10a）位于线 2（代表 $\Delta \tau$=5a）之下，表示产量增长情况下，使用寿命越长，S 的数值越低。图 22.2 中线 6（代表 $\Delta \tau$=10a）位于线 3（代表 $\Delta \tau$=5a）之上，表示产量下降情况下，使用寿命越长，S 越高。

由此推测，与金属产量基本稳定的国家相比，金属产量不断增长的国家的废金属指数相对较低，并且金属制品的使用寿命越短，废金属指数越高；与此相反，金属产量不断下降的国家，其废金属指数相对较高，并且金属制品的使用寿命越长，废金属指数越高。对于金属产量基本稳定的国家，废金属指数主要与循环率有关。无论金属产量如何变化，提高金属循环率，均有利于提高废金属指数。

22.3　实例分析：中国若干金属的废金属指数

22.3.1　估算方法

在中国，废金属回收后，通常直接进入金属再生部门，折旧废金属和加工废金属的数量未列入统计数据，因此无法根据式（22.3）来计算废金属指数。

假设某种废金属的再生收率为 η，再生金属在金属产量中的比率（简称再生金属占有率）为 φ，则在第 τ 年的废金属再生过程中，存在下式：

$$\eta(R_{w\tau} + R_{ni\tau}) = \varphi P_\tau$$

式中，$R_{ni\tau}$ 是第 τ 年某一种废金属的净进口量，它等于废金属的进口量与出口量的差值，t/a。

上式整理后得

$$R_{w\tau} = \frac{\varphi P_\tau}{\eta} - R_{ni\tau} \qquad (22.7)$$

将式（22.7）代入式（22.3），整理后得

$$S = \frac{\varphi}{\eta} - \varepsilon \qquad (22.8)$$

式中，ε 是由于废金属贸易产生的废金属指数的修正值，且 $\varepsilon_\tau = R_{ni\tau}/P_\tau$。

可见，废金属指数还可以通过计算 φ 和 η 的比值，并考虑废金属的进出口来进行修正的方法估算得到。本章中，φ 的数值是再生金属产量与金属总产量的比值，根据（中国有色金属工业年鉴，1990～2001）中的数据计算得到。由于废金属的再生收率（即 η 的数值）按 2000 年冶炼总回收率估算，铜、铝、铅、锌分别取为 96.88%、88.44%、93.52%、92.85%。废金属的进出口数据也取自（中国有色

金属工业年鉴，1990～2001）。

另外，由于金属再生收率在一定时期内基本不变，从式（22.8）还可看出，再生金属占有率可以间接反映废金属资源的状况。特别是在废金属贸易量较小的情况下更是如此。再生金属占有率越高，废金属指数也越高，废金属资源越充足。

22.3.2 估算结果

按照上述方法，可以估算出中国铜、铝、铅、锌等几种金属的废金属指数，见表 22.1。

表 22.1　中国若干废金属指数估算表
Table 22.1　Estimated values of several metal scrap index for China

年份		1990	1991	1992	1993	1994	1995	1996	1997	1998	1999	2000	平均值
再生金属占有率	铜	0.379	0.337	0.363	0.336	0.363	0.433	0.382	0.321	0.281	0.288	0.254	0.340
	铝	0.008	0.008	0.005	0.010	0.024	0.104	0.068	0.066	0.041	0.075	0.065	0.043
	铅	0.095	0.145	0.132	0.108	0.128	0.288	0.203	0.175	0.122	0.106	0.092	0.145
	锌	0.040	0.039	0.028	0.020	0.040	0.089	0.061	0.044	0.010	0.011	0.036	0.038
贸易修正值	铜	NA	NA	0.395	0.754	1.030	1.084	0.627	0.666	0.782	1.439	1.817	0.955
	铝	NA	NA	0.027	0.040	0.061	0.182	0.154	0.152	0.112	0.140	0.267	0.126
	铅	NA	NA	0.012	0.016	0.009	0.008	0.001	0.000	−0.002	−0.001	−0.000	0.005
	锌	NA	NA	0.011	0.012	0.021	0.018	0.015	0.015	0.014	0.021	0.023	0.016
废金属指数	铜	NA	NA	−0.020	−0.407	−0.655	−0.637	−0.233	−0.335	−0.492	−1.142	−1.555	−0.608
	铝	NA	NA	−0.021	−0.029	−0.034	−0.065	−0.077	−0.078	−0.067	−0.055	−0.193	−0.069
	铅	NA	NA	0.129	0.099	0.128	0.300	0.216	0.187	0.132	0.114	0.098	0.156
	锌	NA	NA	0.019	0.009	0.022	0.078	0.051	0.032	−0.003	−0.009	0.015	0.023

表 22.1 可见，1990～2000 年，中国的废金属指数普遍较低，甚至出现了废金属指数为负值的状况，表示铜、铝、铅、锌四种金属的废金属资源十分短缺。其中，铜和铝的废金属指数一直为负值，废铜贸易修正值接近于 1，表示该期间铜的生产几乎完全依赖于进口废铜；废铝净进口修正系数在 0.13 左右，表示该期间铝的生产以消耗铝矿资源为主，并消耗进口废铝来补充缺口。铅的情况稍好，废铅指数平均为 0.156，废铅进出口基本持平，表明该期间铅的生产以消耗铅矿石为主，消耗国内废铅资源为辅。在锌的生产方面，也是以消耗锌矿石为主，其次还消耗了极少量的国内与国外的废锌资源。

为了进一步分析中国废金属指数低下的原因，本节以铅金属为例，估算了美国 1990～2000 年的废铅指数。为了便于对比，将中国与美国的有关数据汇总到表 22.2 中。其中美国的有关数据取自文献（U.S. Department of the interior，2002b；

Gerald，2001）。

表 22.2　中国与美国废铅指数估算表
Table 22.2　Estimated values of lead scrap index for China and USA

年份		1990	1991	1992	1993	1994	1995	1996	1997	1998	1999	2000	平均值
再生铅占有率	中国	0.095	0.145	0.132	0.108	0.128	0.288	0.203	0.175	0.122	0.106	0.092	0.145
	美国	0.693	0.72	0.751	0.726	0.725	0.72	0.764	0.766	0.772	0.76	0.769	0.742
废铅指数	中国	0.102	0.155	0.141	0.115	0.137	0.308	0.217	0.187	0.130	0.113	0.098	0.156
	美国	0.701	0.729	0.760	0.735	0.734	0.729	0.773	0.775	0.781	0.769	0.778	0.751

由表 22.2 可见，1990~2000 年，中国的废铅指数在 0.102~0.308，平均 0.156；而美国同期内的废铅指数为 0.701~0.781，平均 0.751，比中国高 0.6 左右，表示废美国铅资源比较充足。另外，中国的再生铅占有率平均为 0.145，而美国为 0.742。从中也可看出，与美国相比中国废铅资源相对短缺。

22.3.3　讨论

1. 中国废金属指数较低的原因

仍以铅金属为例。

在铅产量方面，2000 年中国的铅产量增长到 1990 年的 3.71 倍，表示 1990~2000 年，中国铅产量处于持续增长状态，而美国在同期内，基本维持不变。如图 22.3 所示。

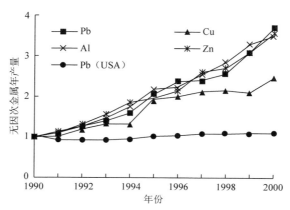

图 22.3　若干金属的年产量变化

Fig. 22.3　The variation of annual production for several metals

在废铅回收方面，文献（毛建素和陆钟武，2003）的研究表明，1900~2000 年，中国的折旧废铅循环率在 0.141~0.536，平均只有 0.271，加工废铅循环率在

0.015～0.046，平均为 0.023；而在同一时期，美国的折旧废铅循环率在 0.686～0.902，平均值高达 0.825，加工废铅循环率在 0.036～0.041，平均值为 0.039。

在铅制品的使用寿命方面，由于各国的铅制品大都以铅酸电池为主，其次是电力电缆、建材等，使得各国铅制品的使用寿命相差不大，对废铅指数的影响也较小。

可见，1990～2000 年中国废铅指数较低的主要原因，一是在该期间中国的铅产量持续高速增长，二是在该期间中国铅的循环率较低。

对于铜、铝、锌等几种金属，中国有色金属工业统计数据表明，2000 年的产量分别增长到 1990 年的 2.454、3.499、3.546 倍，如图 22.3 所示，表示均处于持续高速增长状态。另外，从表 22.1 中的数据可见，这三种金属的再生占有率普遍较低，废金属指数也十分低下，由此可以推测，这几种金属的循环率势必也比较低下。

2. 中国废金属指数低下的后果

由于中国废金属指数低下，中国金属业的快速发展很大程度上以大量消耗金属矿资源为代价，甚至不得不较多地进口金属矿石或者进口废金属资源。例如，目前情况下，中国铜的生产很大程度上依赖于进口铜精矿和废铜资源，而铅业中也开始加大铅精矿的进口量，2000 年已经加大到国内总需求的 1/3 以上（中国有色金属工业年鉴编辑委员会，1990～2001）。同时，国内废金属资源的短缺，还将使得近期内的金属冶炼仍将以原生金属的冶炼为主。

3. 中国废金属资源前景预测

作为发展中国家，中国铜、铝、铅、锌这四种金属的产量增长势头，可能仍会维持一段时间，因此，近期内只有通过提高金属循环率来提高废金属指数。从统计数据（中国有色金属工业年鉴编辑委员会，1990～2001）看，这几种金属产量中有很大一部分是用于出口，如果中国的金属生产主要用于国内需求，那么这几种金属的年产量就可以大大降低，废金属资源的状况也将明显好转，金属生产依赖于国外进口金属矿石和废金属资源的状况也会得到改善。可以预料，伴随中国金属产量进入缓慢增长期，以及废金属回收状况的不断改善，废金属资源不足的局面将会逐步扭转。

22.4 结 论

（1）废金属指数是衡量工业中废金属资源充足程度的判据。废金属指数越高，废金属资源越充足。

（2）研究废金属资源问题，必须考虑金属年产量的变化、金属的生产与金属制品报废之间的时间差以及废金属的回收状况等因素，否则难以得出科学的结论。

（3）获得了废金属指数的变化规律。结果表明，废金属指数的大小主要与金属年产量的变化和金属循环率有关。在金属产量持续增长的情况下，工业中的废金属资源必然相对短缺，而且产量增长愈快，愈是如此。反之，在金属产量缓慢增长、保持不变、逐渐下降的情况下，废金属指数将依次明显高于金属年产量高速增长情况下废金属指数。提高金属循环率，工业中的废金属资源状况将明显改善。

（4）研究表明，1990～2000年，中国的铜、铝、铅、锌的废金属指数十分低下，平均值分别只有−0.608、−0.069、0.156和0.023，表明这几种金属的废金属资源严重不足，甚至不得不依赖于进口废金属。而同一时期美国的废铅指数平均为0.751，表明废铅资源比较充足。进一步分析表明，中国的废金属指数低下的主要原因，一是研究期间内，中国的金属产量高速增长，二是金属的循环率较低。

参 考 文 献

姜松. 2000. 中国再生有色金属资源的开发利用. 中国资源综合利用，（1）：18-21.

陆钟武. 2000. 关于钢铁工业废钢资源的基础研究. 金属学报，36（7）：728-734.

陆钟武. 2002. 论钢铁工业的废钢资源. 钢铁，37（4）：66-70.

陆钟武. 2003. 关于循环经济几个问题的分析研究. 环境科学研究，16（5）：1-10.

毛建素，陆钟武. 2003. 论铅业的废铅资源. 世界有色金属，7：10-14.

余德辉，王金南. 2001. 循环经济21世纪的战略选择. 再生资源研究，5：2-5.

中国有色金属工业年鉴编辑委员会. 1990～2001. 中国有色金属工业年鉴. 北京：中国有色金属工业年鉴出版社.

Bertram M，Graedel T E，Rechberger H，et al. 2002. The contemporary European copper cycle：waste management subsystem. Ecological Economics，42：43-57.

David T A，Nasrin B. 1994. Wastes as Raw Materials. The Greening of Industrial Ecosystems. Washington DC：National Academy Press：69-89.

Frosch R A. 1994. Closing the Loop on Waste Materials. The Greening of industrial Ecosystems. Washington DC：National Academy Press：37-47.

Gerald R S. 2001. Lead Recycling in the United States in 1998. http：//minerals.usgs.gov/minerals/pubs/mcs/2001/mcs2001.pdf.

Graedel T E，Allenby B R. 1995. Industrial ecology. New Jersey：Prentice Hall，110.

Sagar A D，Frosch R A. 2004. A perspective on industrial ecology and its application to a metals-industry ecosystem. Journal of Cleaner Production，12：985-995.

Spatari S，Bertram M，Fuse K.，et al. 2003. The contemporary European zinc cycle：1 year stocks and flows. Resources Conservation and Recycling，39：137-160.

U.S. Department of the interior. 2002a. Recycling Metals，http：//minerals.usgs.gov/minerals /pubs/commodity/.

U.S. Department of the interior. 2002b. U.S. Geological Survey-Mineral Yearbook 1994-2000. http：//minerals.usgs.gov/minerals /pubs/commodity/380494-380400.pdf.

第 23 章 论铅业的废铅资源
On Lead Scrap Resource for Lead Industry*

铅业中铅的来源有两个：一是铅矿石，二是废铅。前者是天然资源，后者是回收的再生资源。如果铅业多用废铅，少用铅矿石，那么将不仅有利于保存铅矿资源，而且还有利于环境保护。但是，要做到这一点，首先需要有充足的废铅资源。

目前，世界各国（地区），废铅资源的实际情况差别很大，有的国家废铅资源较充足，可以大力发展再生铅业。例如，美国就属于这种情况，其再生铅量占铅产量的70%以上（U.S. Department of the Interior，2002）。而有的国家，废铅资源不足，再生铅业难以良性发展。例如，中国就属于这种情况，其再生铅仅占铅产量的百分之十几（中国有色金属工业年鉴编辑委员会，1990～2001）。由此可见，废铅资源决定着一个国家铅业的总体结构。

为什么有的国家废铅资源比较充足，而有的国家废铅资源相对短缺？废铅资源充足的国家，会不会永远充足？什么时候会发生变化？废铅资源短缺的国家，情况又会怎样？

为了清楚地回答以上问题，本章将沿用文献（陆钟武，2002a；2000）中的研究方法，分析铅业中的废铅资源问题。

23.1 分 析 方 法

23.1.1 废铅的来源

对于一个国家（地区）来说，铅业的废铅有以下4种来源（陆钟武，2002a；2000）：

（1）内部废铅。这是产生于铅生产企业内部，同时又作为企业自身的生产原料重新利用的废铅。通常，这部分废铅不进入市场流通，因此不作为本文研究的废铅资源。

（2）加工废铅。这是来自国内制造加工工业的废铅。一般情况下，这部分废

---

* 毛建素，陆钟武. 2003. 论铅业的废铅资源. 世界有色金属，7：10-14.

铅在较短时间内就能进入废铅再生企业，所以又称"短期废铅"。

（3）折旧废铅。这是铅制品完成其使用寿命后报废时形成的废铅。由于从铅的生产，到最后变成折旧废铅，一般要经历较长的时间，所以折旧废铅又称"长期废铅"。

从铅演变成折旧废铅，中间有一个"时间差"。只有引入这个"时间差"概念，才能把铅业的废铅资源问题弄清楚，否则研究工作将毫无收获。

（4）进口废铅。这是从其他国家（地区）进口的废铅。由于进口废铅来自国外，因此不能把它看做是进口国的废铅资源。

以上 4 种不同来源的废铅，只有加工废铅和折旧废铅量两种废铅，是一个国家（地区）的废铅资源量。在数量上，折旧废铅量常常是加工废铅量的好几倍。所以，在研究废铅资源问题时，要特别注重折旧废铅。

23.1.2 废铅指数

废铅资源量的绝对值，不能说明铅业废铅资源的充足程度。这是因为一个国家废铅资源是否充足，是相对于这个国家的铅产量而言的。在铅产量很小的情况下，只要有少量的废铅就显得很充足，反过来也是同样的道理。

为此，将下面的比值定义为一个国家的废铅指数，代表符号为 S，即

$$S = \frac{\text{统计期内国内回收的折旧废铅量与加工废铅量之和}}{\text{统计期内该国的铅产量}}$$

或

$$S = \frac{\text{统计期内国内的废铅资源量}}{\text{统计期内该国的铅产量}}$$

废铅指数 S 是衡量铅业废铅资源充足程度的判据。S 值愈大，铅业的废铅资源愈充足；S 值愈小，废铅资源愈短缺。

23.1.3 铅产量变化与废铅指数之间的关系

一个国家（地区），在一个历史时期内，铅产量随时间的变化，在总体上可划分为以下 3 种情况：①保持不变；②持续增长；③逐渐下降。下面将分别研究这 3 种情况下的废铅指数。

为此，将在完全相同的假设条件下，针对产量变化的 3 种情况，各举 1 个例题。

3 个例题共同的假设条件是：

（1）在铅生产出来的当年就被制成铅制品。在制造加工过程中，每吨铅产生 0.03t 加工废铅，随即返回铅业，作为炼铅原料，重新处理；

（2）铅制品使用 4 年报废（兰兴华和殷建华，2000；姜松，2000），报废后每吨铅形成 0.5t 折旧废铅，随即返回铅业，作为炼铅原料，重新处理；

（3）不考虑铅及铅制品的进、出口。

1. 铅产量保持不变

例 1　已知某国铅产量一直是 100 kt/a，到 2000 年年底，已稳定 4 年以上，见图 23.1。求 2000 年该国的废铅指数 S。

解：2000 年铅业可得废铅量为

加工废铅量　　$100 \times 0.03 = 3$　kt/a

折旧废铅量　　$100 \times 0.5 = 50$　kt/a

故废铅指数为

$$S = \frac{3 + 50}{100} = 0.53$$

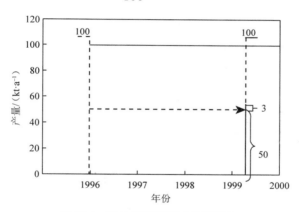

图 23.1　产量不变情况下的废铅量

Fig. 23.1　Quantity of lead scrap in case of constant output of lead

2. 铅产量持续增长

例 2　已知某国铅产量持续增长，1996 年为 100 kt/a，4 年后（2000 年）为 140 kt/a，见图 23.2。求 2000 年该国的废铅指数 S。

解：2000 年铅业可得废铅量为

加工废铅量　　$140 \times 0.03 = 4.2$　kt/a

折旧废铅量　　$100 \times 0.5 = 50$　kt/a

故废铅指数为

$$S = \frac{4.2 + 50}{140} = 0.39$$

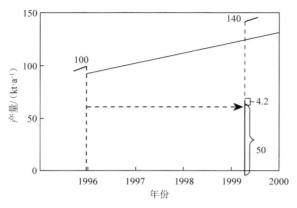

图 23.2 产量增长情况下的废铅量

Fig. 23.2 Quantity of lead scrap in case of increasing annual output of lead

3. 铅产量下降

例 3 已知某国铅产量持续下降，1996 年为 100 kt/a，4 年后为 60 kt/a，见图 23.3。求 2000 年该国的废铅指数 S。

解：2000 年铅业可得废铅量为

加工废铅量　　$60 \times 0.03 = 1.8 \, \text{kt/a}$

折旧废铅量　　$100 \times 0.5 = 50 \, \text{kt/a}$

故废铅指数为

$$S = \frac{1.8 + 50}{60} = 0.86$$

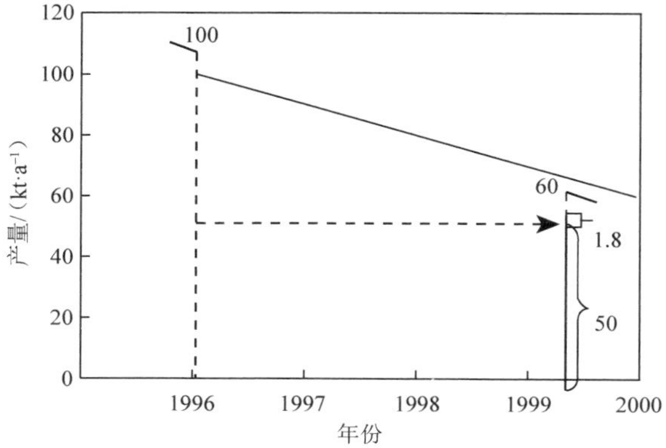

图 23.3 产量下降情况下的废铅量

Fig. 23.3 Quantity of lead scrap in case of decreasing annual output of lead

4. 小结

以上 3 个例题的计算结果分别是：例 1　　$S=0.53$；例 2　　$S=0.39$；例 3　　$S=0.86$。

在 3 个例题的计算结果之间，为什么会有这么大的差别呢？原因就是铅产量变化的情况不同。在例 2 中，铅产量持续增长，所以 S 值较低；而且不难理解，产量增长愈快，S 值愈低。在例 3 中，铅产量持续下降，所以 S 值较高；而且可以推测，产量下降愈快，S 值愈高。而例 1 的情况介于例 2 和例 3 之间。

当然，如果在折旧废铅形成过程中，不存在 23.1.1 节所说的"时间差"，那么铅产量的变化对 S 值的影响也就不存在了。然而，那是不可能的。由此可见，在研究铅业的废铅资源时，引入"时间差"的概念十分重要。

23.1.4　废铅实得率与废铅指数之间的关系

如果改变 1.3 中的计算条件，假设制造加工过程中，每吨铅产生 0.04t 加工废铅，并且铅制品报废后每吨铅形成 0.8t 折旧废铅，那么上述 3 个例题中的计算结果，将大为不同。

比如，例 1 情况下，2000 年铅业可得废铅量为

加工废铅量　　$100 \times 0.04 = 4$　kt/a

折旧废铅量　　$100 \times 0.8 = 80$　kt/a

故废铅指数为

$$S = \frac{4+80}{100} = 0.84$$

较原来的 0.53 提高了 0.31。

不难算出，例 2 情况下，S 值变为 0.61；例 3 情况下，S 值变为 1.37。分别较原来提高了 0.22、0.51，表示废铅资源明显充足。特别是在产量下降（例 3）的情况下，废铅指数超过了 1，出现废铅资源过剩的局面。这时，部分废铅将用于出口或流入社会。

可见，假设条件中的这两个指标，是影响废铅指数的重要因素。文献（陆钟武，2000；陆钟武，2002b）中称每吨铅形成的折旧废铅量，为折旧废铅实得率；每吨铅产生的加工废铅量，为加工废铅实得率，分别用符号 a、b 表示。a、b 的数值愈高，废铅资源愈充足。

由于制品加工过程中，铅的利用率越高越好，而利用率越高，产生的加工废铅越少，相应的加工废铅实得率会越小，因此提高加工废铅实得率，不能作为改善废铅资源状况的途径。只有大力提高折旧废铅实得率，才能有效地改善废铅资源状况。

23.2 实例——中国、美国、瑞典三国废铅指数的估算

23.2.1 铅产量的变化情况

中国、美国、瑞典的铅产量变化情况如图 23.4 所示。由图可见：①近 10 年来，中国的铅产量一直高速增长；②美国的铅产量曾稍有下降，后又缓慢增长，并趋于稳定；③瑞典的铅产量基本保持不变。图中铅产量的数据，中国的取自中国有色金属工业年鉴（1990～2001），美国和瑞典的取自地质调查矿产年鉴（1994～2000）。

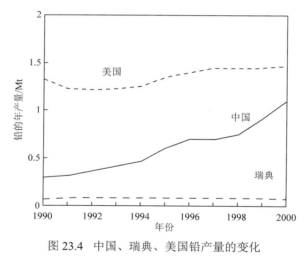

图 23.4　中国、瑞典、美国铅产量的变化

Fig. 23.4　The variation of annual output of lead in China，Sweden and USA

23.2.2 废铅实得率的变化情况

1. 估算方法

由于废铅经回收后，通常直接进入再生铅厂，折旧废铅和加工废铅的数量未列入统计数据，因此无法根据前文中废铅实得率定义，直接计算 a 和 b 的数值。

考虑废铅经过再生过程形成再生铅，而再生铅的数量是列入统计的，因此本节将以再生铅的数据为基础，根据不同国家的铅的再生收率，中国 88%（杨春明和马永刚，2000），美国和瑞典 95%（Sten，1999），用再生铅量除以再生收率，折算出各国不同年份的废铅资源量。然后，按照各国废铅的构成，中国折旧废铅占

86%，美国和瑞典占 95%（U.S. Department of the interior，2002；Gerald，2001），用废铅资源量乘以折旧废铅所占的百分比，得到折旧废铅量；废铅资源量中的其余部分，便是相应年份的加工废铅量。最后，考虑从铅演变成折旧废铅的"时间差"，估计为 4 年，用折旧废铅量除以 4 年前所在年份的铅产量，得到 4 年前所在年份的折旧废铅实得率 a；用各年的加工废铅量除以当年的铅产量，得到加工废铅实得率 b。

2. 估算结果

按照上述方法，估算出中国、美国、瑞典 3 个国家铅业的折旧废铅实得率 a 和加工废铅实得率 b，估算结果列入表 23.1～表 23.3 中。表中铅产量和再生铅产量的数据，中国的取自中国有色金属工业年鉴（1990～2001），美国和瑞典的取自地质调查矿产年鉴（1994～2000）。

由表可见，中国的折旧废铅实得率 a 在 0.141～0.536，加工废铅实得率 b 为 0.015～0.046；同时期内，美国的 a 在 0.686～0.902，b 为 0.036～0.041；瑞典的 a 在 0.409～0.629，b 为 0.017～0.031。通过对比不难发现：中国的废铅实得率较低，美国较高，瑞典介于二者之间。

23.2.3 废铅指数的估算

在 23.2.2 节估算出各年的废铅资源量的基础上，用废铅资源量除以当年的铅产量，得到当年的废铅指数 S。按照这种方法估算的中国、美国、瑞典 3 个国家铅业的废铅指数分别汇总到表 23.1～表 23.3 中。

表 23.1 中国废铅指数估算表

Table 23.1 Estimated values of lead scrap index for China

年份	1990	1991	1992	1993	1994	1995	1996	1997	1998	1999	2000
铅产量/kt	296.5	319.7	366	411.9	467.9	607.9	706.2	707.5	756.9	918.4	1099.9
再生铅产量/kt	28.2	46.3	48.3	44.3	59.9	175.3	143.6	123.7	92.3	97.4	102.
废铅量/kt	32.05	52.61	54.89	50.34	68.07	199.20	163.18	140.57	104.89	110.68	115.91
折旧废铅/kt	27.56	45.24	47.21	43.29	58.54	171.31	140.33	120.89	90.21	95.18	99.68
加工废铅/kt	4.49	7.37	7.68	7.05	9.53	27.89	22.85	19.68	14.68	15.50	16.23
a/(t/t)	0.197	0.536	0.383	0.293	0.193	0.157	0.141	—	—	—	—
b/(t/t)	0.015	0.023	0.021	0.017	0.020	0.046	0.032	0.028	0.019	0.017	0.015
废铅指数	0.11	0.16	0.15	0.12	0.15	0.33	0.23	0.20	0.14	0.12	0.11

注：此表数据仅适用于中国内地。

表 23.2　美国废铅指数估算表

Table 23.2　Estimated values of lead scrap index for USA

年份	1990	1991	1992	1993	1994	1995	1996	1997	1998	1999	2000
铅产量/kt	1330	1230	1220	1230	1260	1350	1400	1450	1450	1460	1470
再生铅产量/kt	922	885	916	893	913	972	1070	1110	1120	1110	1130
废铅量/kt	970	932	964	940	961	1023	1126	1168	1179	1168	1189
折旧废铅/kt	922	885	916	893	913	972	1070	1110	1120	1110	1130
加工废铅/kt	48.53	46.58	48.21	47	48.05	51.16	56.32	58.42	58.95	58.42	59.47
a/(t/t)	0.686	0.790	0.877	0.902	0.889	0.822	0.807	—	—	—	—
b/(t/t)	0.036	0.038	0.040	0.038	0.038	0.038	0.040	0.040	0.041	0.040	0.040
废铅指数	0.73	0.76	0.79	0.76	0.76	0.76	0.80	0.81	0.81	0.80	0.81

表 23.3　瑞典废铅指数估算表

Table 23.3　Estimated values of lead scrap index for Sweden

年份	1990	1991	1992	1993	1994	1995	1996	1997	1998	1999	2000
铅产量/kt	69.6	88.0	91.2	84.6	82.6	82.0	84.1	86.2	92.6	86.00	82.8
再生铅产量/kt	22.1	38.8	37.1	37.8	36.0	36.0	41.9	51.5	52.0	48.0	45.0
废铅量/kt	23.26	40.84	39.05	39.79	37.89	37.89	44.11	54.21	54.74	50.53	47.37
折旧废铅/kt	22.1	38.8	37.1	37.8	36	36	41.9	51.5	52	48	45
加工废铅/kt	1.16	2.04	1.95	1.99	1.89	1.89	2.21	2.71	2.74	2.53	2.37
a/(t/t)	0.517	0.409	0.459	0.609	0.629	0.585	0.535	—	—	—	—
b/(t/t)	0.017	0.023	0.021	0.024	0.023	0.023	0.026	0.031	0.030	0.029	0.029
废铅指数	0.33	0.46	0.43	0.47	0.46	0.46	0.52	0.63	0.59	0.59	0.57

23.2.4　小结

在 1990~2000 年，中国、美国、瑞典 3 个国家废铅指数的估算值在以下范围：中国 $S=0.11$~0.33；美国 $S=0.73$~0.81；瑞典 $S=0.33$~0.63。

3 个国家废铅指数 S 的估算值之间差别如此大，主要原因：一是这三个国家铅产量变化的情况差别很大：中国，持续高速增长；美国，曾稍有下降，然后缓慢增长，并趋于稳定；瑞典，基本不变；二是废铅回收状况差别很大：中国的废铅实得率最低，a 和 b 的平均值分别只有 0.271 和 0.023；同时期内，美国 a 和 b 的平均值分别高达 0.825 和 0.039；瑞典介于中国和美国之间，a 和 b 的平均值分别为 0.535 和 0.025。

23.3 讨　　论

23.3.1　中国

（1）由于铅产量持续高速增长，同时废铅的实得率很低，中国铅业的废铅指数仅为 0.11~0.33，废铅资源十分短缺。在这种情况下，铅业对铅矿石的依赖程度极高，铅业的发展只能以大量消耗铅矿资源为代价，同时铅业结构中，以原生铅的冶炼为主。

（2）如果提高废铅的实得率，同时每年适当进口一些废铅，那么将可有效缓解中国废铅不足的局面。当然，是否进口废铅，还与国际市场上废铅价格的走势等其他因素有关，应慎重考虑。

（3）可以预料，伴随中国的铅产量进入缓慢增长期，以及废铅回收状况的不断改善，废铅不足的局面将逐步好转。到那时，铅业可以较多的使用废铅，铅矿资源会得到一定的节省。

23.3.2　美国

（1）目前情况下，美国的废铅资源较为充足。铅业只需消耗少量的铅矿资源，便可获得较大的铅产量。

（2）可以预料，若美国铅产量进一步增长，则它的 S 值将会开始下降。S 值降低的幅度，决定于铅产量增长的幅度。如果出现铅产量高速增长的情况，那么几年后，就会产生废铅短缺的局面。

23.3.3　瑞典

（1）20 世纪 90 年代以来，瑞典的铅产量基本保持不变，S 值为 0.33~0.63，并大致呈上升趋势，可以推测，这主要是废铅实得率不断提高的结果。

（2）瑞典的 S 值之所以明显低于美国的 S 值，主要是由于美国的 a 值高出瑞典的 0.3，b 值高出 0.014。由于缺乏瑞典废铅资源方面的详细资料，进一步的原因，留待细查。

（3）可以预料，如果进一步改善废铅回收状况，瑞典的废铅指数还会继续上升。并可望达到美国目前的水平，那时瑞典的废铅资源状况将较为充足。

23.4　结　　论

（1）废铅资源状况是决定一个国家（地区）铅业总体结构的主要因素之一。在对铅业进行决策时，要把废铅资源的当前状况与中长期预测结合起来，否则可能出现偏差。

（2）废铅指数是衡量铅业废铅资源充足程度的判据。把它与铅产量的变化联系起来，并且注意到折旧废铅形成的"时间差"问题，是研究铅业废铅资源问题的基本思路。

（3）在铅产量持续增长的情况下，铅业的废铅资源必然短缺，而且产量增长愈快，愈是如此。中国的废铅资源量保持在每年十余万吨，废铅指数只有 0.10～0.33，废铅资源严重不足，是铅产量高速增长的必然结果。

（4）在铅产量缓慢增长、保持不变、逐渐下降的情况下，铅业的废铅指数将依次明显高于铅产量高速增长情况下废铅指数。

（5）在铅产量变化的情况一定的条件下，提高废铅的回收状况，铅业的废铅资源将明显充足。近 10 年来，在废铅指数方面，美国明显高于瑞典，主要原因是废铅回收状况，美国比瑞典好。

致　　谢

研究过程中，得到北京矿冶研究总院孙传尧院长、沈阳蓄电池研究所杨凌主任、北京安泰科技信息中心冯君丛主编等人的帮助。

参 考 文 献

姜松. 2000. 中国再生有色金属资源的开发利用. 中国资源综合利用，（1）：18-21.

兰兴华，殷建华. 2000. 发展中的再生铅工业. 中国资源综合利用，（8）：19-21.

陆钟武. 2000. 关于钢铁工业废钢资源的基础研究. 金属学报，36（7）：728-734.

陆钟武. 2002a. 论钢铁工业的废钢资源钢铁. 钢铁，37（4）：66-70.

陆钟武. 2002b. 钢铁产品生命周期的铁流分析——关于铁排放量源头指标等问题的基础研究. 金属学报，38（1）：58-68.

杨春明，马永刚. 2000. 中国废铅蓄电池回收和再生铅生产.中国电工技术学会铅酸蓄电池专业委员会. 第七届全国铅酸蓄电池学术年会论文全集. 广东南海，198-202.

中国有色金属工业年鉴编辑委员会. 1990～2001. 中国有色金属工业年鉴. 北京：中国有色金属工业年鉴出版社.

Gerald R S. 2001. Lead Recycling in the United States in 1998. http：//minerals.usgs.gov/minerals/pubs/mcs/2001 /mcs-2001.pdf.

Sten K. 1999. Closing the technospheric flow of toxic metals，Modeling lead losses from a lead-acid battery system for Sweden. Journal of Industrial Ecology，3（1）：23-40.

U.S. Department of the interior. 2002. U.S. Geological Survey-Mineral Yearbook 1994-2000. http：//minerals.usgs.gov/minerals /pubs/commodity/380494-380400.pdf.

第24章 关于我国废铅实得率低下的原因的研究
A Study on the Reasons of Low Lead Recycling Rate in China[*]

毛建素和陆钟武（2003）研究了铅业的废铅资源问题，明确了中国废铅资源短缺的主要原因之一是废铅的实得率太低。然而，废铅实得率低下的原因又是什么呢？

本章将通过解析铅的生命周期流动过程，建立精铅生产与所形成的废铅之间的定量关系，找出废铅实得率的影响因数，然后结合中国铅业实际，估算这些影响因数的数值，从而得出对这个问题的答案。

由于在废铅的构成中，折旧废铅通常占主要地位，因此本章只讨论折旧废铅的获得情况。文中的废铅实得率特指折旧废铅实得率。

24.1 废铅实得率

某年废铅实得率定义为

$$a_\tau = \frac{R_{old(\tau+\Delta\tau)}}{P_\tau} \tag{24.1}$$

式中，a_τ——第 τ 年的废铅实得率，(t/t)；

P_τ——第 τ 年的精铅产量，t；

$R_{old(\tau+\Delta\tau)}$——第 $\tau+\Delta\tau$ 年返回到铅生产企业的折旧废铅量，t；其中，$\Delta\tau$ 是从精铅生产到最后形成折旧废铅之间的"时间差"，反映铅的一个生命周期所经历的时间，单位是年。本研究中取 $\Delta\tau = 4$ 年。

根据毛建素和陆钟武（2003）的估算，中国的废铅实得率 a 在 0.115～0.536，平均只有 0.231；而美国的废铅实得率在 0.686～0.902，平均 0.825，如表 24.1。相比之下，中国的废铅实得率十分低下。

---

[*] 毛建素，陆钟武. 2003. 关于我国废铅实得率低下的原因的研究. 世界有色金属，11：24-27.

表 24.1 中国和美国的废铅实得率

Table 24.1 The lead recycling rates of China and USA

年份	1986	1987	1988	1989	1990	1991	1992	1993	1994	1995	1996	平均值
中国	0.115	0.184	0.196	0.143	0.197	0.536	0.383	0.293	0.193	0.157	0.141	0.231
美国	—	—	—	—	0.686	0.790	0.877	0.902	0.889	0.822	0.807	0.825

24.2 铅的生命周期流动示意图

铅的生命周期通常包括铅的采选、提炼，铅制品的制造和使用，以及铅制品使用寿命终了后的回收或抛弃几个阶段，是一个漫长的物质流动过程。在这个过程中，若针对铅元素，绘制出与折旧废铅有关的各股铅的流动方向和数量，则如图 24.1 所示。

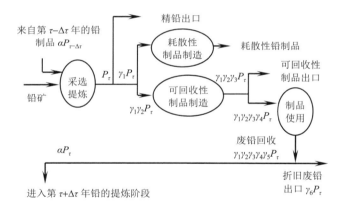

图 24.1 铅的生命周期流动示意图

Fig. 24.1 the diagram of lead flow in its life cycle

图 24.1 中各变量符号含义如下：

P_τ——中国精铅的年产量，t；

γ_1——精铅的国内消费率，即精铅产量中，国内消费部分所占的比率；

γ_2——可回收性铅消费率，即国内精铅消费中可回收性铅消费所占的比率。这里，可回收性铅消费是指可回收性铅制品（Robert，1994），如铅酸电池、铅材等，所消费的铅；其他如颜料等所消费的铅称为耗散性消费；

γ_3——铅的加工利用率，即在铅制品生产阶段，铅制品中的铅量占生产投入铅量的比率；

γ_4——可回收性铅制品的国内消费率，即可回收性铅制品产量中，国内消费部分所占的比率；

γ_5——折旧废铅回收率，即国内回收的折旧废铅量与国内消费的可回收性铅制品量的比值；

γ_6——折旧废铅净出口率，即折旧废铅的净出口量与精铅产量的比值。考虑折旧废铅的形成与精铅生产之间的时间差，第τ年的折旧废铅净出口率表达为

$$\gamma_{6\tau} = \frac{R_{old\tau}}{P_{\tau-\Delta\tau}} \qquad (24.2)$$

式中，$\gamma_{6\tau}$——第τ年的折旧废铅净出口率；

$R_{old\tau}$——第τ年的折旧废铅净出口量，是折旧废铅出口量与进口量的差值，t；

$P_{\tau-\Delta\tau}$——第$\tau-\Delta\tau$年的精铅产量，t。

图24.1中的两点假设是：①精铅在其生产出来的当年就被制成铅制品；②可回收性铅制品的使用寿命为$\Delta\tau$，这类铅制品在其生产年份$\Delta\tau$年以后，全部报废形成折旧废铅，并在报废当年返回铅业，进行再生处理。

不难看出，与以往的产品生命周期物流图（Graedel and Allenby，1995；陆钟武，2002）相比，这张生命周期流动图具有以下几个特点：①侧重分析与折旧废铅有关的铅的流动，而其他流动，如加工废铅的回收，含铅废物、污染物的排放等，隐含在相应阶段的γ数值中，图中并未画出；②考虑了铅制品的可回收性，把铅消费分成了可回收性铅消费和耗散性铅消费两个类型；③考虑了精铅、铅制品以及废铅的贸易问题。

按照图24.1，在精铅年产量为P_τ的情况下，经过$\Delta\tau$年，将在国内形成数量为$\gamma_1\gamma_2\gamma_3\gamma_4\gamma_5 P_\tau$的折旧废铅。在这些折旧废铅中，将有数量为$\alpha_\tau P_\tau$的废铅返回到国内铅的提炼阶段，其余部分出口国外。若用下标符号τ或$\tau+\Delta\tau$区分不同过程发生的时间，不难得出折旧废铅的平衡方程：

$$\alpha_\tau P_\tau = \gamma_{1\tau}\gamma_{2\tau}\gamma_{3\tau}\gamma_{4\tau}\gamma_{5(\tau+\Delta\tau)}P_\tau - \gamma_{6(\tau+\Delta\tau)}P_\tau$$

方程两边同除以P_τ，得到第τ年的废铅实得率：

$$\alpha_\tau = \gamma_{1\tau}\gamma_{2\tau}\gamma_{3\tau}\gamma_{4\tau}\gamma_{5(\tau+\Delta\tau)} - \gamma_{6(\tau+\Delta\tau)} \qquad (24.3)$$

可见，废铅实得率的影响因数有精铅和铅制品的国内消费率γ_1、γ_4，可回收性铅消费率γ_2，加工利用率γ_3，折旧废铅回收率γ_5和折旧废铅净出口率γ_6。

24.3 废铅实得率影响因数的估算

24.3.1 估算方法

γ_1、γ_2、γ_3、γ_4、γ_5和γ_6的数值分别按以下方法估算：

精铅的国内消费率 γ_1，等于各年的精铅国内消费量，除以相应年份的精铅产量，其中精铅产量和消费量的数据取自文献（中国有色金属工业年鉴编辑委员会，1990～2001）。

在铅的消费结构中，把铅酸电池、电缆和铅材作为可回收性铅消费，用这3 种铅制品的铅消费量之和，除以相应年份的国内铅消费量，得到各年的可回收性消费率 γ_2。各年的铅消费结构数据，主要来自文献（蔡显弟等，1996；李富元等，1999）。

在制品生产阶段，普通机械加工过程，如电力电缆、铅材等生产过程，铅的利用率较低，在 60%～70%（姜松，2000）；铅酸电池、化工制品等生产过程，铅的利用率较高，为 85%～98%。本文取 γ_3=90%。

对于可回收性铅制品的国内消费率 γ_4，近似取铅酸电池的国内消费率。铅酸电池的年产量及进出口数据，分别来自文献（中国机械工业年鉴，1985～2001）和文献（中国对外经济贸易年鉴，1990～2001）。

在废铅国际贸易方面，废铅的出口额减去废铅的进口额，得到各年废铅的净出口量。取折旧废铅占废铅总量的 86%，得到各年的折旧废铅净出口量。然后，按式（24.2）计算得到各年的折旧废铅净出口率。各年的废铅进出口数量，来自文献（杨春明和马永刚，2000）。

最后，在算得 γ_1、γ_2、γ_3、γ_4、γ_6 数值的基础上，按下式算出折旧废铅的回收率：

$$\gamma_{5(\tau+\Delta\tau)} = \frac{\alpha_\tau + \gamma_{6(\tau+\Delta\tau)}}{\gamma_{1\tau}\gamma_{2\tau}\gamma_{3\tau}\gamma_{4\tau}} \qquad (24.4)$$

24.3.2 估算结果

按照上面的方法，估算得到中国 1986～1996 年 γ_1、γ_2、γ_4 的数值，以及 1990～2000 年 γ_6、γ_5 的数值，估算结果分别汇总到表 24.2～表 24.6 中。

表 24.2　γ_1 数值的估算表
Table 24.2　The estimated values of γ_1

年份	1986	1987	1988	1989	1990	1991	1992	1993	1994	1995	1996	平均值
铅产量/kt	239.6	246.4	241.4	302	296.5	319.7	366.0	411.9	467.9	607.9	706.2	
铅消费量/kt	249	256	250	250.6	244	249.9	259.3	299.6	298	447.7	464.3	
γ_1	1.039	1.039	1.036	0.830	0.23	0.782	0.708	0.727	0.637	0.736	0.657	0.819

表 24.3 γ_2 数值的估算表

Table 24.3　The estimated values of γ_2

年份	1986	1987	1988	1989	1990	1991	1992	1993	1994	1995	1996	平均值
铅消费量/kt	249	256	250	250.6	244	249.9	259.3	299.6	298	447.7	464.3	
铅酸电池/kt	56.9	70.8	90.9	110.2	134	NA	186	205.8	192.8	209	317	
电缆护套/kt	50.0	29.0	42.7	49.2	45.8	NA	27.5	17.08	15.79	NA	NA	
建材/kt	10.1	9.3	5.9	7.3	7.0	NA	15.3	10.49	11.03	NA	NA	
可回收性铅消费量/kt	117	109.1	139.5	166.7	186.8	NA	228.8	233.4	219.6	NA	NA	
γ_2	0.470	0.426	0.558	0.665	0.766	NA	0.882	0.779	0.737	NA	NA	0.660

表 24.4 γ_4 数值的估算表

Table 24.4　The estimated values of γ_4

年份	1986	1987	1988	1989	1990	1991	1992	1993	1994	1995	1996	平均值
电池产量/GWh	3.220	5.072	4.550	NA	6.980	5.146	6.837	7.773	NA	7.080	9.487	
电池出口量/GWh	0.158	0.477	0.007	NA	0	1.968	0.020	0.014	0.022	0.011	0	
γ_4	0.951	0.906	0.998	NA	1	0.618	0.997	0.998	NA	0.998	1	0.941

表 24.5 三种算法下 γ_5 的估算值

Table 24.5　The estimated values of γ_5 in three methods

年份	1990	1991	1992	1993	1994	1995	1996	1997	1998	1999	2000	平均值
方法 A	0.657	1.003	0.088	NA	0.129	NA	0.656	0.571	NA	NA	NA	0.517
方法 B	0.220	0.445	0.332	NA	0.308	NA	0.643	0.531	NA	NA	NA	0.413
方法 C	0.275	0.509	0.377	NA	0.348	NA	0.684	0.577	NA	NA	NA	0.461

注：方法 A、B、C 分别指按照公式（24.4）、(24.5)、(24.6) 计算 γ_5。

表 24.6 γ_6 数值的估算表

Table 24.6　The estimated values of γ_6

年份	1990	1991	1992	1993	1994	1995	1996	1997	1998	1999	2000	平均值
废铅出口量/kt	6.132	5.250	1.500	0.621	1.510	0.589	0.152	0.061	1.751	1.060	0.037	
废铅进口量/kt	1.690	0.140	5.712	7.322	5.793	5.690	0.820	0.204	0.007	—	0.05	
γ_6	0.159	0.178	−0.150	−0.191	−0.124	−0.137	−0.016	−0.003	0.032	0	0.	−0.023

可见，精铅的国内消费量占总产量的 65.7%～103.9%，平均 81.9%，并且逐年呈下降趋势。其中，前几年精铅的国内消费率大于 1，是由于国内消费量大于精铅产量造成的。在国内的铅消费中，可回收性铅消费占国内精铅总消费的 42.6%～88.2%，平均 66%。铅制品的国内消费率为 61.8%～100%，平均 94.1%。

在废铅贸易方面，近 10 年来，中国经历了从废铅净出口，到废铅净进口，再到进出口平衡的过渡，平均折旧废铅净出口率–0.023。

在废铅回收方面，按照式（24.4）计算的废铅回收率，具有较大的波动范围，平均值为 0.517。为避免得出错误的结论，研究中还分别采用了另外两种算法，来估算 γ_5 的数值。一种是，取近 10 年的平均折旧废铅出口率，即 $\gamma_6 = -0.023$，这时：

$$\gamma_{5(\tau+\Delta\tau)} = \frac{\alpha_\tau - 0.023}{\gamma_{1\tau}\gamma_{2\tau}\gamma_{3\tau}\gamma_{4\tau}} \tag{24.5}$$

另一种是，不考虑废铅的出口，式（24.4）变为

$$\gamma_{5(\tau+\Delta\tau)} = \frac{\alpha_\tau}{\gamma_{1\tau}\gamma_{2\tau}\gamma_{3\tau}\gamma_{4\tau}} \tag{24.6}$$

按以上两种方法估算的各年的 γ_5 数值，也汇总到表 24.4 中。可见，这两种算法下，γ_5 的数值较为稳定，并且平均值分别为 0.413 和 0.461。

总之，在可回收性铅制品中，只有约一半的铅得到了回收，并形成了折旧废铅。

24.3.3 讨论

（1）近 10 年来，由于中国从精铅进口国变成了精铅出口国，精铅的国内消费率只有 80%左右，并逐年下降，同时还有部分铅制品用于出口，导致了中国各年的精铅产量中，约有 25%的铅不能返回到中国铅业，是造成中国废铅实得率低下的重要原因之一。

（2）在铅的消费结构中，可回收性铅消费仅占国内总消费的 66%，其余部分为耗散性消费，这部分铅将随着铅制品的使用，永久地损失到环境之中，是造成中国废铅实得率低下的又一重要原因。

（3）近 10 年来，国内折旧废铅的回收率太低，只有 50%左右，是造成中国废铅实得率低下的又一重要原因。根据杨春明和马永刚的研究（2000），目前中国约有 300 家再生铅厂，平均年产量在几百吨，生产集中度很低，由此推测可能有部分再生铅的产量没有列入统计数据。由此可能造成计算结果的部分偏差。

（4）可以预料，如果加强宏观调控力度，适当降低精铅和铅制品的出口，减少不可回收性铅消费，提高废铅的回收水平，那么废铅的实得率可望得到较大的改善。

24.4 改善建议

改善中国废铅实得率的主要对策是：

（1）改善铅的消费结构，增大可回收性铅消费，减少并逐步淘汰铅的耗散性消费。实际上，中国已经开始了这方面的工作，例如铅酸电池在铅消费中的比例逐年上升，并且在 2000 年，中国颁布法令，取消了铅在汽油添加剂中的使用。

（2）借鉴国外先进的管理经验，建立废铅回收法规，完善回收机制，促进废铅的回收。

（3）加强宏观调控力度，适当降低铅品的出口。

（4）加强铅制品加工企业的管理，淘汰落后的工艺、技术和设备，提高铅的加工利用率。

（5）在铅业的统计指标体系中，补充再生铅业的相关数据，主要包括各年废铅资源量、废铅的构成、废铅再生企业生产技术数据等。

24.5 结　　论

（1）通过铅的生命周期流动图，来分析废铅实得率的影响因数，并建立它们之间的定量关系，是研究废铅实得率有关问题的可行方法。研究过程中，要注意到精铅、铅制品和废铅的贸易，铅消费的结构与类型，以及精铅生产与铅制品报废之间的时间差，否则难以奏效。

（2）近 10 年来，中国废铅实得率只有 0.115～0.536，平均 0.231。造成中国废铅实得率低下的主要原因，一是精铅及铅制品的出口比率太高，二是国内铅消费中，耗散性铅消费所占的比率较高，三是折旧废铅的回收状况不好。

（3）可以预料，如果加强宏观调控力度，适当降低精铅和铅制品的出口，减少耗散性铅消费，提高废铅的回收水平，那么废铅实得率可望得到较大的提高。

致　　谢

研究过程中，得到北京矿冶研究总院孙传尧院长、沈阳蓄电池研究所杨凌主任、北京安泰科技信息中心冯君丛主编等人的帮助。

参 考 文 献

蔡显弟，冯君丛，江达，等. 1996. 中国铅的消费及回收. 世界有色金属，(6)：37-40.
姜松. 2000. 中国再生有色金属资源的开发利用. 中国资源综合利用，(1)：18-21.
李富元，李世双，王进. 1999. 国内外再生铅生产现状及发展趋势. 世界有色金属，(5)：26-30.
陆钟武. 2002. 钢铁产品生命周期的铁流分析——关于铁排放量源头指标等问题的基础研究. 金属学报，38（1）：58-68.

毛建素，陆钟武. 2003. 论铅业的废铅资源. 世界有色金属，(7)：10-14.

杨春明，马永刚. 2000. 中国废铅蓄电池回收和再生铅生产. 中国电工技术学会铅酸蓄电池专业委员会. 第七届全国铅酸蓄电池学术年会论文全集. 广东南海，198-202.

中国对外经济贸易年鉴编辑委员会.1990～2001. 中国对外经济贸易年鉴. 北京：中国经济出版社.

中国机械工业年鉴编委会编. 1985～2001. 中国机械工业年鉴. 北京：机械工业出版社.

中国有色金属工业年鉴编辑委员会. 1990-2001. 中国有色金属工业年鉴. 北京：中国有色金属工业年鉴出版社.

Graedel T E，Allenby B R. 1995. Industrial ecology. New Jersey：Prentice Hall.

Robert.U A. 1994. Industrial metabolism：Theory and policy. The greening of industrial ecosystems. Washington D.C.：National Academy Press，23-37.

第 25 章 中国铅流变化的定量分析
Quantitative Analysis on the Changes in Anthropogenic Lead Flows of China[*]

近 10 年以来，我国铅生产量和消费量大幅度上升，2010 年的产量和消费量与 2000 年的相比，分别增加了 2.78 倍和 6.52 倍，两者都约占世界的 1/3（中国有色金属工业协会，2001；2011）。随着我国汽车、电动自行车等铅酸电池需求产品产量的增长，我国铅酸电池产量也愈来愈高，到 2010 年，铅酸电池总产量已达到约 14470 万 kVAh，超过世界总产量的 1/4（王金良等，2011）。金属铅和铅产品需求量的迅速增加，使得涉铅企业数量也迅速增加，如 2005 年，我国有规模以上铅锌冶炼企业 466 家，而 2006 年就增加了 55 家（李卫锋等，2010）。可是，由于我国有些涉铅企业工艺设备落后，企业布点分散，地方保护以及人们环保意识淡薄，一些企业违规操作等铅金属使用不良现象（张正浩等，2005；王敬忠等，2012），导致我国一些地方铅污染比较严重，类似陕西凤翔县和湖南嘉禾县等儿童血铅超标以及中毒事件频发（Ji et al.，2011；林星杰，2011）。因此，为了减少铅矿资源的大量消耗，减少铅污染，遏制铅污染事件频发，我国政府采取了一系列措施，包括颁布了一系列铅金属使用政策，推动铅行业清洁生产，关停大量中小企业，促进产业结构升级等（Mao and Ma，2012；Meyer et al，2008；沈越等，2011）。这些措施涵盖管理、技术、政策、产品替代等多个方面，从而从多个角度来引导我国铅金属的使用。

铅金属使用状况的变化和铅金属使用措施的执行将使我国铅流发生改变。研究铅流的改变情况，能够检验铅流改善措施执行的有效性。目前为止，已有学者对我国 1999 年、2000 年、2004 年和 2006 年铅流现状进行了分析（Mao et al，2007；郭学益等，2009；Guo et al.，2009；Wang et al.，2008），但尚未有研究者对我国铅流改变进行研究。因此，亟需对近些年我国铅流改变进行研究，从而更好地指导我国铅金属的使用，以及为我国铅矿资源环境优化管理提供定量参考。

---

[*] 马兰，毛建素. 2014. 中国铅流变化的定量分析. 环境科学，35（7）：2829-2833.

25.1 研究方法

25.1.1 铅元素人为流动分析

Mao 等（Mao et al., 2008a；Mao et al., 2008b）在 2008 年运用"STAF"物流分析法构建了铅元素人为流动基本框架，并运用此框架分析了 2000 年包括我国在内的全球 52 个国家的铅流现状。本研究仍采用该方法，所用铅元素人为流动基本框架如图 25.1 所示。

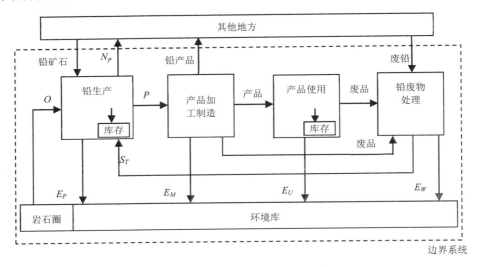

图 25.1 铅元素人为流动基本框架

Fig. 25.1 Framework of anthropogenic lead cycle

注：O：国内铅矿石消耗量；P：国内精铅消费量；E_P：铅生产阶段环境释放量；E_M：产品加工制造阶段环境释放量；E_U：产品使用阶段环境释放量；E_W：铅废物处理阶段环境释放量；S_T：铅循环回收量；N_P：精铅净出口量。

铅元素人为流动主要由铅生产、铅产品加工制造和使用、铅废物处理等四个环节组成，分别代表铅产品生命周期过程中所经历的四个阶段。框架中流类型主要包括：铅矿石进入铅生产阶段的铅矿资源流；前一阶段进入后一阶段的含铅物流；由于产业代谢，耗散型铅产品使用等现象（Mao et al., 2008a；Mao et al., 2008b），在生产、加工、使用和废弃处理过程中向环境释放的铅流；由于资源、产业布局方面的地域差异，研究区域和其他区域之间进行贸易活动的铅贸易净进出口流。在框架中，铅流在每个阶段遵循物质守恒原理，即物质的输入量等于输出量。

本章中，除特别说明外，所有数据均指含铅元素量。

25.1.2 对比研究

为了解我国铅元素人为循环流动的改变，研究中采用对比法。采用相同的铅流分析方法，并以已完成的中国 2000 年的铅流结果（Mao et al.，2008a；Mao et al.，2008b）为基准，选择若干重要指标，对比 2010 年与 2000 年的铅流结果。借此，从对比分析中找到节省铅矿石资源，减少铅污染的途径。

考虑本研究关注铅人为流动对外部资源环境的影响，因此选择与铅矿石资源消耗，铅环境释放和循环回收等方面相关的若干指标。这些指标又进一步分为绝对指标和相对指标。

1. 绝对指标

在铅元素人为流动过程中，国内铅矿石消耗量 O 能直接反映国内铅矿石资源消耗状况；各个阶段的环境释放量 E_P、E_M、E_U 和 E_W 能直接反映铅各个阶段的环境释放状况，而 4 个阶释放量总和，即环境释放量（用字母 E 表示），能直接反映铅环境释放总体状况（Mao et al.，2008a）。因此，本研究将选取这些量作为本研究的绝对指标。

2. 相对指标

在反映天然资源利用方面，资源效率是衡量工业过程中资源利用效率的最重要指标（Halada and Yamamoto，2001；Eco-Efficient 2002 Fair，2002），因此本研究采用铅资源效率来反映国内铅矿资源的消耗状况。环境效率是衡量工业过程中环境释放物对环境影响程度的主要指标（Mao et al.，2007；毛建素和陆钟武，2004），本研究采用铅环境效率来反映铅的释放对环境的影响状况。循环率是直接反映物质循环回收状况的最常用指标，因此本研究采用铅循环率来反映我国铅资源回收利用状况。分别采用字母 r、q 和 α 分别代表铅资源效率，环境效率和循环率，3 者的定义及计算公式如下。

铅资源效率 r，指国内投入单位铅矿石所能产出的铅量（Halada and Yamamoto，2001；Eco-Efficient 2002 Fair，2002）。资源效率愈高，获得同等数量的铅所消耗的国内铅矿资源愈少（毛建素和陆钟武，2004）。其计算公式为

$$r = \frac{P + N_P}{O} \tag{25.1}$$

因为生产出的铅主要用于铅消费和精铅净出口，所以本章中铅产量用铅消费量 P 和精铅净出口量 N_P 之和（即 $P+N_P$）代替。

铅环境效率 q，指整个铅产品生命周期中单位铅释放。环境效率愈高，获得同等数量的精铅所释放的铅愈少，对环境的影响程度也愈小（Mao et al.，2007）。其

计算公式为

$$q = \frac{P + N_P}{E_P + E_M + E_U + E_W} \qquad (25.2)$$

铅循环率 α，指循环回收的铅与铅产量比值。铅循环率愈高，铅循环回收的愈多，再生铅产量在铅产量中所占比重也愈大。其计算公式为

$$\alpha = \frac{S_T}{P + N_P} \qquad (25.3)$$

在公式（25.3）中，S_T 和图 25.1 所指代的量一致，指铅循环回收量。

25.1.3 数据来源和计算说明

2010 年我国精铅消费量、矿山产铅量、再生铅产量、铅精矿进出口量、除铅蓄电池之外的其他铅产品的进出口量，铅选矿实际回收率和铅冶炼总回收率均来自中国有色金属工业年鉴（2011）。精炼铅消费结构由北京矿冶研究总院整理（王晔，2011）。产品加工制造阶段所使用的铅酸电池、铅合金和铅材的加工系数，该阶段铅的排放率等的数据估算均参考相关生产企业环评报告；其他铅产品加工系数采用 Mao 等（2008a）的数据。在铅酸电池进出口方面，铅酸电池主要以两种形式进出口：电池的形式和铅酸电池作为商品的部件形式（Mao et al.，2008a），铅酸电池净进出口是两种形式净进出口的铅酸电池量之和。估算以电池形式进出口的铅酸电池量的相关数据来自王金良等（2011）的研究和文献（中国汽车工业年鉴，2011）。在我国，铅酸电池主要用于汽车和电动车领域，而在进出口方面，电动车以电动自行车进出口为主（王金良等，2011）。因此，本研究在计算以部件形式进出口的铅酸电池量时，直接以进出口汽车和电动车的铅酸电池量近似代替。而用于估算进出口汽车铅酸电池量的数据来自文献（柴静，2011），王金良等（2011）的研究结果和文献（中国轻工业年鉴，2011），用于估算进出口电动自行铅酸电池的数据来自柴静（2011），王金良等（2011）的研究结果和中国汽车工业年鉴（2011）。废铅回收率取值参考柴静（2011）的研究结果，铅酸电池回收率取值参考冯涛（2009）的研究结果，其他铅产品回收率数据采用 Mao 等（2008a）研究中的数据。来自废铅酸电池的废杂铅的比例占废杂铅总量的比例来自李敏等（2012）的研究结果。

25.2 结果与讨论

25.2.1 2010 年我国铅流分析结果

根据相关统计数据，参照 Mao 等（2008a）的铅流计算方法，分析了 2010 年

我国铅流现状,结果如图 25.2 所示:

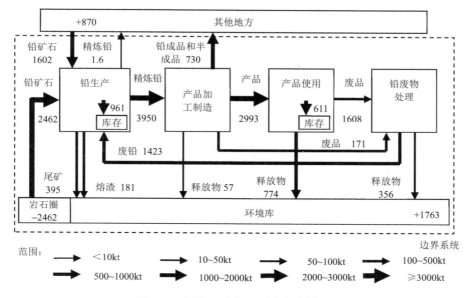

图 25.2　中国 2010 年铅元素人为循环

Fig. 25.2　Anthropogenic lead cycles of China in 2010

由图 25.2 可知,2010 年我国铅矿石消耗量为 2462kt,铅环境释放量为 1763kt,其中铅生产阶段为 576kt,铅产品加工制造阶段为 57kt,产品使用阶段为 774kt,废品处理阶段为 356kt。根据公式（25.1）~公式（25.3）计算得,2010 年我国铅资源效率为 1.61t/t,环境效率为 2.24t/t,循环率为 36.01%。这些结果表明,我国 2010 年铅矿资源的消耗和铅污染还很严重。

25.2.2　指标对比分析

1. 指标对比

根据我国 2000 年铅元素人为流动图[图 25.3,本图引自 Wang 等（2008）的研究结果,将图文改成中文形式,并对应图 25.2 修改了箭头的粗细和虚实],整理出 2000 年铅流的绝对指标,并计算了相对指标。

将 2000 年和 2010 年的指标进行对比,结果见表 25.1。

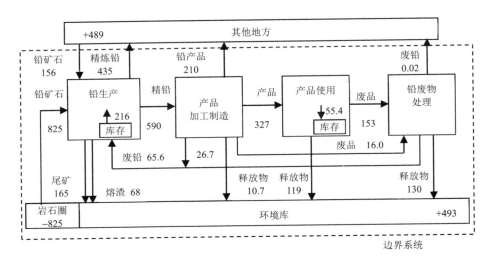

图 25.3　中国 2000 年铅元素人为循环图（单位：kt）

Fig. 25.3　Anthropogenic lead cycles of China in 2000（Unit：kt）

表 25.1　2000 年和 2010 年我国铅流指标对比

Table 25.1　Comparison of the indexes for lead flows in 2010 and 2000

指标类别	指标	2000 年	2010 年	变化倍数*
绝对指标	国内铅矿石消耗量 O/kt	825	2462	2.97
	铅环境释放物量 E/kt	493	1763	3.57
	铅生产阶段铅释放量 E_P/kt	233	576	2.47
	产品加工制造阶段铅释放量 E_M/kt	11	57	5.18
	产品使用阶段铅释放量 E_U/kt	119	774	6.50
	铅废品处理阶段铅释放量 E_W/kt	130	356	2.74
相对指标	铅资源效率 r/(t/t)	1.24	1.61	1.30
	环境效率 q/(t/t)	2.08	2.24	1.07
	铅循环率 α/%	6.44	36.01	29.57**

*为 2010 年指标值与 2000 年相应指标值的比值；**为 2010 年指标值减去 2000 年相应指标值的差值。

2. 分析

从表 25.1 可以看出，与 2000 年相比，2010 年我国铅矿石消耗量提高了 1.97 倍，铅环境释放物量提高了 2.57 倍，而且各个阶段的释放量均大于 2000 年的，表明我国铅矿资源的消耗量和污染物排放量呈增大趋势，我国铅资源环境状况并未改善。从表面上看，造成我国铅矿石消耗量增大的主要原因是我国铅需求量逐年增大（柴静，2011）；造成我国铅环境释放量增大的主要原因：在铅生产、铅产品加工制造等各个阶段，我国目前所采取的有关铅流改善的管理、技术、政策等措施还不够先进、健全、完善（张正浩，2005；Mao and Ma，2012；任立明等，2013）。因此，我国还需进一步采取一些措施来改善我国铅流，一方面限制汽车等铅终端产品产量的增长，采用其他品种电池来替代铅酸电池的使用，适当减少铅产品的出口等，从而减少铅消费，最终减少铅需求量；另一方面继续加强相关部门的监督管理力度，加强铅行业技术革新，完善政策等，从而优化铅资源环境现状。引起以上铅流变化的深层原因将在下一章中进行讨论（马兰和毛建素，2014）。

表25.1也显示，2010年我国铅资源效率是2000年的1.3倍，环境效率是2000年的1.07倍，表明近年来我国在节约铅矿资源，减少铅污染方面措施执行后有一定成效；2010年我国铅循环率比2000年升高了29.57%，表明近年来我国在资源回收利用，以及限制铅出口方面的措施执行结果比较有效（柴静，2011）。

25.3 结　　论

（1）2010年我国消耗了2462kt铅矿石，同时释放了1763kt铅。铅资源效率为1.61t/t，环境效率为2.24t/t，循环率为36.01%。高消耗量和释放量，低效率和循环率，表明我国铅工业对矿石资源的依赖度还很高，铅污染还很严重。

（2）与2000年相比，2010年我国铅资源效率提高到2000年的1.3倍，环境效率提高到2000年1.07倍，循环率升高了29.57%。这说明近年来我国铅流改善措施执行后有一定成效。

（3）与2000年相比，2010年我国铅矿石消耗量提高了1.97倍，铅环境释放量提高了2.57倍。这表明我国铅矿石资源消耗和铅污染仍在加重，铅资源环境状况改善任重道远。

参 考 文 献

柴静. 2011. 铅行业的基本特征. 中国金属通报，（11）：19-21.

冯涛. 2009. 废铅回收面临的问题及对策. 中国资源综合利用，20（9）：10-11.

郭学益，钟菊芽，宋瑜，等. 2009. 我国铅物质流分析研究. 北京工业大学学报，35（11）：1554-1560.

李卫锋，张晓国，郭学益，等. 2010. 我国铅冶炼的技术现状及进展. 中国有色冶金，4（2）：29-33.

李敏，刘毅，朱东方，等. 2012. 废旧蓄电池中再生铅资源的回收利用. 河南化工，29（4）：25-27.

林星杰. 2011. 铅冶炼行业重金属污染现状及防治对策. 有色金属工程，1（4）：23-27.

马兰，毛建素. 2014. 中国铅流改变原因分析. 环境科学，35（8）. 3219-3224.

毛建素，陆钟武. 2004. 关于中国铅的资源效率研究. 环境科学研究，17（3）：78-80.

任立明，王志国，郑磊. 2013. 我国社会源危险废物产生、回收及处置现状与管理对策. 中国环境管理，5（2）：

59-64.

沈越, 陈扬, 孙阳昭, 等. 2011. 我国废铅酸蓄电池污染防治技术及政策探讨. 中国环保产业, (4): 49-52.

王金良, 孟良荣, 胡信国. 2011. 我国铅蓄电池产业现状与发展趋势——铅蓄电池用于电动汽车的可行性分析 (1). 电池工业, (2): 111-116.

王敬忠, 曹国庆, 张蓉. 2012. 铅蓄电池发展与环境影响. 中国金属通报, (21): 16-21.

王晔. 2011. 中国铅行业及展望. 有色金属工程, (1): 27-29.

张正洁, 李东红, 许增贵. 2005. 我国铅污染现状、原因及对策. 环境保护科学, (4): 41-42.

中国轻工业联合会. 2011. 中国轻工业年鉴 (2011). 北京: 中国轻工业年鉴社, 150-415.

中国有色金属工业协会. 2001. 中国有色金属工业年鉴 (2001). 北京: 有色金属工业协会, 434-435.

中国有色金属工业协会. 2011. 中国有色金属工业年鉴 (2011). 北京: 有色金属工业协会, 597-622.

中国汽车工业协会. 2011. 中国汽车工业年鉴 (2011). 北京: 中国汽车工业协会, 8-18.

Eco-Efficient 2002 Fair. 2002. Eco-efficiency-Creating well-being and business by increasing resource productivity. http://www.eco-efficient.net/teh.html.

Guo X Y, Zhong J Y, Song Y. 2009. Substance flow analysis of lead in China. Epd congress 2009, San Francisco: Minerals, Metals & Materials Soc, 889-898.

Halada K, Yamamoto R. 2001. A review of the development of resource productivity concept and eco-efficient technology. Northeastern University. International symposium on industry and ecology 2001. Shenyang: Northeastern University, PRC, 14-36.

Ji A L, Wang F, Luo W J, et al. 2011. Lead poisoning in China: a nightmare from industrialization. Lancet. 377 (9776): 1474-1476.

Mao J S, Dong J, Graedel T E. 2008a. The multilevel cycle of anthropogenic lead: I. Methodology. Resources Conservation and Recycling, 52 (8-9): 1058-1064.

Mao J S, Dong J, Graedel T E. 2008b. The multilevel cycle of anthropogenic lead: II. results and discussion. Resource Conservation & Recycling, 52 (8-9): 1050-1057.

Mao J S, Ma L. 2012. Analysis of current policies on lead usage in China. International Journal of Biological Sciences and Engineering, 3 (4): 234-245.

Mao J S, Yang Z F, Lu Z W. 2007. Industrial flow of lead in China. Transactions of Nonferrous Metals Society of China, 17 (2): 400-411.

Meyer P A, Brown M J, Falk H. 2008. Global approach to reducing lead exposure and poisoning. Mutation Research-Reviews in Mutation Research, 659 (1-2): 166-175.

Wang T, Mao J, Johnson J, et al. 2008. Anthropogenic metal cycles in China. Journal of Material Cycles and Waste Management, 10: 188-197.

第 26 章　中国铅流改变原因分析
The Reasons for the Changes in Anthropogenic Lead Flows of China[*]

人类对铅金属的使用，既消耗岩石圈中的铅矿资源，又因为工业代谢和耗散型铅制品的使用等原因，增加了环境中的含铅污染物（Ayres，1994；Mao et al.，2008a）。如何降低资源消耗和减少铅污染这个有关铅流改善的问题已成为人类铅金属使用过程中迫切需要解决的问题。尽管各个国家纷纷采取技术的研发革新、管理方法的改进、材料和产品等不同措施来解决这个问题（Genaidy et al.，2009；Baker et al.，2010；Rai，2012），但是由于各国经济发展需求、法律法规健全、社会发达等程度不同，最终呈现的结果不同：有些国家的铅流得到改善，有些国家的铅流还在进一步恶化（Mao et al.，2008b）。对于我国，一方面，近年来国内铅产量和铅消费量仍在快速增加（王晔，2011）；另一方面，铅污染严重，儿童血铅超标以及铅中毒事件频发（王敬忠等，2012；Ji et al.，2011）。为了解决这些问题，有些学者曾针对我国进行过铅流分析（Mao et al.，2007），政府也从管理、技术、政策等多个角度着手，采取了一系列措施来改善我国铅流（Mao and Ma，2012；任立明等，2013；沈越等，2011）。这些使我国铅流发生了改变。那么，引起我国铅流改变的原因是什么呢？Wang 等（2008）针对 2004 年我国在铅废品处理阶段释放的铅比 2000 年少的状况，将产生原因归结为我国提高了铅回收率并控制铅释放。但仍需更全面了解产生这些变化的深层原因，从而为进一步铅资源环境改善提供依据。

26.1　影响铅流的因素分析

26.1.1　铅流分析框架

一般来说，铅流在社会经济系统中流经 4 个阶段：铅生产、制品加工制造、制品使用和废品处理。铅流分为 4 种：从资源投入到废物处理的正向流，废铅循

---

[*] 马兰，毛建素. 2014. 中国铅流变化原因分析. 环境科学，35（8）. 3219-3224.

环回收利用流，铅释放入环境中的铅释放物流，与其他地区的进出口贸易流（Mao et al.，2008a）。在整个系统的开始，研究区域内铅矿石、净进口的铅矿石和回收的废铅，经过铅生产阶段从而生产出精铅。根据需求不同，这些精铅一部分运往不同企业用于制品的加工制造，另一部分通过贸易净出口的形式进入其他区域。铅制品也分两部分：供本地区使用和通过贸易净出口的形式进入其他区域使用。铅制品经过一定时间段的使用后寿命终止，变为铅废品，经过集中收集、拆除等处理过程最终生成废铅。最后，回收的废铅作为铅生产的原料重新进入铅生产阶段，进行再生铅生产。整个铅流流动过程中，因为工业代谢和耗散型铅制品的使用等原因，一定量的铅没有利用而直接释放入环境中（Ayres，1994；Mao et al.，2008a）。图 26.1 为铅流分析基本框架（Mao et al.，2008a）。在框架中，铅流在每个阶段遵循物质守恒原理，即物质的输入量等于输出量。本文中，除特别说明外，所有数据均指含铅量。

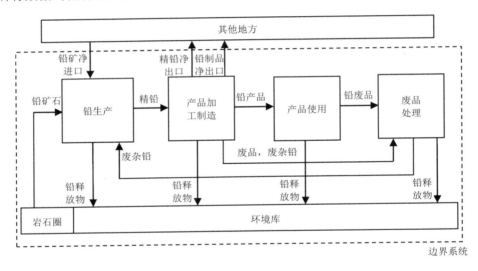

图 26.1 铅流分析框架

Fig. 26.1 Framework of the anthropogenic lead flows

26.1.2 影响指标

从铅流的分析框架（图 26.1）中可以看出，铅流的改变包括铅流规模大小的改变和相关效率系数的改变。因此，影响铅流的因素包括两方面：因铅流规模改变而使铅流改变的规模性因素，因相关效率系数改变而使铅流改变的技术性因素。最能体现这两种因素的是与之对应的指标。故本章中，为了表征影响铅流的这两种因素，将影响铅流的因素所对应的指标分为规模性指标和技术性指标。规模性指标是指自身规模改变，能引起其他铅流发生改变，且其值为绝对量的指标；技术性指标是指自身大小改变，能引起铅资源环境效率改变，且其值为相对系数的指标。

1. 规模性指标

依据铅流分析框架（图 26.1），辨析出影响铅流的规模性指标包括：铅产量、铅矿石净进口量、国内铅消费量、精铅净出口量、铅制品净出口量、国内终端制品消费量、废铅回收量和国内铅矿石消耗量。

铅产量指生产阶段所生成的精铅量。铅产量影响铅资源的消耗，铅产量越大，消耗的铅资源就越多。在整个社会系统中，铅产量的多少主要由社会对精铅的需求量来决定。在实际的统计数据中，需求的多少通过铅消费量来反映。因此，铅消费量决定铅产量。此外，铅产量还受市场价格、国内资源状况等其他因素的影响（柴静，2011）。

铅矿石净进口量指用于铅生产的净进口的铅矿石量。在国内资源有限的情况下，铅矿石净进口量影响铅产量的大小，铅矿石净进口越多，铅产量就越大。因此，铅矿石净进口量受铅生产的需求影响，生产的需求越多，需要净进口的铅矿石就越多。

国内铅消费量指国内精铅消费量。国内精铅消费影响铅产量，国内精铅消费量越大，铅产量就越大。因此，国内铅消费量的变化会拉动铅产量的变化。国内铅消费量受终端铅制品的需求影响，终端制品消费量变大，国内铅消费量也变大（柴静，2011）。

精铅净出口量指净出口的精铅量，其在一定程度能够影响铅产量。当精铅净出口量变大时，为了满足精铅净出口量，铅的生产量就会加大。因此，精铅净出口量的变化拉动铅产量的变化。精铅净出口量主要受国外铅消费需求的影响，消费需求越大，精铅出口量也就越大。此外，在对资源和环境备受关注的当今世界，铅作为资源耗竭型和环境污染型资源，一般不鼓励其产品出口，因此精铅净出口量很大程度受宏观政策因素的制约。除以上两者之外，精铅净出口量还受市场价格等其他因素影响（柴静，2011）。

铅制品净出口量指净出口的铅制品量，其在一定程度能够影响铅消费量，铅制品净出口量变大时，铅消费量就会加大。铅制品净出口量主要受国外铅制品消费需求的影响，消费需求越大，铅制品净出口量也就越大（柴静，2011）。此外，作为铅产品，同样受政策约束。除以上两者之外，它还受市场价格等其他因素影响（柴静，2011）。

铅制品国内消费量指最终用于国内消费的终端铅制品量，它影响废铅回收量。

在技术和管理等回收因素不变的前提下，国内终端制品量越大，意味着若干年后将形成的废铅资源数量越大，从而可回收的废铅也就越大。铅制品国内消费量主要受产品应用领域发展需求的影响，如铅酸电池相关汽车业、通信业的发展，这些末端用户需求越大，铅制品的国内消费量就越多。

废铅回收量指铅制品使用寿命终结后，经过收集、拆分和处理后所得到的废铅量。在铅矿资源有限的情况下，废铅回收量影响铅产量，废铅回收量越多，铅产量越大。废铅回收量主要受国内可回收铅制品消费量的影响。国内可回收铅制品消费量越大，就有越多的铅废品报废回收，废铅回收量就越大（冯涛，2009）。

国内铅矿石消耗量指铅生产消耗的国内铅矿石量，它影响铅产量的大小，国内铅矿石消耗量越大，铅产量就越大。因此，国内铅矿石消耗量主要受铅生产需求影响，需求越大，国内铅矿石消耗量就越大。

2. 技术性指标

依据铅流分析框架（图26.1），辨析出影响铅流技术性指标包括：铅循环率、铅环境释放率和国内终端制品消费比率。

铅循环率指废铅回收量与铅产量的比值。铅循环率一方面影响废铅资源和铅矿石资源的消耗状况。铅循环率变大，废铅资源消耗量（即废铅回收量）就增大，铅资源消耗量就减少。另一方面，铅循环率改变影响铅资源环境效率的改变。通过文献（Mao et al.，2007）中铅资源效率和环境效率的间接计算公式可以得到，在其他量不变的情况下，资源效率和环境效率随着铅循环回收率增大而增大，减小而减小。铅循环率主要受区域内终端制品消费比率、回收技术和管理力度等因素影响（Mao et al.，2007；Mao and Ma，2012；任立明等，2013）。

铅环境释放率指铅环境释放量与铅产量的比值。铅环境释放率影响铅资源环境效率的改变。通过文献（Mao et al.，2007）中铅资源效率和环境效率的间接计算公式可以得到，在其他量不变的情况下，资源效率和环境效率随着铅环境释放率减小而增大，增大而减小。铅环境释放率受生产技术、回收技术、管理力度等宏观因素影响（Mao et al.，2007；Mao and Ma，2012；任立明等，2013）。

铅制品国内消费率指铅制品国内消费量与铅制品总产量（即国内消费量与净出口量之和）的比值，铅制品国内消费率会影响到循环率，当它变大时，说明国内将有更多的折旧废铅，便于国内就地回收，在回收技术和管理因素不变的前提下，铅循环率变大；相反，如果该国内消费率较小，则意味着生产的铅制品将随贸易进入其他国家或地区，在我国现有废铅贸易限制条例下，将不能返回到再生产阶段，造成国内铅循环率较低的局面。

影响铅流的指标整理在如下表26.1中。

表 26.1 影响铅流的指标

Table 26.1 Indices influencing anthropogenic lead flows

类别	指标名称	2000 年 数值	2000 年 来源	2010 年 数值	2010 年 来源	变化倍数 (2010/2000)
规模性指标	铅产量/10kt	110	中国有色金属工业年鉴，2001	415.8	中国有色金属工业年鉴，2011	3.78
	铅矿石净进口量/10kt	15.6	同上	160.2	同上	10.27
	国内铅消费量/10kt	66.0	同上	395.0	同上	5.98
	精铅净出口量/10kt	43.5	同上	0.2	同上	0.005
	铅制品净出口量/10kt	21	Wang 等，2008	73	马兰和毛建素，2014	3.48
	铅制品国内消费量/10kt	32.7	同上	299.3	同上	9.15
	废铅回收量/10kt	6.56	同上	142.3	同上	21.69
	国内铅矿石消耗量/10kt	82.5	同上	246.2	同上	2.98
技术性指标	铅循环率/%	6.44	按 Wang 等人（2008）研究数据估算	36	同上	5.59
	铅环境释放率/%	44.8	同上	42.4	按马兰和毛建素（2014）研究数据估算	0.95
	铅制品国内消费率/%	60.9	同上	80.4	按马兰和毛建素（2014）研究数据估算	1.32

26.2 原 因 分 析

26.2.1 终端消费对铅消费的拉动作用

在铅流流动过程中，铅流规模主要表现在铅生产量和消费量两个方面，其中铅消费主要用于国内铅制品生产，而这些铅制品又进一步用于国内终端消费和出口国外，国内终端消费又主要体现在铅酸电池等铅制品在汽车、电动车、通信等领域的使用（王晔，2011），因此近些年来，这些领域的扩展和铅制品的净出口将对国内铅消费起到拉动作用。

我国主要净出口铅制品是铅酸电池，而铅酸电池主要以两种形式净出口：电池的形式和铅酸电池作为汽车、电动自行车等主要净出口商品部件的形式（中国有色金属工业协会，2002~2011；吴浩亮等，2011）。根据《中国汽车工业年鉴（2002~2010）》（中国汽车工业年鉴，2002~2010）中有关汽车进出口数据，《中国轻工业年鉴（2002~2010）》（中国轻工业年鉴，2002~2010）中有关电动自行车进出口数据，《中国有色金属工业年鉴（2002~2010）》中有关铅酸电池制品以电池形式进出口的进出口量，Mao 等（2008）的研究相关系数的取值和计算方法，计算出 2001~2009 年的铅酸电池净出口量；根据《中国有色金属工业年鉴（2002~2010）》中其他制品进出口量的统计数据，计算出其他制品净出口量；根据相关生产企业环评报告估算出不同制品的加工系数，参考 Mao 等（2008a）的计算方法，计算出 2001~2009 年铅制品国内消费量。最终，根据表 26.1 的相关数据，《中国有色金属工业年鉴（2002~2010）》中 2002~2009 年铅消费量的统计数据，估算出的 2001~2009 铅制品净出口量和铅制品国内消费量，绘制了 2000~2010 年铅制品消费对铅消费的拉动图（图 26.2）。

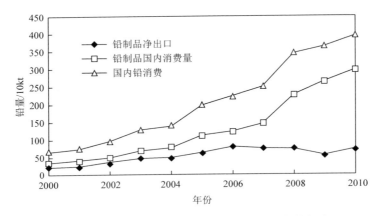

图 26.2 2000~2010 年中国终端消费对铅消费的拉动

Fig. 26.2 The final consumption of lead products pulled the growth of domestic lead consumption

从图 26.2 可以看出近年来，2000~2010 年，我国终端制品消费量以年均 24.7% 的速度增长；铅制品净出口量在 2000~2006 年呈缓慢增长趋势，2006~2010 年均值呈下降趋势；国内铅消费逐年快速增长。因此，近年来国内终端铅消费是拉动铅消费快速增长的主要原因。

26.2.2 铅消费对铅生产的拉动作用

据统计，我国生产的精炼铅主要用于国内铅消费和出口（王晔，2011），因此这两方面的数量对铅生产及产量增长起拉动作用；从另一方面看，生产铅的原料主要来自铅矿石和回收的废铅，而铅矿石又进一步分成国产铅矿和从国外净进口的铅矿两部分，因此这些年铅产量的增加又拉动了这三部分原料性资源消耗量的增加。

根据《中国有色金属工业年鉴（2002~2011）》中有关铅生产的铅产量相关数据，参考 Mao 等（2008）相关数据的计算方法，估算出 2001~2009 年国内铅矿石消耗量和废铅回收量。最后，结合《中国有色金属工业年鉴（2002~2010）》精铅进出口量和铅矿石进出口量的统计数据,以及表 26.1 相关数据,绘制了 2000～2010 年中国对铅生产的拉动图(图 26.3)。

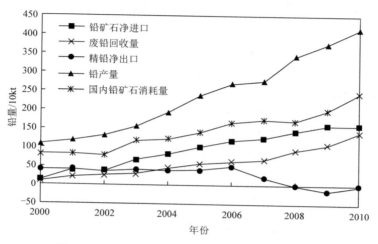

图 26.3 2000～2010 年中国铅消费对铅生产的拉动

Fig. 26.3 The growth of lead consumption further increased the growth of lead production

从图 26.3 可以看出,2000～2010 年,我国铅产量以年均 14.2%的速度快速增长;在 2006 年之前,精铅净出口量趋于稳定,2006 年之后,精铅净出口量急剧下降,2008 年之后甚至出现负值。因此,对于铅生产来说,国内铅消费是拉动铅产量增大的主要原因。图中也显示,铅矿净进口量、废铅回收量和国内铅矿石净进口量都呈增大趋势,说明由于铅生产的增大,使国内铅矿消耗量、废铅回收量和铅矿净进口量均迅速增大。

26.2.3 铅制品国内消费率增大便于铅循环率提高

由表 26.1 可知,2010 年铅制品国内消费率是 2000 年的 1.32 倍,说明铅制品国内消费率增大,根据前面的分析,将便于铅循环率的提高。而表中 2010 年的铅循环率是 2000 年的 5.59 倍,证明了铅制品国内消费率的增长促进了国内铅循环率的提高。

26.2.4 先进技术的采用

近年来,在我国铅矿开采行业,与新技术采用有关的主要包括:选矿采用浮选工艺;适当提高自动化水平;新建大中型铅矿山采用适合矿床开采技术条件的先进采矿方法和大型设备;成功研发高碱工艺等先进工艺并加以利用;采用尾矿再选和综合利用技术（雷力等，2007；秦江波等，2011）。这些比较先进的技术的采用，降低了开采难度，有利于铅资源的开采利用。在铅冶炼行业，大部分企业淘汰了土烧结盘、简易高炉、烧结锅、烧结盘等落后的炼铅工艺及设备，采用了国内先进的具有自主知识产权的富氧底吹强化熔炼和富氧顶吹强化熔炼技术，以及国外新的冶炼铅技术（李卫锋等，2010）。在铅产品加工制造行业，一些铅酸电池生产企业相继采用了机械化程度比较高的工艺和设备，从而提高了资源利用率（肖勇军和王昶，2007）。在三废处理方面，也采用一些比较先进的技术进行铅废物处理，降低了三废中的含铅率（林星杰，2011）。最值一提的是，在废品回收时，摒弃了人工破碎和露天环境下进行破碎作业的落后处理方式，采用自动破碎或机械化破碎分选技术，从而对铅循环率的增大有一定的贡献（李敏等，2012）。

在耗散型铅制品制造方面，一方面，我国对耗散型铅制品铅含量规定了比较严格的技术限值（曾润和毛建素，2010）。为达到这些限值，制造耗散型铅制品时必须采用相对先进的技术，这样就降低了铅在许多应用上的用量。另一方面，采用先进技术研发生产出铅制品替代品，如研发生产出有机颜料来代替含有铅化合物的颜料（曾润和毛建素，2010），从而改变了国内制品的消费结构，使国内可回收铅制品在制品消费中占的比例增大，从而使更多的铅回收使用。

以上先进技术的采用，提高了资源利用效率，使我国铅环境释放率降低。

26.2.5　宏观管理加强

近年来，我国对铅行业从市场准入，技术革新，产业结构升级，法规政策的制定等多方面加强了管理。如对市场准入规模的规定，在 2007 年颁布的《铅行业准入条件》中明确规定：新建、改扩建再生铅项目单系列生产能力必须在 5 万 t/a 以上，现有再生铅企业单系列生产能力不低于 3 万 t/a；并且必须严格按照这个规定对企业进行排查，对不合格企业进行整顿甚至关停（Mao and Ma，2012）。

在管理方面最显著的是对废旧铅酸电池加强了管理。主要体现在：在技术政策和技术标准方面，发布了一系列政策和标准，规范废电池处理处置和资源再生行为，推广使用新技术，鼓励以新型电池替代铅酸电池等；在科研支撑方面，大力支持相关研究；管理行动方面，由国家环保部开展的环保专项行动进行彻查，在彻查过程中对不合格企业或项目作出严格处理（Mao and Ma，2012）。

宏观管理加强最显著的结果是：一些规模小、水平落后的企业被大量关停、如 2011 年 11 月 30 日，中国环保部网站公布的《铅蓄电池生产、组装及回收（再生铅）企业名单》显示，全国关停铅酸电池生产企业达 649 家，关停比例超过 80%。

正是国家宏观管理的加强，关停了规模小、技术落后的涉铅企业，同时推动先进技术采用，从而整体提高了资源利用率，使 2010 年的铅环境释放率降低，其值是 2000 年的 0.95 倍。

26.3 结　　论

通过以上对影响我国铅流因素的分析，发现引起我国铅流变化的原因是多方面的，主要表现在以下方面。

（1）影响铅流的规模性指标主要包括铅产量、铅消费量、贸易量。近年来产生这些指标间变化的主要原因是：国内铅制品消费量以年均 24.7%的速度急剧增长，拉动铅消费量增长；铅消费拉动铅产量以年均 14.2%的速度快速增长，最终拉动了铅矿资源消耗量、废铅回收量和铅矿净进口量增长。

（2）影响铅流的技术性指标主要包括铅循环率、铅环境释放率和铅制品国内消费率，影响这些指标变化的原因是：铅制品国内消费率增大便于铅循环率提高；国家宏观管理中关停规模小、技术落后的涉铅企业，并推进先进技术应用，整体上提高了资源利用率，降低了铅环境释放率。

参 考 文 献

柴静. 2011. 铅行业的基本特征. 中国金属通报，(11)：19-21.

冯涛. 2009. 废铅回收面临的问题及对策. 中国资源综合利用，20（9）：10-11.

雷力, 周兴龙, 文书明, 等. 2007. 我国铅锌矿资源特点及开发利用现状. 矿业快报，(9)：1-4.

李敏, 刘毅, 朱东方, 等. 2012. 废旧蓄电池中再生铅资源的回收利用. 河南化工，29（4）：25-27.

李卫锋, 张晓国, 郭学益, 等. 2010. 我国铅冶炼的技术现状及进展. 中国有色冶金，4（2）：29-33.

林星杰. 2011. 铅冶炼行业重金属污染现状及防治对策. 有色金属工程，1（4）：23-27.

马兰, 毛建素. 2014. 中国铅流改变定量分析. 环境科学，(7)：2829-2833.

秦江波, 于冬梅, 孙永波. 2011. 中国矿产资源现状与可持续发展研究. 经济研究导刊，(22)：11-12.

任立明, 王志国, 郑磊. 2013. 我国社会源危险废物产生、回收及处置现状及管理对策. 中国环境管理，5（2）：59-64.

沈越, 陈扬, 孙阳昭, 等. 2011. 我国废铅酸蓄电池污染防治技术及政策探讨. 中国环保产业，(4)：49-52.

王敬忠, 曹国庆, 张蓉. 2012. 铅蓄电池发展与环境影响. 中国金属通报，(21)：16-21.

王晔. 2011. 中国铅行业发展及展望. 有色金属工程，1（1）：27-29.

吴浩亮, 张明锋, 陈波, 等. 2011. 中国铅酸蓄电池行业现状与展望. 工程建设与设计，(7)：122-125.

肖勇军, 王昶. 2007. 影响我国铅业全面可持续发展的主要因素分析. 生态经济，(10)：216-219.

曾润, 毛建素. 2010. 我国耗散型铅使用的变化及趋势分析. 环境科学与技术，33（2）：192-195.

中国汽车工业协会. 2002~2010. 中国汽车工业年鉴（2002~2010）. 北京：中国汽车工业协会.

中国轻工业联合会. 2002~2010. 中国轻工业年鉴（2002~2010）. 北京：中国轻工业年鉴社.

中国有色金属工业协会. 2001~2011. 中国有色金属工业年鉴（2001~2011）. 北京：有色金属工业协会.

Ayres R U. 1994. Industrial metabolism: theory and policy. The greening of industrial ecosystems. Washington DC: National Academy Press, 23-37.

Baker E, Chon H, Keisler J. 2010. Battery technology for electric and hybrid vehicles: Expert views about prospects for advancement.Technological Forecasting and Social Change, 77 (7): 1139-1146.

Genaidy A M, Sequeira R, Tolaymat T, et al. 2009. Evidence-based integrated environmental solutions for secondary lead smelters: Pollution prevention and waste minimization technologies and practices. Science of the Total Environment, 407: 3239-3268.

Ji A L, Wang F, Luo W J, et al. 2011. Lead poisoning in China: a nightmare from industrialization. Lancet, 377 (9776): 1474-1476.

Mao J S, Dong J, Graedel T E. 2008a. The multilevel cycle of anthropogenic lead: Ⅰ. Methodology. Resources Conservation and Recycling, 52 (8-9): 1058-1064.

Mao J S, Dong J, Graedel T E. 2008b. The multilevel cycle of anthropogenic lead: Ⅱ. Results and discussion. Resources conservation and recycling, 52 (8-9): 1050-1057.

Mao J S, Ma L. 2012. Analysis of current policies on lead usage in China. International journal of biological sciences and engineering, 3 (4): 234-245.

Mao J S, Yang Z F, Lu Z W. 2007. Industrial flow of lead in China. Transactions of Nonferrous Metals Society of China, 17 (2): 400-411.

Rai P K. 2012. An eco-sustainable green approach for heavy metals management: two case studies of developing industrial region. Environmental monitoring and assessment, 184: 421-448.

Wang T, Mao J, Johnson J, et al. 2008. Anthropogenic metal cycles in China. Journal of Material Cycles and Waste Management, 10: 188-197.

第 27 章 中国铅的使用政策现状分析
Analysis of Current Policies on Lead Usage in China*

27.1 Introduction

Lead, an important industrial material, is used mainly in the manufacture of lead-acid batteries, cable sheathing, lead alloys, glass manufacturing, gasoline, and paint (Ye and Wong, 2006; Zhang et al., 2005). The long-term heavy use of lead can lead to high environmental risks and threats to human health (Zheng et al., 2010). China is currently the world's largest producer of primary and refined lead, and the largest lead-consuming country (Chen et al., 2009). Unfortunately, the bulk production and consumption of lead products in China has resulted in many lead pollution incidents in recent years. For example, More than 4000 children were exposed to lead hazards in several Chinese provinces from 2009 to the first half of 2011 (Ji et al., 2011). Accordingly, the government has formulated a series of policies on lead usage that aim to mitigate lead pollution in the environment. The introduction of lead usage policies can minimize lead pollution incidents, reduce the emissions of lead-containing pollutants, encourage the security operations of enterprises, and strictly control the unnecessary usage of lead (Wang and Lin, 2010; Meyer et al., 2008).

To date, studies of the policies of lead usage in China have been narrowly focused on areas such as lead resources development and control (Li and Ma, 2010); lead metal smelting (Wang et al., 2011); lead-containing product usage (Chen et al., 2009; Zeng and Mao, 2010); lead-containing contaminant control (Shen et al., 2011), electronic wastes (e-wastes) management and recycling (Yu et al., 2010; Zhou and Xu, 2012); and laws, policies, and standards related to lead pollution accidents and the control of lead exposure (Xue and Zeng, 2010). To address the lack of an overall analysis of the lead usage policies promulgated by China, we performed a

---

* Mao J S, Ma L. 2012. Analysis of current policies on lead usage in China. International Journal of Biological Sciences and Engineering, 3 (4): 234-245.

systematic analysis of the lead-related policies released by China in the last 10 years (2003-2012), based on the life cycle framework of lead product. Our analysis of the core regulatory basis of these policies enabled us to provide recommendations for China's policymakers.

27.2　Methodology

27.2.1　Lead Product Life Cycle

The product life cycle of lead includes production, fabrication and manufacturing, use, waste management, and recycling stages (Fig. 27.1). The production stage, which includes primary and secondary lead production, is followed by the fabrication and manufacturing of items such as lead-acid batteries, lead sheet and lead pipe, cable sheathing, and lead alloy and other products. Once these products have passed through the use stage and reached the end of their useful lives, the lead is reclaimed and recycled, becoming a secondary resource material, while the wastes remaining from the recycling process are often incinerated or land filled (Mao et al., 2008) At each stage, lead is emitted into the environment by industrial processes (Ayres, 1994) in the form of tailings, slag, dust or other debris (Mao et al., 2009). In addition, the product life cycle may be influenced by lead trading activities among regions and countries that occur because of geographical differences in resource availability and industrial capabilities.

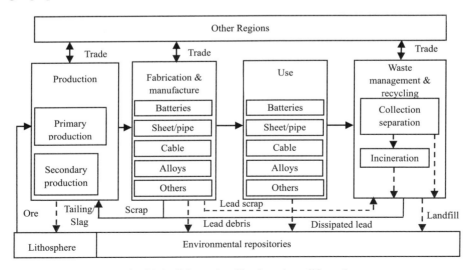

Fig. 27.1　Schematic of lead products life cycle

27.2.2 Policy Classification

We classified the lead usage policies issued by China in recent years into policies for lead production, fabrication and manufacturing, use, waste management and recycling, and trade. Additional sub-categories are shown in Table 27.1.

Table 27.1 Classification of lead-related policies

Main category	Sub-category
Lead production policy	Primary lead production policy
	Secondary lead production policy
Lead products fabrication and manufacturing policy	Dissipative lead products fabrication and manufacturing policy
	Recyclable lead products fabrication and manufacturing policy
Lead products use policy	Dissipative lead products use policy
	Recyclable lead products use policy
Waste management and recycling policy	Lead-acid batteries-related policy
	E-wastes-related policy
Lead trade policy	Lead materials trade policy
	Lead products trade policy
	Lead scrap trade policy
	Other relevant trade policies

27.3 Existing Policies and Analysis

Lead-related policies promulgated in China can be defined as rules and regulations, or as standards. Both types of policies are related to the technology used (such as the principles and approaches to pollution prevention or the requirements for cleaner production), scale of enterprises (size of enterprises permitted to undertake a process), and lead emission limits. In general, China has 33 sets of rules and regulations (Table 27.2), and about 38 standards (Table 27.3).

Table 27.2 Recent rules and regulations issued for lead usage in China

Category	Name of policy	Year of issuance	Issued by
Lead production policy	Interim measures on the management of lead and zinc smelting enterprise announcement	2007	National Development and Reform Commission (NDRC)
	Interim measures on the management of lead and zinc smelting enterprise access announcement	2010	Ministry of Industry and Information Technology (MIIT)

Continued

Category	Name of policy	Year of issuance	Issued by
Lead production policy	Technical policy on pollution prevention and control for lead-zinc smelting industry	2012	Ministry of Environmental Protection (MEP)
	Cleaner production evaluation indicator system for their lead and zinc industry (on trial)	2007	NDRC
	Guidance on best available technologies of pollution prevention and control for lead smelting (on trial)	2011	MEP
	Lead and zinc industry access conditions	2007	NDRC
	Secondary lead industry access conditions (exposure draft)	2011	MIIT
Lead products fabrication and manufacture policy	Rules for their implementation of lead-acid battery production license	2007	National Industrial Product Manufacture Licensing Certificate Office (NIPMLCO)
	Rules for their implementation of lead-acid battery production license (revised)	2011	General Administration of Quality Supervision, Inspection and Quarantine (AQSIQ)
	Guidelines on the lead-acid battery industry site environmental monitoring (on trial)	2011	MEP
	Lead-acid battery industry access conditions	2012	MIIT, MEP
	Implementation program for their cleaner production of battery industry	2011	MIIT
	Cleaner production evaluation indicator system for battery industry (on trial)	2006	NDRC
	Notice on strengthening the pollution prevention and control of lead-acid batteries and secondary lead industry	2011	MEP
	Hygienic standard for cosmetics	2007	Ministry of Health (MOH)
	Heavy metal pollution comprehensive prevention planning for battery industry (exposure draft)	2010	MIIT
Waste management and recycling policy	Technical policy for the pollution prevention of discarded batteries	2003	State Environmental Protection Administration (SEPA)
	The circular on strengthening environmental management of waste electrical and electronic equipment	2003	SEPA
	Document on environmental management of WEEE	2003	SEPA
	Management of recycling home appliances and electronic equipment	2004	NDRC
	The ordinance on the management of waste household electrical and electronic products recycling and disposal (draft)	2005	NDRC
	Technical policy on pollution prevention of discarded appliances and electronic product	2006	MIIT, Ministry of Science and Technology (MOST) MEP
	The ordinance on management of pollution control from electronic information products	2007	MIIT

Continued

Category	Name of policy	Year of issuance	Issued by
Waste management and recycling policy	Administrative measure on pollution prevention of waste electrical and electronic equipments	2008	MEP
	Regulation on management of waste electrical and electronic equipments recycling and disposal	2009	NDRC, MIIT
Lead trade policy	Notice of the Ministry of Finance and the State Administration of Taxation on adjusting the tax refund rates for exported goods	2003	Ministry of Finance (MOF), State Administration of Taxation of China (SATC)
	Notice of the Ministry of Finance, National Development and Reform Commission, Ministry of Commerce, General Administration of Customs, State Administration of Taxation on adjusting the tax refund rates of certain commodities and supplementing the catalogue of prohibited commodities in processing trade	2006	MOF, NDRC, Ministry of Commerce (MOFCOM), General Administration of Customs (GAC), SATC
	Notice of the Customs Tariff Commission of the State Council on the tentative tax rate of import and export of some commodities	2006	Customs Tariff Commission (CTC)
	Notice of the Ministry of Finance and the State Administration of Taxation on lowering the export refund rates for some commodities	2007	MOF, SATC
	Notice of cancellation of export tax refund rates for some of the goods	2010	MOF, SATC
	Prohibited goods catalog for processing and Trade	2004	MOFCOM, GAC, MEP
	Catalogue of solid waste forbidden to import	2009	MEP, MOFCOM, NDRC, GAC, AQSIQ
	Administrative measures on the import of solid waste	2011	MEP, MOFCOM, NDRC, GAC, AQSIQ

Table 27.3　Recent standards related lead metal usage in China

Category	Name of standard	Year of Issuance	Issued by
Lead production standard	Secondary Lead and Lead Alloy Ingots (GB/T 21181—2007)	2007	AQSIQ, Standardization Administration of the People's Republic of China (SAC)
	Cleaner production standard-Lead smelting industry (HJ 512-2009)	2009	MEP
	Cleaner production standard-Lead electrorefining industry (HJ 513-2009)	2009	MEP
	Emission standards of pollution for secondary lead industry (exposure draft) [a]	2010	MEP, AQSIQ
	Emission standard of pollutants for lead and zinc industry (GB 25466—2010)	2010	MEP, AQSIQ

Continued

Category	Name of standard	Year of Issuance	Issued by
Lead products fabrication and manufacture standard	Standard permissible limits and testing method for release of lead or cadmium from ceramic cookware (GB 8058—2003)	2003	AQSIQ
	Standard permissible limits of lead and cadmium release from ceramic ware in contact with food (GB 12651—2003)	2003	AQSIQ
	The technical requirement for environmental labeling products-Ceramic ware, glass-ceramic ware and glass dinnerware (HJ/T 312—2006)	2006	SEPA
	Lead and cadmium control specifications for manufacturer of export domestic ceramicware (SN/T 2509—2010)	2010	AQSIQ
	Packaging glass containers-Release of lead cadmium arsenic and antimony-Permissible limits (GB 19778—2005)	2005	AQSIQ, SAC
	Technical requirement for environmental labeling products-Water soluble coatings (HJ/T 201—2005)	2005	SEPA
	Indoor decorating and refurbishing materials-Limit of harmful substances of interior architectural coatings (GB 18582—2008)	2008	AQSIQ, SAC
	Technical requirement for environmental labeling products - Gravure and flexible printing ink (HJ/T 371—2007)	2007	SEPA
	Technical requirement for environmental labeling products - Offset printing ink (HJ/T 370—2007)	2007	SEPA
	Limits and determination method of certain toxic elements in printing ink (QB 2930.1—2008)	2008	NDRC
	Technical requirement for environmental labeling products–Solvent-based wood coatings for indoor decoration and refurbishing (HJ/T 414—2007)	2007	SEPA
	Indoor decorating and refurbishing materials - Limit of harmful substances of solvent based coatings for woodenware (GB18581—2009)	2009	AQSIQ, SAC
	Water-soluble fertilizers-content-limit and determination of mercury, arsenic, cadmium, lead and chromium content (NY1110—2006)	2006	Ministry of Agriculture (MOA)
	Water-soluble fertilizers-Content-limit of mercury arsenic cadmium lead and chromium content (NY 1110—2010) (revised)	2010	MOA
	National safety technical code for toys (GB 6675—2003)	2003	AQSIQ
	Gasoline for motor vehicles (GB 17930—2006) (revised)	2006	AQSIQ, SAC
	Gasoline for motor vehicles (GB 17930—2011) (revised)	2011	AQSIQ, SAC
	Limit of lead extraction in faucets (JC/T 1043—2007)	2007	NDRC
	General rules for compound food additives (GB26687—2011)	2011	MOH
	Cleaner production standard-Lead acid battery industry (HJ447—2008)	2008	MEP

Continued

Category	Name of standard	Year of Issuance	Issued by
Lead products use Standard	Lead-acid batteries for electric vehicles (QC742—2006)	2006	NDRC
	Lead-acid batteries for motorcycles (JB/T4282—2007)	2007	NDRC
	Lead-acid batteries for stationary valve-regulated (GB/T 19638.2—2005)	2005	AQSIQ, SAC
	Lead-acid storage batteries used for energy storage (GB/T 22473—2008)	2008	AQSIQ, SAC
	Valve regulated lead mesh lead acid battery for communications (YD 1715—2007YD-T)	2007	MIIT
	Electric bicycles-cell or battery and chargers (QB/T2947.1—2008)	2008	NDRCn
	Lead-acid battery used for passenger trains (GB13281—2008)	2008	AQSIQ, SAC
	Lead-acid storage batteries for motorcycles (GB/T 23638—2009)	2009	AQSIQ
	Lead-acid batteries used for electric road vehicles (GB/T 18332.1—2009)	2009	AQSIQ
Waste management and recycling standard	Cleaner production Standard-Waste lead acid battery recycling industry (HJ 510—2009)	2009	MEP
	Technical Specifications of Pollution Control for Treatment of Lead-acid Battery (HJ 519—2009)	2009	MEP
	Lead and lead alloy scrap (GB/T 13588—2006)	2006	AQSIQ, SAC
	Lead and lead alloy scrap (GB/T 13588—2006)	2006	AQSIQ, SAC

a Standard is exposure draft and has no standard number.

27.3.1 Lead Production

China has promulgated 12 policies for lead production, most of which are concerned with market access conditions for lead manufacturers, clean production requirements, and/or pollutant emission standards. These policies appear to emphasize secondary lead production. The lead production industry policies generally include provisions for access management of lead production enterprises, production technology, pollution control technology, and pollutant limits. For example, the recently released *Technical policy on pollution prevention and control for lead-zinc smelting industry* requires enterprises to pay more attention to clean production and also provides comprehensive regulations on the control of raw materials, productive technology and equipment selection, and production process pollution control. This policy stipulates the sizes of enterprises permitted to undertake smelting operations. Many smaller-scale and private enterprises will be required to cease operations, which will play contribute to improving the overall technological level of the industry. Table 27.4 presents the minimum allowable production levels for industries. The lowest lead-production capacity permitted in the lead mining industry is 30 000 t/a (*Secondary lead industry access conditions*, issued 2011) for an existing enterprise, in contrast to the minimum capacity of 10, 000 t/a that was allowed in the 2007 *Lead and zinc industry access conditions* for the same type of enterprise. When the newer regulation is implemented, many small-scale enterprises will be required to cease operation.

Table 27.4 Requirements of production scale

Name of policy[a]	Scope of application	Requirement of production scale	
		Type of project	Scale
Lead and zinc industry access conditions (2007)	Lead smelting industry	New project[b]	>50 000 t/a
		Existing enterprise[b]	>50 000 t/a (after enterprise improvement)
	Secondary lead Industry	Existing enterprise	>10 000 t/a
		Reconstruction project and expansion project	>20 000 t/a
		New project	>50 000 t/a
	Lead mining industry	Minimum production scale[c]	≥30 000 t/a
		Medium-sized mining[c]	>300 000 t/a
		Beneficiation enterprise using flotation process	>1 000t/d
Secondary lead industry access conditions (exposure draft) (2011)	Secondary lead industry	New project, reconstruction project and expansion project[b]	>50 000 t/a
		Existing enterprise[b]	≥30 000t/a

a Sources of corresponding policies are shown in Table 27.2.
b Applies production of lead only, other metals are not considered.
c Single-unit equipment for lead production.

The standards seem to be especially focused on limiting lead emissions in the lead smelting industry (Table 27.5). For instance in *Emission standards of pollution for secondary lead industry*, lead concentrations in atmospheric emissions for secondary lead refineries are limited to 6 mg/m^3 and 4 mg/m^3 for existing and new enterprises, respectively; whereas in *Emission standard of pollutants for lead and zinc industry* these values are 10 mg/m^3 and 8 mg/m^3, respectively. This indicates that the requirements for lead emissions are stricter in secondary lead refineries than in general lead and zinc refineries.

Table 27.5 Emission limits for lead smelting

Pollutant	Scope of application	Name of standard	Objective type	Limit[a]
Lead and its compounds in atmospheric pollutants	Lead smelting industry	Emission standard of pollutants for lead and zinc industry	Existing enterprise	10 mg/m^3
			New enterprise	8 mg/m^3
	Secondary lead industry	Emission standard of pollutants for lead and zinc industry	Existing enterprise	10 mg/m^3
			New enterprise	8 mg/m^3
		Emission standards of pollution for secondary lead industry (exposure draft)	Existing enterprise	6 mg/m^3
			New enterprise	4 mg/m^3
Total lead in wastewater discharge	Lead smelting industry	Emission standard of pollutants for lead and zinc industry	Existing enterprise	1.0 mg/L
			New enterprise	0.5 mg/L
			Special emissions limits	0.2 mg/L
	Secondary lead industry	Emission standard of pollutants for lead and zinc industry	Existing enterprise	1.0 mg/L
			New enterprise	0.5 mg/L
			Special emissions limits	0.2 mg/L
		Emission standards of pollution for secondary lead industry (exposure draft)	Existing enterprise	1.0 mg/L
			New enterprise	0.5 mg/L
			Special emissions limits	0.2 mg/L

a Sources of standards shown in Table 27.3.

These new policies have promoted remediation in the lead smelting industry. Ministry of Environmental Protection reported more than 10 thousand lead-related enterprises were investigated in 2011, and based on that many small scale enterprises with old-dated technology were banned and shut down, some enterprises were suspended operation pending consolidation (Luo, 2012). Similarly, the introduction of advanced technologies has promoted the upgrading of the industrial structure. For instance, in Guiyang County of Hunan Province, a 100,000 t/a oxygen-rich lead smelting production line has been established, which has significantly improved the company's technological processes (Mepprc, 2011).

27.3.2 Lead Products Fabrication and Manufacturing

Some lead-containing products such as paints, additives, and so on, are dissipative; the lead in these products cannot be recycled (Mao and Graedel, 2009). Although the lead in these products represents a small amount of the total lead consumption, it is easily accessible in general usage and human contact is common, resulting in hazards to human health (Zeng and Mao, 2010). China's recently introduced policies focus on dissipative lead products (Table 27.2 and Table 27.3).

The dissipative lead products manufacturing policy issued by China establishes lead limits. For instance, the *Limit of lead extraction in faucets* specifies that lead levels shall not exceed 11 μg/L, and the *General rules for compound food additives* stipulates a maximum limit for lead of 2.0 mg/kg. Some of the policies have been amended. For instance, the maximum lead content in the *Gasoline for motor vehicles* 2011)is 0.005g/L, which amends and reduces the lead levels in the 2006 version of this policy. The publication *Lead and cadmium control specifications for manufacturer of export and domestic ceramic ware* is the industrial standard for importation, exportation, and quarantine procedures, and prescribes the quality management system requirements for lead and cadmium control for the manufacturer of export and domestic ceramic ware. This standard should reduce lead-related trade conflicts.

Table 27.6 Limits for five types of dissipative lead products in China

Type of product	Name of standard [a]	Year of issuance	Upper limit
Toys	National safety technical code for toys	2003	Migration limit 90 mg/kg
Water soluble coatings	Technical requirement for environmental labeling products-Water soluble coatings products	2005	Migration limit 90 mg/kg
	Indoor decorating and refurbishing materials-Limit of harmful substances of interior architectural coatings	2008	Migration limit 90 mg/kg
Printing ink	Technical requirement for environmental labeling products-Gravure and flexible printing ink	2007	Migration limit 90 mg/kg
	Technical requirement for environmental labeling products-Offset printing ink	2007	Migration limit 90 mg/kg
	Limits and determination method of certain toxic elements in printing ink	2008	Migration limit 90 mg/kg
Solvent coatings	Technical requirement for environmental labeling products- Solvent-based wood coatings for indoor decoration and refurbishing	2007	Migration limit 90 mg/kg
	Indoor decorating and refurbishing materials-Limit of harmful substances of solvent based coatings for woodenware	2009	Migration limit 90 mg/kg
Gasoline for motor vehicles	Gasoline for motor vehicles	2006	0.005 g/L
	Gasoline for motor vehicles	2011	0.005 g/L

a Sources of standards shown in Table 30.3.

Dissipative lead uses play an important role in lead emissions (Mao et al., 2009). Product lead contents included in this policy include toys, water soluble coatings, printing ink, solvent coatings, and gasoline for motor vehicles. The lead limits for these products are shown in Table 27.6. In a comparison of Chinese standards to corresponding

US and EU standards, we found that the upper limits of migratable lead in water soluble coatings, printing ink, solvent coatings are similar, and that the upper limit of lead in gasoline is the same as the EU. As such, some lead limits of China are comparable to those of developed countries.

Lead-acid batteries are the primary lead-containing product that undergoes recycling. In recent years, the increased use of electric bicycles and electric motorcycles has led to a corresponding increase in the demand for lead-acid batteries, which has resulted in heightened production (Cherry et al., 2009; Weinert et al., 2008). Chen et al. reported that specific lead pollution incidents in China are closely related to the production of lead-acid batteries (Chen et al., 2012). Issues associated with lead-acid battery manufacturing in China include a general lack of clean production, high lead consumption, poor corporate structure and layout, serious low-level redundant construction, high "three wastes" (solid, liquid, and gases) emissions, and low recycling rates (Zhang et al., 2005; WU et al., 2008). To address these problems China has introduced a number of policies for the manufacturing of lead-acid batteries, giving special attention to: ①requirements for cleaner production, particularly as specified in several policies, such as *Implementation program for the cleaner production of battery industry*, *Cleaner production evaluation indicator system for battery industry (on trial)*, and so on; ②production capacity of lead-acid battery manufacturers, as specified in *Lead-acid battery industry access conditions* (2012); and ③establishment of a licensing system, as specified in *Rules for their implementation of lead-acid battery production license*. These polices have strengthened the management and monitoring of lead-acid battery manufacturers.

The promulgation and implementation of these policies has played a key role in lead pollution control. For example, He et al. showed that the lead levels in children's blood has decreased as a result of prohibiting the use of lead-containing gasoline and the cessation of lead product manufacturing operations, especially enterprises making lead-acid batteries (He et al., 2009). Luo noted that 81% of lead-acid battery companies were closed in 2011 (Luo, 2012), which greatly reduced industrial pollution from lead-acid storage battery production.

27.3.3 Lead Products Use

Lead-acid batteries are the most commonly used lead product in China (Chen et al., 2009). Accordingly, recent policies focus primarily on the specification and classification of lead-acid batteries, and include provisions for testing and transporting lead-acid batteries in an effort to standardize quality requirements and management. Most policies relate to valve regulated lead-acid batteries, which are widely used in motors, vehicles, and telecommunications. Because these applications require well-sealed batteries, liquid and the acid fumes from these batteries can be prevented from spilling into the environment (Wang, 1998; Chang et al., 2009).

27.3.4 Waste Management and Recycling

Mao et al. indicated that almost half of China's lead pollutants come from outdated secondary lead refineries (Mao et al., 2007), and that obsolete lead-acid batteries are a primary source of lead scrap in secondary lead production (Mao et al., 2006). Hence, some waste management and recycling policies relate to the collection and recycling of lead-acid batteries. These policies mainly regulate the recycling technology and pollution prevention during collection of lead acid batteries (Table 27.2 and Table 27.3).

Waste management and recycling policies have established technical specifications for the collection, disassembly, pollution prevention, and control of used batteries. The policies emphasize cleaner production and the elimination of primitive types of production. Specifically, the publication *Technical Specifications of Pollution Control for Treatment of Lead-acid Battery* not only stipulates that obsolete lead-acid batteries are hazardous waste, but also extend the producer's responsibility from production to recycling of discarded batteries (extended producer responsibility).

The release of these policies helped promote the recycling of used lead-acid batteries and compelled the establishment of scrap recovery of lead-acid battery systems in China. Most of the scrap lead-acid batteries in China are handled by small traders (Mao et al., 2006); as a consequence, lead recovery and lead pollution prevention has not been effective. There is an urgent need for the release of the policy.

The quantity of e-wastes has been increasing rapidly due to rapid growth in the consumption of electric products and some illegal import of e-wastes (Yang et al., 2008). Lead emissions occur during the treatment of e-wastes. This is especially true for small-scale private enterprises with simple, outdated technology, and can result in substantial harm to staff and local residents (Tsydenova and Bengtsson, 2011). Huo et al. cite a notorious example of unregulated e-waste disposal activities occurring in Guiyu (Huo et al., 2007). Many recent policies have been promulgated specifically for e-wastes and are used to strengthen the management of e-wastes. Such management includes the collection, storage, disassembly, recycling, and final disposal of e-wastes; it also extends to producer responsibility, catalog management, and licensing schemes for waste electrical and electronic equipment recycling enterprises. The issuance of these policies also reflects the trend from inactive "waste disposal" to active "pollution source control and resource recycling." For instance, the publication *Technical policy on pollution prevention of discarded appliances and electronic product* (2006) focuses mainly for waste disposal, whereas the *Ordinance on management of pollution control from electronic information products* (2007) promotes pollution source control and recycling. Two policies, *Ordinance on management of pollution control from electronic information products* and *Regulation on management of waste electrical and electronic equipments recycling and disposal* are similar to the EU's *Restriction of the use of certain hazardous substances directive* (EU RoHS directive 2002/95/EC) (EP and EC, 2002) and the *European Union directive on waste from electrical and electronic equipment* (EU WEEE directive) (Yu et al., 2010; Liu, 2006), and have played key roles in the use of hazardous substances and e-wastes management, respectively.

27.3.5 Lead Trade

As a result of rapid growth in domestic demand for lead products and common resource bottlenecks in China, the government no longer encourages exports of energy-intensive smelting products. Constant trade friction with other countries is the result of different requirements regarding the lead content in products and the fact that some foreign solid wastes transferred to China have caused environment pollution. Consequently, China has introduced a series of lead trade policies to lower the tax refund rates for exported lead goods and reduce the number of concerned species of lead goods (policies issued in 2003, 2006); this was followed by the cancellation of the tax refund for all lead-containing goods (policy issued in 2010). These policies resulted in the reduction of exports of lead-containing goods (Liu, 2004). Some regulations relate to the import of lead-containing products: *Prohibited goods catalog for processing and trade*, the *Catalogue of solid waste forbidden to import*, and the *Administrative measures on the import of solid waste* (Table 27.2).

27.4 Discussion and Recommendations

27.4.1 Discussion

Our analysis indicates an increase in the release of lead-related policies in recent years as well as an extension in the scope and depth of the policies. Although these policies may have a positive effect for environmental improvement in China, they are still inferior to those of other developed countries because of the following:

(1) Lack of law-level policies for some special lead-containing products. For example, the *Consumer Product Safety Improvement Act* (CPSIA) was enacted in the United States of America in 2008 for children's products based on their total lead content (USCPSC, 2008). In China, however, only the standard *National safety technical code for toys* has been issued; no law or regulation has yet been formally promulgated. Because of the high potential risks to human health posed by these products (Guney and Zagury, 2012), laws similar to the *National safety technical code for toys* standard should be enacted in near future.

(2) Lower limitations for environmental lead emissions. For dissipative lead products, China limits only migratable lead content, not the total lead contents of the products. For instance, the limit for the migration lead in paints is 90 mg/kg (*Indoor decorating and refurbishing materials - Limit of harmful substances of interior architectural coatings*), but no limit is stated for total lead content. In contrast, the maximum permitted lead concentration in the EU is 1000 mg/kg (see the EU's *Restriction of the use of certain Hazardous Substances directive* (RoHS Directive 2002/95/EC) (EP and EC, 2002) and the limit for migratable lead is 90 mg/kg in the *European standard EN71-3: 2002* (CEN, 2002).

(3) Infrequent updating of lead-related policies. *The American Society for Testing and Materials standard, ASTM F963: Standard Consumer Safety Specification for Toy Safety*, was updated in 2003, 2007, 2008, and 2011 since 2000, based on results of current research (ASTMI, 2012). In China the corresponding standard, the *National safety technical code for toys*, has not be revised since its promulgation in 2003.

(4) Weak implementation of lead-related policies and standards. The process of adoption for most policies in China is subject to steps that include versions for drafts, exposure drafts, trials, and the formal final version. These steps allow enterprises to improve their technology incrementally. The process does, however, slow down and

weaken policy implementation. As a result, policies that are compulsory and mandatory in more-developed countries are provided only as recommendations for enterprises in China. This situation seriously affects policy enforcement.

Questions that require further investigation include whether or not the existing policies influence the use of lead in China, and how these policies affect lead emissions and overall environmental quality.

27.4.2 Recommendations

Lead-related environmental quality can be improved by emphasizing and improving the current regulatory environment and investigating the environmental regulations of other countries.

The current regulatory environment in China can be improved through establishing new policies in areas where none currently exist; reducing the steps required from the draft to final stages; clarifying lead emission limits and enterprise access conditions by amending existing policies; and sharing the responsibility of prevention of lead pollution among government, enterprises and other social entities, so that the implementation of policies can be strengthened.

The differences between policies and standards in China and those in other developed countries should be thoroughly investigated. The potential influence and results of these policies needs to be better understood. Results can provide Chinese policymakers with guidance and the basis for providing more advanced policies for China.

Acknowledgements

This project was financially supported by the National Natural Science Foundation of China (General Program) under grant 41171361.

References

ASTMI (The American Society for Testing and Material standards International). 2012. ASTM F963-11 Standard Consumer Safety Specification for Toy Safety. Show F963 historical versions and work items. http://www.astm.org/search/standards-search.html?query=Standard Consumer Safety Specification for Toy Safety, reskin=true. (Access 2012-08-10)

Ayres R U. 1994. Industrial Metabolism: Theory an d Policy, the Greening of Industrial Ecosystems. Washington DC:

National Academy Press, 23-37.

CEN (European Committee for Standardization). 2002. The European standard: Safety of toys-part 3: Migration of certain elements (EN-71-3: 1994/AC: 2002) (Directive 88/378/EEC (C 188, 2005-08-02). http: //esearch.cen.eu/ esearch/Details.aspx?id=6829713. (Access 2012-08-06)

Chang Y, Mao X X, Zhao Y F, et al. 2009. Lead-acid battery use in the development of renewable energy systems in China Journal of Power Sources, 191 (1): 176-183.

Chen H Y, Li A J, Finlow D E. 2009. The lead and lead-acid battery industries during 2002 and 2007 in China. Journal of Power Sources. 191 (1): 22-27.

Chen L G, Xu Z C, Liu M, et al. 2012. Lead exposure assessment from study near a lead-acid battery factory in China. Science of the Total Environment, 429: 191-198.

Cherry C R, Weinert J X, Yang X M. 2009. Comparative environmental impacts of electric bikes in China. Transportation Research Part D: Transport and Environment, 14 (5): 281-290.

EP and EC (The European Parliament and the Council of the European Union). Directive 2002/95/EC on the restriction of the use of certain hazardous substances in electrical and electronic equipment. Official Journal L 037, 13/02/2003 p. 0019–0023. http: //eur-lex.europa.eu/LexUriServ/LexUriServ.do?uri=CELEX: 32002L0095: en: HTML.

Guney M, Zagury G J. 2012. Heavy metals in toys and low-cost jewelry: critical review of U.S. and Canadian legislations and recommendations for testing. Environmental Science & Technology, 46 (8): 4265-4274.

He K M, Wang S Q, Zhang J L, et al. 2009. Blood lead levels of children and its trend in China. Science of the Total Environment, 407 (13): 3986-3993.

Huo X, Peng L, Xu X J, et al. 2007. Elevated blood lead levels of children in Guiyu, an electronic waste recycling town in China. Environmental Health Perspectives, 115 (7): 1113-1117.

Ji A L, Wang F, Luo W J, et al. 2011. Lead poisoning in China: a nightmare from industrialization. Lancet, 377 (9776): 1474-1476.

Li S L, Ma Y Y 2010. Comprehensive management and policy research on lead-zinc resources development–a case of lead-zinc resources in Kazi Town of Shaanxi Province. Safety and Environmental Engineering, 17 (6): 88-93. (In Chinese with English abstract)

Liu J W. 2004. Research on tax refund policy for export. Beijing: Peking University Press, 1-4. (In Chinese)

Liu X B. 2006. Electrical and electronic waste management in China: progress and the barriers to overcome. Waste Management & Research, 24 (1): 92-101.

Luo J W. 2012. The national VideoPhone conference on special action of enviornmental protection in 2012. State council VideoPhone conference on special action of enviornmental protection. China environmental news. http: //www.ce-news.com.cn/xwzx/ zhxw/ybyw/201203/ t20120320_ 714337.html (In Chinese) (Access 2012-03-23).

Mao J S, Cao J, Graedel T E. 2009. Losses to the environment from the multilevel cycle of anthropogenic lead. Environmental Pollution, 157 (10): 2670-2677.

Mao J S, Dong J, Graedel T E. 2008. The multilevel cycle of anthropogenic lead: I. Methodology. Resources, Conservation and Recycling, 52 (8-9): 1058-1064.

Mao J S, Graedel T E. 2009. Lead In-Use Stock. Journal of Industrial Ecology, 13 (1): 112-126.

Mao J S, Lu Z W, Yang Z F. 2006. The eco-efficiency of lead in China's lead-acid battery system. Journal of Industrial Ecology, 10 (1-2): 185-197.

Mao J S, Yang Z F, Lu Z W. 2007. Industrial flow of lead in China. Transactions of Nonferrous Metals Society of China, 17 (2): 400-411.

MEPPRC (Ministry of Environmental Protection of the People's Republic of China). 2011. The transition and upgrading of crude lead smelters at Guiyang county of Hunan province. http: //www.mep.gov.cn/zhxx/gzdt/201112/ t20111215_221371.htm. (In Chinese) (Access 2012-08-11).

Meyer P A, Brown M J, Falk H. 2008. Global approach to reducing lead exposure and poisoning. Mutation Research-Reviews In Mutation Research, 659 (1-2): 166-175.

Shen Y, Chen Y, Sun Y Z, et al. 2011. Study on Technology and Policies of Pollution Prevention and Control for Waste Lead-acid Battery Recycling. China Environmental Protection Industry. 4: 49-52 (In Chinese with English abstract).

Tsydenova O, Bengtsson M. 2011. Chemical hazards associated with treatment of waste electrical and electronic equipment. Waste Management, 31 (1): 45-58.

USCPSC (US Consumer Product Safety Commission). 2008. Consumer Product Safety Improvement Act (CPSIA) of 2008 (Public Law110–314. Access 2008-08-14. http: //www. cpsc.gov/ cpsia.pdf.

Wang B, Sun Q H, Hu X W, et al. 2011. Screening of best available techniques for lead smelting pollution prevention and control. Journal of Environmental Engineering Technology, 1 (6): 526-532 (in Chinese with English abstract).

Wang C M, Lin Z L. 2010. Environmental policies in China over the past 10 years: progress, problems and prospects. Procedia Environmental Sciences, 2: 1701-1712.

Wang Z. 1998. Manufacture and application of valve-regulated lead/acid batteries in China. Journal of Power Sources, 73 (1): 93-97.

Weinert J, Ogden J, Sperling D, et al. 2008. The future of electric two-wheelers and electric vehicles in China. Energy Policy, 36 (7): 2544-2555.

Wu M, Tian L Y, Fan H P. 2008. The production, use and recycle of lead-acid battery in China. Automobile & Parts, 3: 37-39 (In Chinese with English abstract).

Xue P L, Zeng W H. 2010. Policy issues on the control of environmental accident hazards in China and their implementation. 2: 440-445.

Yang J X, Lu B, Xu C. 2008. WEEE flow and mitigating measures in China. Waste Management, 28 (9): 1589-1597.

Ye X B, Wong X. 2006. Lead exposure, lead poisoning, and lead regulatory standards in China, 1990–2005. Regulatory Toxicology and Pharmacology, 46 (2): 157-162.

Yu J L, Williams E, Ju M T, et al.2010. Managing e-waste in China: policies, pilot projects and alternative approaches. Resources Conservation and Recycling, 54 (11): 991-999.

Zeng R, Mao J S. 2010. Trend Analysis on Dissipative Uses of Pb in China. Environmental Science & Technology,

33（2）：192-195（in Chinese with English abstract）.

Zhang Z J，Li D H，Xu Z G.2005. Present conditions，reasons and measures of lead pollution in China. Environmental Protection Science，31（4）：41-47.

Zheng N，Liu J S，Wang Q C，et al. 2010. Heavy metals exposure of children from stairway and sidewalk dust in the smelting district, northeast of China. Atmospheric Environment，44（27）：3239-3245.

Zhou L，Xu Z M. 2012. Response to waste electrical and electronic equipments in China: legislation, recycling system, and advanced integrated process. Environmental Science & Technology，46（9）：4713-4724.